The Making of the

BRITISH
LANDSCAPE

Nicholas Crane is an author, geographer, cartographic expert and recipient of the Royal Scottish Geographical Society's Mungo Park Medal in recognition of outstanding contributions to geographical knowledge, and of the Royal Geographical Society's Ness Award for popularising geography and the understanding of Britain. He has presented several acclaimed series on BBC2, among them *Map Men*, *Town*, *Britannia* and *Coast*. He was elected President of the Royal Geographical Society in 2015.

By Nicholas Crane

The Making of the

BRITISH
LANDSCAPE

From the Ice Age to the Present

NICHOLAS CRANE

WEIDENFELD & NICOLSON

A W&N PAPERBACK

First published in Great Britain in 2016
This paperback edition first published in 2017
by Weidenfeld & Nicolson
an imprint of The Orion Publishing Group Ltd
Carmelite House, 50 Victoria Embankment
London EC4Y 0DZ

An Hachette UK Company

1 3 5 7 9 10 8 6 4 2

A CIP catalogue record for this book is
available from the British Library.

ISBN 978 0 7538 2667 6

Typeset by Input Data Services Ltd, Somerset

Printed and bound by CPI Group (UK) Ltd, Croydon, CR0 4YY

MIX
Paper from
responsible sources
FSC® C104740

www.orionbooks.co.uk

To care about a place,
you must know its story

CONTENTS

BOOK TWO

BOOK THREE

LIST OF ILLUSTRATIONS

Section One

1. Rannoch Moor. (Frame Focus Capture Photography/Alamy Stock Photo)
2. Cheddar Gorge in the Mendip Hills. (James Osmond/Getty Images)
3. The rump of Doggerland. (Malcolm English)
4. The chalk cliffs of a new island. (James Osmond/Getty Images)
5. An impression of 'peak wildwood'. (Loop Images Ltd/Alamy Stock Photo)
6. Pioneering flint mines. (Historic England Archive)
7. The portal dolmen or tomb of Pentre Ifan. (John Fletcher/Dreamstime. com)
8. Long barrows. (John Henshall/Alamy Stock Photo)
9. Henges, defined by an earthen enclosure. (robertharding/Alamy Stock Photo)
10. Houses at Skara Brae, Orkney. (Mat Ladley/Alamy Stock Photo)
11. Silbury Hill. (Chris Lock/Alamy Stock Photo)
12. Round barrows. (English Heritage/Heritage Images/Getty Images)

Section Two

1. Field systems defined by banks – or reaves. (Historic England Archive)
2. The White Horse of Uffington. (robertharding/Alamy Stock Photo)
3. The broch on Mousa island. (David Robertson/Alamy Stock Photo)
4. Interior of Mousa broch. (Vincent Lowe/Alamy Stock Photo)
5. The hillfort known as Maiden Castle. (Last Refuge robertharding/Getty Images)
6. The Roman fort on Hardknott Pass in the Lake District. (David Lyons/ Alamy Stock Photo)
7. The Roman baths (with subsequent additions) in Bath. (eye35.pix/ Alamy Stock Photo)

MAPS

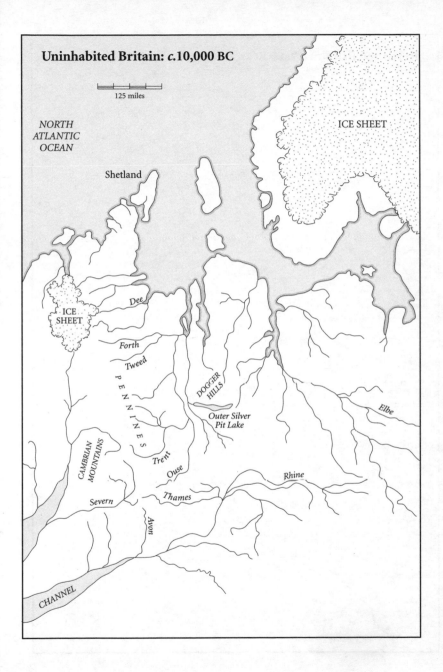

Uninhabited Britain: *c.*10,000 BC

125 miles

NORTH
ATLANTIC
OCEAN

ICE SHEET

Shetland

ICE
SHEET

Dee

Forth

Tweed

P E N N I N E S

DOGGER
HILLS

Outer Silver
Pit Lake

Elbe

CAMBRIAN MOUNTAINS

Trent

Ouse

Rhine

Severn

Thames

Avon

CHANNEL

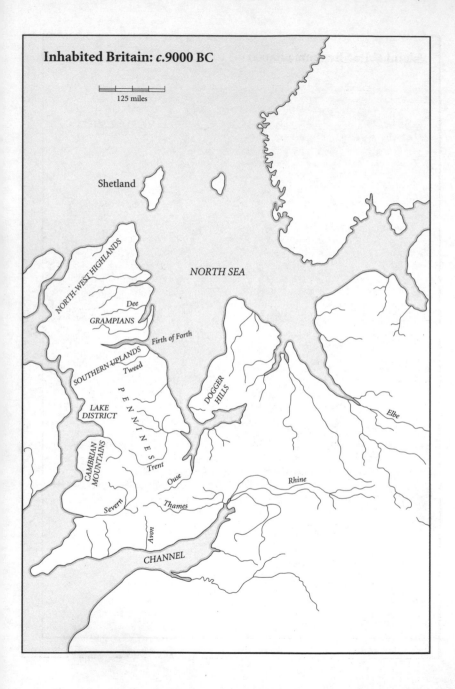

Inhabited Britain: *c.*9000 BC

125 miles

Shetland

NORTH SEA

NORTH-WEST HIGHLANDS

Dee

GRAMPIANS

Firth of Forth

SOUTHERN UPLANDS

Tweed

LAKE DISTRICT

P E N N I N E S

DOGGER HILLS

Elbe

CAMBRIAN MOUNTAINS

Trent

Ouse

Rhine

Severn

Thames

Avon

CHANNEL

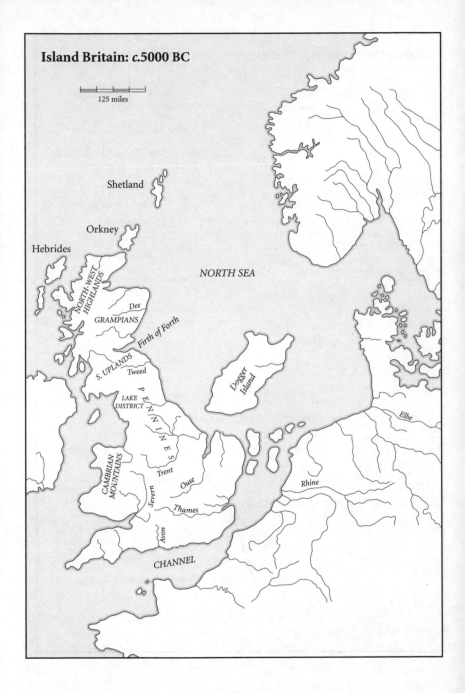

Island Britain: *c.*5000 BC

125 miles

Shetland

Orkney

Hebrides

NORTH-WEST HIGHLANDS

Dee

GRAMPIANS

Firth of Forth

NORTH SEA

S. UPLANDS

Tweed

LAKE DISTRICT

P E N N I N E S

Dogger Island

Elbe

CAMBRIAN MOUNTAINS

Trent

Ouse

Severn

Thames

Rhine

Avon

CHANNEL

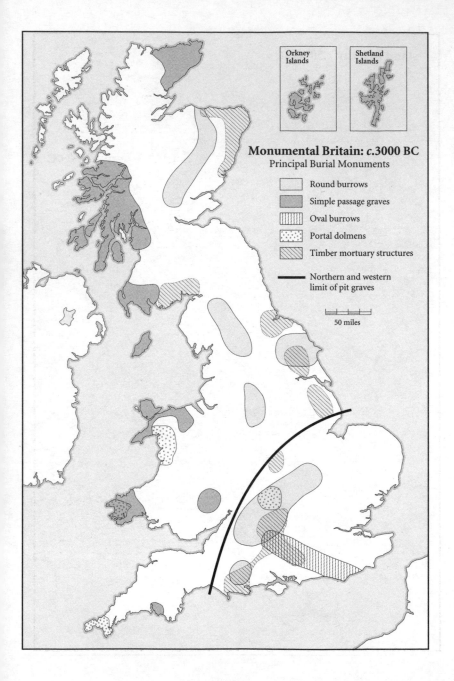

Orkney
Islands

Shetland
Islands

Monumental Britain: *c.*3000 BC
Principal Burial Monuments

Round burrows

Simple passage graves

Oval burrows

Portal dolmens

Timber mortuary structures

Northern and western
limit of pit graves

50 miles

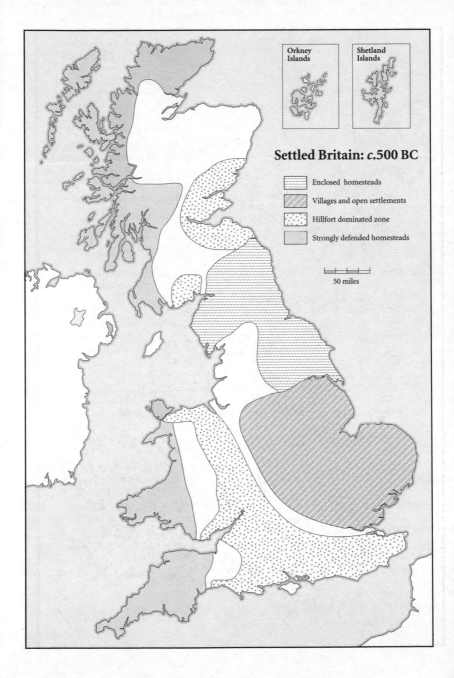

Orkney
Islands

Shetland
Islands

Settled Britain: *c.*500 BC

	Enclosed homesteads
	Villages and open settlements
	Hillfort dominated zone
	Strongly defended homesteads

50 miles

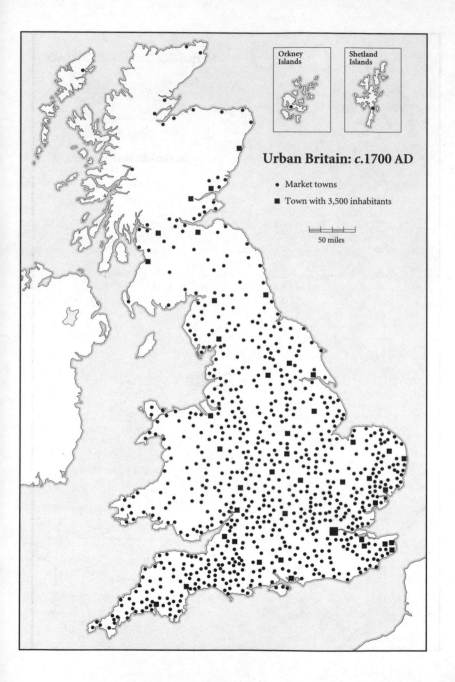

Urban Britain: *c.*1700 AD

- • Market towns
- ■ Town with 3,500 inhabitants

Orkney
Islands

Shetland
Islands

50 miles

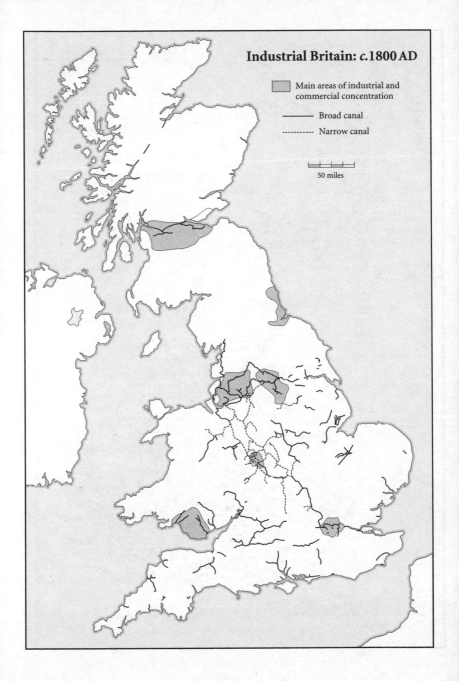

Industrial Britain: *c.*1800 AD

Main areas of industrial and
commercial concentration

——— Broad canal

----------- Narrow canal

50 miles

Orkney
Islands

Shetland
Islands

Inverness

Aberdeen

Dundee

Connected Britain: *c.*1950
The railway system before
widespread closure

Glasgow
Edinburgh

50 miles

Carlisle

Newcastle

Leeds   Hull

Liverpool   Manchester

Holyhead

Derby

Norwich

Birmingham

Cardiff   Bristol

LONDON

Southampton

Exeter

Orkney
Islands

Shetland
Islands

Modern British National Parks

50 miles

1. Cairngorms
2. Loch Lomond and the Trossachs
3. Northumberland
4. North York Moors
5. Lake District
6. Yorkshire Dales
7. Snowdonia
8. Peak District
9. Broads
10. Brecon Beacons
11. Pembrokeshire Coast
12. Exmoor
13. Dartmoor
14. New Forest
15. South Downs

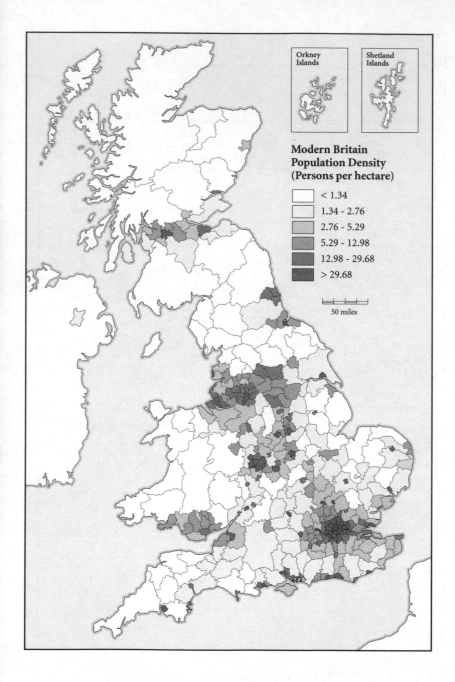

Orkney
Islands

Shetland
Islands

**Modern Britain
Population Density
(Persons per hectare)**

< 1.34

1.34 - 2.76

2.76 - 5.29

5.29 - 12.98

12.98 - 29.68

> 29.68

50 miles

BOOK ONE

PART ONE

Ice Age to Island

ONE

Edge Land

10,000–9000 BC

Several winters back an overladen northerly smothered the county of Norfolk with snow, blocking roads and burying hedges. A few days into the freeze, I decided to ski to my sister's village. In the half-light of late afternoon the undulating drifts and whorls merged with the burdened sky and I had to use a compass to find my way. The blizzards had been followed by an exhausted calm. Through a window in the overcast ceiling, a planet blinked. Every few minutes I paused to check my bearing and to stare into the bleak snowscape. Nothing moved and I could see no electric lights. It was as if Britain had returned to the Ice Age. Yet beneath my skis I knew there was a field where I'd once met a man with a pocketful of Roman coins he'd picked from the freshly tilled soil. Over to my left was a rearing *arête* I knew to be a railway embankment from the age of steam. And ahead of me, entirely lost beneath the Arctic blanket, was a meandering stream that had been deep enough to take Iron Age trading vessels. The village itself, unseen in its hollow, had been named a thousand years ago after 'a farmstead where broom grows'. For one transformative evening, the ancient mosaic of human trace-marks had been wiped clean by an Arctic wind; returned to the state it had known before temperatures rose.

Only twelve thousand years ago, 'Britain' was the colourless, glacial extremity of a continuous landmass which reached east to Kamchatka and south to Table Mountain. A low, undulating plain – Doggerland – linked the higher ground of south-eastern Britain with rising land between the Rhine and Elbe. Glaciers mantled the peaks of northern Britain and filled the Great Glen to a depth of 600 metres. Westerly winds hurled spindrift cross the frozen North Atlantic, bringing mean winter temperatures that fell to minus 17°C. Pack ice washed up on beaches. South of the blinding ice stretched the tundra, a cold, arid land of mosses and lichens, sedges, grasses and stunted shrubs that were snow-covered in winter and blasted by frigid, dust-laden winds in summer. Vegetation struggled to reach half a metre in height. Pale clumps of creeping willow, juniper and dwarf birch clung to pockets of soil which had been turned slightly saline by desiccative winds and meagre rainfall. Rivers ran in weakened braids through valleys choked with barren tracts of rubble, gravel and sand. Few animals could tolerate this impoverished continental margin. On the upper Thames, Arctic beetles hung in meltwater pools. Burrows beneath the snow hid colonies of Northern collared lemmings and steppe pika. Not much moved. Above ground, occasional ptarmigan croaked as they took to the wing. Infrequent foxes, bears and wolves patrolled the open permafrost. In summer, when grass rose from the tundra, herds of horse and reindeer came up from Doggerland to calve and graze, then they would return east to the shelter of the lowlands.

But in around 9700 BC, the climate changed. Mean July temperatures may have leapt by as much as 7°C in as little as fifty years. Rain fell. In the far north, ice creaked and sheared. Glens trembled as meltwater cataracts thundered along glacial scars. Along valley-sides, down-wasting glaciers left lines dividing frost-shattered crags and cliffs from emerging levels of bedrock smoothed and scoured by the passage of ice. Rock walls collapsed. Hundreds of millions of tons of loose stone was water-worked into cones and fans and gullies. Lakes accumulated behind barricades of moraine. In the south, the undulating tundra changed hue. Bogbean, sea thrift and sedge that had endured glaciation in frigid, saline marshes and moister nooks began to spread in livid green mats. Floodplains pricked with creamy meadowsweet, blood-drops of great burnet and pink tongues of ragged-robin; hill slopes gathered carpets of yellow-flowered sea plantain, greater knapweed and mountain avens. And into this waking land walked some hominids.

Britain's portal, the Thames, led them inland. Beyond the tidal limit they reached a fork in the flow where a smaller river joined from the north (later, this confluence became known as 'The Stones', or 'Staines'). Here, they turned to follow the tributary, perhaps because it led to a fording-point. It's not known how long the group spent in the valley of the Colne (a name so old that its roots are lost), or whether they also climbed the gravel terraces to look for herds on the levels to the east (where the tundra has been concreted with Heathrow's runways). But they did kill at least one reindeer and carry its severed limbs to a low rib of land projecting from the valley-side into the Colne's floodplain. Settling on this dry vantage point, they worked river pebbles into blades and scrapers and cut meat from bone and fixed tools and weapons. The hunters stayed a couple of days, maybe a bit longer, eating, resting, making and mending, and then they moved on. In this harsh, open land on the edge of Europe, they were subsisting at the limit of humanity's range. Britain was a biological and cultural death-zone. These people were survivalists, Edgelanders, and they migrated in small, cohesive bands, seldom staying anywhere for more than a few nights, following the herds of horse and reindeer, moving, moving. Wherever they paused to process a kill, they left bones, scatters of stone flakes and sometimes hearths. Thirty or so sites are known in southern Britain. These modest landmarks open Britain's history of continuous occupation.

Humans had been here before. For at least 700,000 years small groups had been making forays to the land that would become Britain. But for this entire period, the habitability of northern Europe had been dictated by an unpredictable Ice Age. There had been times when Britain had been buried beneath a mile of ice, and then warm-ups which had seen the land retaken by lush woodland. It was during these interglacial episodes that migrating hunters had ventured north from the continent's kinder latitudes, foraging and preying on the wandering herds. Then the ice would return, sometimes suddenly. On one occasion, temperatures fell by as much as 15°C in a single human generation. Survivors trekked south in search of shelter from the creeping freeze. Some sought refuge in caves. During at least five interglacials, humans had occupied sites in Britain, only to be driven south by another snap of the climate. Had the pioneers of the tenth millennium BC known where to look, they would have found traces of their predecessors; hunters who had left tools and teeth and the scored bones of slaughtered mammoths, and carvings of horses and reindeer, and human skulls that had been turned into drinking vessels. But the climatic spasm of 9700 BC was different. This time the thermal window would admit 63 million humans.

Britain's destiny was directed by the Sun and southerners. The former tended to dictate how habitable Britain was at any time, and the latter the manner of that habitation. The Sun was God of all things. If the amount of solar warmth – insolation – reaching the upper atmosphere of the Earth had been constant through time, the course of global history would have been entirely different. But insolation changes according to cyclic fluctuations in the way Earth orbits the Sun. One of these cycles relates to slight alterations in the tilt of the Earth's axis. Another of the cycles is caused by variations in the shape of the Earth's orbit. A third cycle arises from the wobble of the Earth on its axis, and by the way the shape of the orbit actually rotates. The cycles operate on periodicities ranging from hundreds of thousands to tens of years, and they interact in complex ways with each other and with a variety of 'ground-effects' such as the amount of incoming light being reflected by the Earth's surfaces, and sources of greenhouse gases like volcanoes and the Earth's flora and fauna. Apart from the chilly reversal between 10,900 and 9700 BC, the overriding pattern since around 20,000 BC had been one of global warming driven in the main by the rising arc of solar insolation.

The second forcing mechanism was entirely terrestrial. Two thousand miles from Britain, where the planet was much warmer, a combination of topography and weather systems delivered around 300 millimetres of rain a year and created an ecosystem that was near-perfect for human existence. Mapped by rivers, lakes, woodland and grassland, it described a fertile crescent of abundance over a thousand miles in length. Cereals grew on grasslands grazed by herds of game and people had taken to living in circular dwellings – houses – up to four metres in diameter and gathering in great number to participate in feasting and cult rituals, practices that obliged them to find ever more ingenious ways of increasing yields from their lands. The Fertile Crescent was incubating an ideology with a long reach.

With every warming decade, Britain became greener. Ecological zones were unlocked from latitudes and altitudes. The tundra retreated northward and upward. Where permafrost had once stunted growth, stands of birch and pine, juniper and willow shaded carpets of year-round grasses. The herds of reindeer and horse changed their migration routes. Hunters and the hunted faced unprecedented difficulties: rivers were quicker and deeper to cross; spring growth impeded sightlines; rain-drenched ground froze and sealed vegetation beneath an unbreakable carapace of ice; trees poached grazing from roaming herds. By 9500 BC, the northern glaciers had gone and the Great Glen had emptied of ice. Ben Nevis had emerged from its stilled, white coverlet as a great, fissured creature from the deep, shaking its precipitous, unbuttressed flanks and sending rocks tumbling to spreading screes. On the great plateau of Rannoch, a 400-metre dome of ice had disappeared and exposed a puddled, grey waste. Relieved of its burden, the earth's crust sprang slowly upward in the far north where the ice had been thickest, while southern coasts were subjected to accelerated sea-level rise as vast quantities of glacial meltwater increased the volume of the oceans. In this dynamic land, shorelines drowned, lakes shrank and grasslands were colonised by trees. As the sources of snowmelt diminished, rivers adopted more regular flows and abandoned their braided threads for single, meandering courses between banks stabilised by vegetation.

Greenhouse gases surged. Fed by the tropics and thawing tundra, the amount of methane being locked away in the ice of Greenland leapt from a low of 450 parts per billion to over 700 ppb by 9600 BC. Atmospheric carbon dioxide crept up from an Ice Age minimum of less than 200 parts per million, to 260 ppm by 9400 BC. By 9300 BC, maximum annual temperatures in Britain had bounced back to their pre-glacial high. Cold-tolerant beetles (one of the indicator species of climate change) were replaced in lowland Britain by warmth-loving species. The fate of the reindeer hunters is unclear. By around 9200 BC, horses and reindeer had disappeared from southern Britain and been replaced by elk, roe deer, red deer, boar and the aurochs – a huge, bovine herbivore. The adult aurochs loomed two metres above the ground. With its keen senses, curved horns, slender body and long legs, the aurochs was quick, agile and a match for hesitant wolves. Bulls were a deep-brown colour or black, and cows a reddish brown, but both had lightly coloured mouths and were countershaded with a white stripe down the spine which made it more difficult for their contours to be recognised by their human predators. Their favoured habitat was level, low-lying, fertile and open; woodland would not have supported the rich grassland they depended upon. Congregating in herds on floodplains and valley floors, they were the biggest beasts in Britain.

Amid the burgeoning biomass were people of the trees. They advanced in fits and starts, their passing marked by lost harpoons, tanged missile heads and scatters of stone flakes. Their relationship with the hunters of horse and reindeer is masked by Britain's blooming postglacial vegetation. Culturally, the woodland people were newcomers to Britain, with subsistence strategies and tool sets (and presumably beliefs and traditions, too) suited to a warming land scattered with trees – mainly birch and pine – and populated by a diverse flora and fauna. They shared tools used on the plains of northern Europe. While many must have walked to Britain across Doggerland, it's possible that some descendants of the big-game hunters had managed to adapt over successive generations to the ways of the wood. Whatever the cultural crossover between people of tundra and wood, both were resourceful. As an edgeland, Britain pushed the limits. Many core aptitudes were carried over from the days of the tundra, to be honed, faceted and polished in temperate woodland.

The woodland people had a more complex, intense relationship with land than their migratory, cold-climate, open-tundra, herd-shadowing forebears. They lived in larger groups and stayed for prolonged periods at select sites, where they exploited the multiplicity of resources that came with a temperate climate and trees. They collected berries and tubers, they netted, trapped and hunted a wide variety of animals, from beavers to boar and deer. Their prey was less gregarious and confined to more localised habitats than the herds of horse and reindeer. They killed aurochs, whose preference for valleys and floodplains put the herds in conflict with humans who used these landscapes for routeways, foraging and hunting missions. They lived in fluid groupings or bands, and understood the need to cooperate for tasks that might require extra hands. Women may have specialised in foraging and men in hunting, and it's likely that an egalitarian ethos prevailed, with food being shared. The dogs they kept were valued for companionship and for their roles in hunting, transport and security. Their stone tools included specialised adzes and axes for working with earth and timber. They could craft containers from birch bark and fabricate multi-piece, sewn clothing from hides. They could make ropes from plant fibres, and turn twine into baskets, mats and nets. They knew how to dry, smoke and cache food, and how to render grease by boiling. They played flutes and set traps. They carried the bow and arrow, the throwing spear and stabbing spear. They crafted multi-barbed harpoon heads from bone and had mastered the right-angled joint, boring holes in shaped antlers to accept handles of trimmed timber. They knew the plants for dyes and potions, and how to make flames with a wooden fire-drill or by striking particular rocks.

Resources multiplied. Wood, glade and warming seas turned the peninsula into a gigantic storehouse. An adult female red deer weighed between 70 and 150 kilograms, and an adult male between 100 and 250 kilograms. Some 50–60 per cent of a deer was edible meat; the hide could be used for lashings, containers, boat hulls, tents and clothing; parts of antler could be used for fabricating bows, projectile points, tools, pegs, toggles and gaming pieces. Bones were whittled and split into points, needles and even ice-skates; the guts were handy for containers; hooves for glue and sinew for reinforcing bows, bindings and sewing thread. Killing an animal was a process of transformation; the cosmos was indivisible from self. People co-existed with the plants and animals they foraged and managed. They related to the patterns of stars, to the stutter of the capercaillie and to the glare of the elk. The wild was them.

This ragged, greening peninsula, this Britain-to-be, was a land of extraordinary variety. That diversity came from its latitudinal reach; its range of weather-systems; its exposure to the Atlantic in the west and to the continent in the south-east; its immensely long coastline; its topographic extremes; its rivers and its stone. The distance from the Shetland Islands to the Lizard Peninsula was nearly 800 miles, far greater than the distance from the Thames to the Mediterranean. It was a land of many regions. There were lower temperatures in the north than the south; more sun in the south-east than in the north-west; more rain in the west than the east. Coasts were characterised according to their resistance to erosion, susceptibility for deposition, and rates of sea-level rise. Atlantic capes were fanged with reefs and defended with cliffs which parted to disclose fjord-like inlets formed where river valleys had flooded. The shores of the south-east were softer and opened to estuarine gulfs and skeins of welcoming creeks. So contorted was peninsula-Britain's coast that at no point could anyone be further than two or three days from the sea. Inland, there were parts of the west where ice-shattered *arêtes* and implacable cliffs walled one land from another. In the south, there were wooded uplands topping 1,500 feet and extensive lowlands threaded by meandering streams and reflective fens where boats were the only practical means of travel. Cutting across Britain were physical borders: much of the peninsula was divided up its central axis by a line of hills separating west from east, and a north–south divide existed between the cooler, shrinking tundra of the far north and the advancing tree-clad regions of the south. When it came to topography, Britain had a bit of everything common to the northern hemisphere. Fed by rainfall and gradient, one feature was particularly present: darting and dreaming their way through the peninsula's virescent slopes were over 100,000 miles of watercourses, grouped into over 1,000 discrete river systems. None of these was a 'river' in the continental sense. The Douro and Garonne, the Loire, Seine, Meuse and Rhine dwarfed watercourses like the Severn and the Thames. The longest river in Britain was just one eighth the length of the Danube. But Britain's prolific rainfall and tilted topographies meant that you were seldom more than a mile from running water.

Britain was built on spectacular foundations: beneath this modestly proportioned extension of the European mainland were rocks spanning three billion years, half the age of the planet. During various eras in the deep past, this part of the earth's crust had been seared by desert heat, scoured by freshwater torrents, blasted by volcanoes, submerged beneath oceans, and subject to shifts in tectonic plates. The result was an extravagant geology ranging from black basalt to glittering quartz, red sandstone to blue-grey slate and bone-white chalk. There were rocks decorated with fossils of extinct creatures, cliffs walled in octagonal columns and caves hung with stone tendrils. More recently, the episodic freezes of a 2.5 million-year Ice Age had created a family of strange features that included symmetrical hummocks, parallel trenches, pits, causeways and curious, freestanding boulders. Britain was abundant in landforms that invited human association. And it was replete with hardware.

Scattered across this well-endowed land were siliceous stones suitable for tool-making. They could be found on beaches and in valley gravels and in the earth of hillslopes. In the east, they could be picked from cliffs. Formed within sedimentary rocks, huge quantities had been transported far from their original source by glaciers or rivers or sea currents. These frequently odd-shaped nodules varied in hardness, texture and colour. They also differed in the way they fractured when struck, qualitative distinctions that were no doubt identified by name. Coarser, opaque chert produced flat fractures, while flint – which was fine-grained and even – fractured into conchoidal faces. Knapping these nodules into edges and points for scraping, cutting and piercing was a skill common to everyone engaged in foraging and hunting. The stones came with their own sound: a rhythmic knocking accompanied by high-pitched tinkling. You could hear it by night and by day, in woodland and in grassland. It was the sound of people knapping, nodule in one hand, hammerstone or antler-stub in the other, striking, looking, striking, the flakes tumbling to create abstract mosaics on the ground that would survive for thousands of years.

Everything moved. In this climatically charged, geographically complex world, food sources were not distributed evenly, or throughout the year. Grassland, upland, lowland, moorland, woodland, glades, estuaries, lakes and coasts all had their own species and seasons of abundance. Continuously shifting ecosystems driven by rapid global warming meant that people lived in a state of relentless adaptation. The greatest security came with a diverse mix of food and material sources such as those found where two different communities of plants and animals met, or overlapped. These boundary or transition zones between biomes offered the best of two worlds. Examples of these transition zones – or ecotones – included wetlands, the edges of estuaries and coastlines. The value of ecotones soared if they also offered ready access to stone-beds or beaches or riverbeds where workable chert or flint could be collected. The geographically advantaged could reduce the range of their roaming and live on the fruits of a locality but the hungry were beckoned by need, sourcing fresh water, food, stone and shelter along familiar routes on seasonal cycles. Britain was not so large or intimidating that a single person could not sample its extremities in a lifetime: a young adult could cover ten or twenty miles a day, for days on end. In the course of a year, it was feasible to trek five thousand miles. In lightweight, shallow-draught boats, people fished, foraged and hunted along coasts and estuaries. During the winter months, they took to snowshoes and sledges. Compared to Europe, the topographies of Britain were of a manageable scale; no rivers were too wide to cross.

A long time ago, I found myself working in unfamiliar mountains where I knew nobody. Within days of arriving, I came across a lone rowan by a rock and for the following months, this became 'my place'. We're all topophiliacs. We have a predisposition to invest locations with attachments. We should perhaps look at that extended tract of blue and green and grey that was Doggerland and Britain in 9000 BC and see most of it as 'space', an abstract, unknown entity. Faintly sketched onto this space were the 'places' that were known to foragers and hunters. The story of Britain is a contest between space and place, between the unknown and known, the insecure and secure, the unconfined and confined. Space was imagined from afar; place was experienced from within.

There were many kinds of place. Associations were cemented with seemingly ordinary locations through events that became archived in the memories of individuals or groups. The bloodied soil that marked the terminus of a long, hard chase, or the place where a lone wolf was seen howling to the moon might be memorialised and handed down by generations of storytellers. The places where family members were born – or sighed their last breath – became dynastic coordinates. Encounters with neighbouring groups or unfamiliar animals assumed a limitary meaning. Places revered for their tool-stone were visited repeatedly and their locations physically memorialised by accretions of flake-scatters.

And there were places venerated for the effect they had on human imagination. There's no reason to suppose that our ancestors weren't moved by some of the sights they came upon. Did they not gaze with pleasure at the play of sunlight on a chuckling stream? Or watch wide-eyed as storm waves pounded sea-girt granite? Or laugh at the feel of sand? The dissonant cosmos nurtured and murdered. Earth systems, the winds and waves, the currents, storms, droughts and freezes, were centrifuges for the imagination, spinning fears and hopes into vortices of expectation; beliefs that things were *meant*. Inherited fictions told of supernatural powers that controlled the changeable world. Faith could alter a perceived outcome. The lifting of eyes was both a conscious quest to escape the reality of life at ground level and an appeal to the clarity of the sky or the physical dominance of a lofty landmark: a hilltop, a treetop, a rocky crag. Caves could be portals to the unlit world; dwelling-places of subterranean forces. Restless, cleansing, life-giving water was both ubiquitous and revered. Springs had miraculous presence; practical yet communing with inner earth. Trees, too, either singly or in groves or canopied woods, were invested with qualities beyond firewood or axe-haft. There were also places that removed the mind from the enveloping cosmos: a peak, a peninsula, an island; viewpoints. And places that spoke, like cataracts and waterfalls and wave-beaten headlands. And rocks or mountain buttresses that looked in profile like birds, or animals or humans.

Fear was a potent place-maker. The world beyond, that space without, hid many threats. Probing a wilderness on foot was an intense experience; a misplaced camp, a navigational error or a failure to read the weather could turn a routine outing into a trial by stress and exhaustion. Method rather than muscle separated the living from the lost. Geography was an instinctive discipline. Noting features was both a means of accumulating coordinates and an act of faith. Every remembered rock and tree was invested with belief that it would be seen again. In a world of constant mobility, route-finding was a spatial preoccupation. Features were picked from tangled topographies and hoarded as waymarks; places indicating bifurcations or dead-ends; route options separating safety and danger. By the second passing, a waymark became a source of reassurance; a place with meaning. The urge to repeat sets of actions at particular locations was both a survival strategy and a way of deriving social reassurance from a potentially chaotic and dangerous habitat. With the passage of time, it formalised the relationship between people and the land. Waymarks became landmarks. Landmarks were lifelines. Teenage graduates of Edgeland knew that death at the hands of another human – from within the group or from clashes with neighbouring groups – ranked with disease as being the most likely way to meet an inconvenient end. The most effective defence was the control of space. Viewpoints and dogs were a form of early warning; numbers and weapons were a means of deterrence. So were suspicion or hostility to strangers and negotiated alliances with other groups. Habitual vigilance was the most effective form of preventative care. Collectively, these various forms of locative paranoia intensified people's association with places they valued the most.

Sleep was a place-maker, too. Lying down with eyes closed for several hours in a night-world of predators demanded faith in a location. The act itself was intimate, bonding person to place in a dreamtime embrace. It was in these nocturnal sanctuaries that the cosmos made contact through its animate constellations, its weeping stars and meteor blizzards and bright-eye moon, and through the warmth of the sun that was radiated to earth and then conducted to shoulder, hip and thigh. People didn't sleep beneath the sky, they slept within it. Terrestrially, these sleep-places summarised the bare essentials of existence for they were usually accompanied by food, fresh water and shelter. The food may have been processed earlier and brought to the site; the water glistened in a nearby stream, river or lake; the shelter may have been the lee of a rock or a swaddling bank of grass or an oak grove or a windbreak of poles and hide. It took less than an hour to raise a windproof lean-to, or a free-standing shelter from materials like stone, branches, turf and hide. East-facing slopes were the warmest at dawn, south-facing slopes held the heat through the day. The body-sized template on the ground was the physical zone of contact, but the sanctuary extended to a wider radius. The relative positions of surrounding furniture – the trees, the rocks and water – were known and a perimeter may have been trodden, ear to the wind, as the final ritual before a night beneath the stars. Come morning, a camp may have been recorded on the land as little more than a panel of compressed grass. Many were accompanied by a small hearth, a scatter of stone chips and discarded bones. A one-night camp was a brief encounter with a bare-necessity location, underpinned by trust in the cosmos. For the inhabitant, it was a special place, remembered for its security – or lack of it. Habitual use of the same camp altered its locative status, promoting it to a place that might have been labelled later as a 'home' or a 'home-base'. At its most expansive, home was an area of land associated with an individual, a family, an extended kin group or an otherwise-bonded social unit. It's conceivable that in areas of lesser sustenance, home-lands were measured in hundreds of square miles.

Places lived through movement; by being connected; by the beating of bounds with foot-soles; by the creation of desire-paths. Paths were the means of recording the passage through time and space of both individual and band; routes that described where you'd come from, where you were, and where you'd be in the future. Imprinted by repetitive footfall, each path was an innate negotiation between human desire and the lie of the land. Their course was not expressly premeditated and their form was not constructed. Desire-paths were memories, retrodden year after year, refined with shortcuts and easements. Up close, a path was a physical trough in the vegetative carpet; viewed mid-frame it appeared as a slender thread meandering like a stream through the wilderness. Every turn had a reason: a boulder avoided; a bog bypassed; a gradient cheated; a ford sought. Home-paths radiated out to water source and firewood, fruit tree, hunting ground, vantage point and crap patch. Paths connected with other paths, evolving into networks. Paths were expressions of movement through space, and so were the discontinuities, the places where people paused to craft tools, where they divided an animal, where they rested or slept, looked at a view or meditated upon the kinds of cosmic issues that fixed the place in a way they felt was timeless.

These landmarks-with-meaning were bundled into landscapes.

Appreciating or imagining tracts of land as coherent spatial entities was a potent means of controlling space; of coming to an accommodation with the wilderness. It was a process of reduction and coding. The interminable could be divided into memorable tracts: categoric spatial entities. Landscapes were variable and powerful. They could be constructed by individuals and viewed collectively through social and cultural prisms. They had no fixed scale. They described physical, measurable space and also imagined topographies. They could be repositories for belief and settings for action. A landscape was like a personalised – or localised – dialect in which every feature was uniquely interpreted. *Homo sapiens* was a maker of landscapes.

Landscapes were mapped in the mind. Views from high-points revealed shapes in miniature and made it possible to collate places into three-dimensional models; cognitive maps that could be recalled and decoded at will. In such a way, it was possible to build a neural database of topographical information. The camp, whether it was used for one night or one month, was the hub of a web of satellite features which comprised a known tract of wilderness; a mappable landscape. A valley, an odd-looking rock, an isolated tree, a parting in the stream were all logged and mentally archived, then more accurately placed with each repassing. Some features were already human constructs. Paths, scatters of stone flakes, the carbonised discs of hearths, the piercings of post holes, were all coded for recall. Sounds and smells had their own coordinates: the spot where a waterfall became audible was a mappable place, and so was the olfactory perimeter of a pine wood, its proximity mapped by the sweet scent of resin. Cognitive maps were a primary tool. The dots on these mental maps were the features that bonded humans to their landscapes. They were the places that mattered; the places invested with meaning. Cognitive maps bound trails and camps, views and waymarks into unitary areas and hafted attachments to homelands. Mind-maps prepared the way for domesticating the wilderness.

Better mapped than Britain was the great plain to the east, but the broad lowland leading back to the continent was diminishing.

Melting ice and the thermal expansion of the oceans were raising sea levels season-by-season, attacking Doggerland's coastlines and flooding its productive wetlands. Sea levels had begun to rise as long ago as 18,000 BC, when the last Ice Age had finally come to an end. The big freeze that had emptied Britain of humans from around 10,700 BC to 9,700 BC had not been enough to arrest the rise of the oceans and by around 9000 BC, Ireland was severed from Britain, and the plain linking Britain to the continent had lost half of its land area. There had been a time when Doggerland had been bigger than Britain, but now it formed a low watershed connecting continental uplands between the Rhine and the Elbe, with hills between the Thames and the Ouse. You could walk from the source of the Rhine to the source of the Thames without crossing sea. At its eastern end, Doggerland may have been a neck of land no more than sixty or so miles wide. In the south, Doggerland tapered to meet the head of a single, gigantic gulf – the Channel – that led west to the Atlantic. In the north, Doggerland's coast was long, convoluted and soft, dominated in its central section by a range of hills – the so-called Doggerland Hills – and by great embayments to each side. One side of the western embayment – the Outer Silver Pit – formed part of Britain's east coast. The rivers, marshes, lakes and estuaries surrounding the Outer Silver Pit were munificent sources of food. The earth system that was destroying Doggerland tide-by-tide was also the main source of its biodiversity: because Doggerland functioned as a dam between two seas, the tides surging in and out of the Channel and the southern North Sea twice every twenty-four hours had an enormous vertical range and so supported a rich and diverse intertidal zone. The Doggerland people probably enjoyed greater returns from their foraging, hunting and fishing than in any area of equivalent size in Britain.

This great plain led – if you were walking west – to chalklands and the Thames. The Thames and its tributaries acted as conduits for incomers, while the chalk offered spectacle, flint and familiarity. Chalk is almost entirely composed of fossils, the shells of minute foraminiferans that died between 135 and 65 million years ago in a shallow sea.

Fossilised as calcium carbonate, they compacted into a soft, brilliant-white stone. Chalk, being young, lay close to the surface. On coasts, it was exposed by erosion as gleaming cliffs, sea-stacks and wave-cut platforms. Inland, its signature was more subtle: hills underlain by chalk tended to curve and swell with verdant, seductive gradients. Chalk caught the eye. It was also the sedimentary rock that bore flint, which lay in thin black seams below the surface and erupted in scatters where seams were exposed by features like sea-cliffs or valley-sides. Chalk was laid like a path all the way from the continent to Britain. It stretched in a subterranean bed beneath the lands of the Seine and the Somme, the Schelde and the Rhine, it lay beneath Doggerland and it lifted into Britain to form the skeleton of the south-east. Chalk under-pinned a peripheral, eastern band of hills that reached their greatest height in the Wolds north of the Humber. Another four ranges of linear, chalk uplands stretched across the south of Britain, converging on a 300-square-mile heartland: the Plain, the most extensive chalk plateau in Britain. Of these four limbs, the longest and widest rolled diagonally across south-east Britain from Doggerland, by way of the Chiltern Hills and Marlborough Downs, to the Plain. Two thinner, but more distinctive bands of chalk – the North and the South Downs – approached as steep-flanked ridges from the east. The stubbiest of the four limbs began near the southern coastal promontory of Portland and reached the Plain by way of the Dorset Downs and Cranborne Chase. The chalklands lived through their rivers. There were many, but three in particular were crucial. One of the rivers had its mouth on the east coast of Britain, another on the south coast, and the third on the west coast. The Thames and the two Avons each led inland to the chalk heartland, the Plain, which formed a tripartite watershed between the three coasts. While ecotonal locations promised the greatest security for a home-base, rivers guided movement. Rivers were ready-made route maps that remained constant from one generation to the next. And they always provided two of life's essentials: drinking water and timber.

From the start, Britain was tilted to the south-east. Land was lower in the south-east. The climate was kinder in the south-east. Rivers were slower in the south east. The causeway from the continent came to the south-east. Chalk-with-flint was a geology of the south-east. Half a year's walking from Britain, Beth Nahrain, lay to the south-east, too. The part of the Fertile Crescent closest to Europe was the region described by the mid and upper sections of the rivers Euphrates and Tigris, and the lower slopes of the Taurus Mountains. The passing of time has given the region a succession of labels: Beth Nahrain ('house of the rivers'); Al Jazirah ('the island'); Mesopotamia ('between the rivers').

The House of the Rivers had lurched from climatic disruption to ecological fruitfulness. The people who roamed the well-watered grasslands and woods in the upper Tigris and Euphrates had moved far beyond mere subsistence and were now gathering periodically at a spectacular hilltop centre to feast and to dance and to raise great stones. Göbekli Tepe (it's unlikely to have been known as 'Potbelly Hill' back in 9000 BC) was no more than a two-day walk from the Euphrates as it described a long bend around the hill. Across the top of Potbelly Hill, work-squads had hacked and carved from the limestone roughly circular enclosures with T-shaped pillars linked by walls and stone benches. In the centre of each enclosure stood a pair of taller T-pillars. One pair topped five metres and was carved with hands, fingers, loincloths and belts. Some of the enclosures had concentric circuits of pillars and walls nesting within each other. In various parts of the sanctuary, an exotic menagerie of scorpions, snakes, aurochs, foxes, boar, ducks, asses and gazelle stared from sculpted limestone. In an age before writing, this was a language shared by all. As many as twenty enclosures may have been constructed across the hill. The scale was literally staggering: the largest eight-metre T-pillars weighed 50 tons and had a volume of 20 cubic metres. Hauling one of these pillars from its quarry across the hill would have required hundreds of people. The organisation, design, measuring and devotion of labour, was on an unprecedented scale. One of the traditions at Potbelly Hill was to abandon an enclosure after use, backfilling it with rubble from the quarries and with vast amounts of animal bones – mainly gazelle, aurochs and ass – that had been smashed during marrow-extraction.

If Britain was the Edgeland, the lands of the Euphrates and Tigris were the Centreland. Potbelly Hill was part of a bigger idea. There were other enclosure sites with T-shaped pillars. It's possible that their primary purpose was to assemble labour from surrounding areas in order that communal structures could be raised. The feasting, dancing, stone-raising, beer-swilling foragers of this fertile region were in an altered state. This was a cult in transition. The people who flocked periodically to Göbekli Tepe were already cropping wild cereals like einkorn and barley and their cooking-ware included pestles, mortars and grinders. Feasting strengthened social bonds and provided an opportunity for pushy individuals to further their social standing. But feasting required food surpluses: wider sources of food, new methods of processing, more intensive production and new techniques of storage. It's possible that places like Potbelly Hill were propagative hubs whose pilgrims played a role in distributing seeds suitable for cultivation.

Beyond Potbelly Hill, the transition was occurring at various other points in the Fertile Crescent. Tools like the axe and the adze for clearing vegetation were in wide use, and people may even have been encouraging the growth and circulation of cereal seeds. In round-houses they ground the grain of cereals between stones and where there was a lack of stone, they built walls from bricks of clay and straw. In places, people gathered together to dwell in communal hubs, or villages. Some buildings were invested with belief. Some had bodies buried beneath their floors, others had the skulls of game hung on their walls. At the spring of Ein as-Sultan in the valley of the Jordan a community of five hundred or more fixed their place in the cosmos by constructing a massive stone wall and an eight-metre-high conical stone tower, plastered inside and out. Every summer solstice the setting sun shared its azimuth with an axis formed by the tower, the angle of its internal staircase and the mountaintop above the village. Lurking in the solstitial shadow of Jericho's tower may have been fear and coercion. Or tower and village may have arisen through a shared belief in communal enterprise. Whatever their founding force, these were physical communities unlike anything in Europe.

Britain, meanwhile, was hardly marked. Seven hundred years after global warming began to uprate northern Europe from Arctic to temperate, human occupation had barely impressed itself on the landscape. Yet the pace of climate change modified habitats with every generation. There was no sense of environmental permanence. Sea levels were surging upward at an average rate of one centimetre a year. Storms at sea could take out tracts of land that had never seen salt water. Birch and pine were spreading towards Britain's furthest regions. As the biogeographical belts migrated, the tundra was colonised by advancing tree-lines. Surviving herds of horse were driven into extinction. And in the south, global warming invited the continent's most inventive species to create the first settlements.

TWO

Wood Land

9000–8000 BC

Seven hundred years or so after the climate began to warm, a group of people reached a level river terrace beneath the folds of southern Britain's chalk heartland. The river is now known as the Kennet and it shares its valley with a main road, a railway, a canal and a national cycle route, all heading west by the easiest route. The Kennet's early travellers may also have been following topographies of least resistance, along the Thames through lands rich in food. Two days or so upstream of the estuary, the waters parted and one river swung north through a notch between two ranges of chalk downlands. The other river continued rising westwards.

In the centuries since the thaw, the Kennet had settled into its bed. The streams of calcium-rich water pouring down from the surrounding chalklands had thickened the valley floor with calcareous marl, and instead of scurrying through braided, gravel channels, the river now flowed steadily between banks of soil. Each side of the water grew birch and pine and aspen. Where the valley widened to about a mile, a lake and fen created the kind of ecotone they were looking for: adjacent biomes of wetland and woodland, with flint to hand. On and off, this stretch of the Kennet was used by people for many generations. Slain aurochs were brought here, and red deer, roe deer and boar. The

people of the Kennet also took pike and mallard, goldeneye duck and
crane. Beavers and wildcats were hunted for furs. On foot-smoothed
earth, they cut wood and scraped hides; bored and whittled bone.
And they prepared mastic, a resin-based product that could be used
as a gum for hafting stone blades to wooden handles. These were thoroughfare sites, where the various limbs of chalk upland converged on
a major cross-Britain, east–west through-route: up the Thames and
Kennet, then over the watershed and down the Avon to the largest
river in western Britain, the Severn. Repeated use formalised the
riverside sites and requisitioned them as 'places'. At their busiest, the
glades of the Kennet were suffused with woodsmoke, chatter and
the knock-tinkle of knapping. But evidence of their presence on the
land would be lost to the human eye (and buried beneath a sewage
works). The activities of so many for so long would be remembered
as a layer of hearths and bones together with 16,000 flakes and spalls,
1,200 blade-like flakes, 280 cores, 285 microliths, 17 axe-adzes, 130
scrapers, 15 awls, 6 hammerstones and a variety of other flint implements; a grand total of 18,402 pieces of stone, of which just 3.5 per cent
were 'finished'. Countless more artefacts lie undiscovered beneath the
Kennet's alluvial carpet.

By the beginning of the ninth millennium BC, the climate had
warmed so much that summer and winter temperatures were comparable to those of Britain in the early twenty-first century. Hearths,
flint-scatters and 'places' were multiplying. Foraging had never been
better. Trees and shrubs could be harvested for nuts and berries,
tubers and roots, while the sea's fringe offered eggs and seaweed, birds
and shellfish. Inlets teemed with fish and wildfowl. A new tree was becoming increasingly common. It thrived on chalk and limestone and
was hardy enough to survive on exposed hillsides and at altitudes of
2,000 feet. Each autumn its branches were hung with golden tails and
clusters of nuts that could be eaten raw or roasted in a fire. Hazel nuts
were a woodland superfood, high in monounsaturated fat, minerals
like magnesium, copper and iron, and in vitamin E. They reduced
blood pressure and inflammation. When stored cool and dry they

could last for two or three months. Britain was an ideal habitat for the hazel. Large mammals such as aurochs and elk, deer and boar were sources of red meat and raw materials like bone and hide. Extreme climate change had created one of the most propitious environmental niches the planet has ever known.

Numbers in Britain swelled while Doggerland shrank. Sea levels were rising at their fastest rate in 20,000 years and vast tracts of what had once been land now lay beneath 70 metres of salt water. Generations of Doggerlanders had been climate-change refugees. Many trekked west in search of new lands. To immigrants, Britain appeared on the horizon as a ridge of upland; high, dry and safe, and as it turned out, thickly scattered with flint. One of the bands that approached Britain across the northern wetlands of Doggerland was drawn by a notch in the upland wall between Scarborough and Filey. Behind the notch was a vale some four miles wide and partially occupied by a lake, the remnant of a much larger body of water that had formed when the ice sheets of the previous glacial maximum had bulldozed a barricade of moraine across the mouth of the vale, blocking the rivers that used to flow east. A huge reservoir had accumulated behind the moraine until the meltwater burst out, southwards, along the valley of the Derwent. A shallow, residual wetland remained. By the beginning of the ninth millennium BC, the lake had shrunk to three miles or so in length, with a ragged shore of peninsulas and embayments. A couple of small islands floated out near the centre of the lake which was fringed with a waterlogged 'carr' shoreline of reeds and sedges, thickened with willow and aspen. Water lilies grew further from shore. The deepest part of the lake was a little over seven metres but in many places a man could wade to his waist. Back from the lake on drier ground spread woods of birch and pine, with an undercroft of waving ferns. Spaces in the tree cover were bright with light-filled grassland. Visible in the distance to the north of the lake, rugged ribs of hard limestone clambered to reach 1,000 feet. To the south, a wave of chalk swelled towards the 800-foot ceiling of the Wolds. For those who had come from the east, this place was reassuringly familiar: safely raised above the rising seas and swaddled by embracing uplands, its fresh-water mere, wetland fringe and accessible woodlands were not unlike the landscapes disappearing on Doggerland.

Boar could be heard in the woods and scores of waterbirds fed at the lake, among them the common crane, red-breasted merganser, common scoter, dabchick, red-throated diver, great crested grebe and brent goose. In the grasslands above the lake grazed elk and aurochs. The Carrs people lived well enough to have had the time to whittle pieces of shale into disc-shaped beads, which could be bored for threading. In their self-sufficient world, tool-making was one of many home-based rituals that solidified bonds between person and place. The crafting of a pointed barb commenced at the water's edge because the antler had to be continuously wetted. Using a flint blade, two parallel grooves were cut in the antler, and then a splinter extracted from between the grooves. The remainder of the process did not require wetting and could be undertaken at an alternative 'dry' location, where the splinter was worked with flint into a tapered point and then serrated along one side with angled barbs. The barbs prevented the point from withdrawing from flesh, which increased blood-loss and hastened death. Barbed points like these could be fitted to arrows, spears and harpoons. They were as effective on pike as they were on elk. Alongside tools for taking birds, land-animals and fish were others for working with materials. Scrapers for cleaning hides could be made from knapped flint or from the bones of aurochs; pointed bodkins for piercing leather were made from the foot-bones of elk.

The Carrs people were also equipped with weapons of mass construction: tools for working timber, soil and stone. Crafted from elk antler, the pick was pointed at one end and pierced with a hole at the other end to take a handle. Swung with force, it could penetrate the hardest of soils and even chalk bedrock. Also formed from elk antler, the mattock had a cutting-end rather than a point, and could be used for breaking softer ground and for soil shifting. It would have been ideal for digging pits, hollows and postholes. The lake people were also carpenters. The typical timber-working axe was made by striking a slice from one end of an oval piece of flint. Axe-heads could break or blunt fairly quickly, but it took a few seconds to resharpen the stone by striking off a flake to form a new blade. Among the wooden implements fashioned here was a long-handled piece of birch with a flat blade that could be used for propelling a raft or canoe. And at some point during the occupation of the lake shore, people felled trees and then split the trunks and branches by hammering wedges into the grain. At least one piece had been split twice, forming a versatile building component, the plank.

Britain's earliest known house stood back from the lakeside. It was roughly circular and had been constructed around a scooped hollow in the ground, with a frame of embedded posts. The above-ground structure is conjectural. Thinner poles had probably been bound with twine – possibly made from nettles – to the main structure, and the whole dwelling might have been clad in reeds from the lakeside, or with hides. Inside, it measured no more than a couple of metres or so. In terms of scale, it was the kind of dwelling that might have been used for sleeping, at a site where most tasks were undertaken in the open. It's unlikely to have been the only house at the lake. People probably came to the dwellings on the lake repeatedly, at various times of the year. It's not possible to say whether they were occupied continuously or episodically. The idea of 'settlement' was becoming more defined.

The Carrs was as good as it got in ninth-millennium, mid-latitude Britain. It offered open water and wetland, edgeland, scrub and woodland. Doggerland's aquatic plain was close by and the adjacent hills provided upland hunting, foraging and flint. Mosaic habitats like this produced the highest densities of game animals and edible plants. Various sites around the lake shore were visited repeatedly over a period of 300 years or so and the formalised human landscape of clearings, camps, hearths and tracks was developed and reiterated through repetitive stays. But the Carrs was more than a coveted ecotone. Physically, the lake was progressively removed from the wilderness and woven into the traditions and beliefs of the people living around its shores. It became part of their society, the bonds between people and place strengthened through acts and rituals of association. People here had worn decorative beads and had left a triangular pendant crafted from a thin piece of shale, which had been incised with short lines running tangentially to longer lines – a pattern also used on the far side of Doggerland. Head-dresses had been fashioned from the skull frontlets of red deer; shamanistic ornamentation that might have been used during waterside ceremonies at the lake – or at the lake's outflow where the water underwent a transformation from still to living at the start of its journey to the sea. Gatherings such as this would have been central to sustaining shared beliefs. Tools worked from animals were returned to the waters of life, the wooden hafts and handles removed and the scrapers and axes of bone and antler cast into the lake. Votive offerings nurtured the value of the lake as a place, and also the value of the artefacts. For the successive generations who settled around the lake shore between the hills, these were much more than the bivouacs of passing foragers. Lives were played out at a location imbued with familiarity. People fell in love here. To many, the Carrs was the place that mattered most.

The lake people, with their children, elders and domesticated dogs, left more than footprints. They may have been no more than periodic callers, staying for a few days or weeks each visit, but their use of the Carrs created an artificial opening in the vegetation on a landscape-scale. Not only did the to-and-fro of daily routines flatten an area of the lake shore, but they were felling trees and setting fire to the lakeside reeds. The burning occurred every year, improving access to the water's edge, opening sight lines for hunting and encouraging fresh plant growth. The building of homes and hearths left a mark too, and so did the construction of a timber walkway. This had been a considerable project, and must have required the combined enterprise of many people. Split timbers were laid with their flat side uppermost on the waterlogged ground of the lake edge to form a dry platform (or several walkways) that extended for as much as 30 metres. Timbered surfaces like these could be used for improving access for canoes, for fishing or for rituals. For the time being, however, the human imprint was temporary.

Today, the lake is farmland. From the edge of a lay-by on the thunderous A64 between Scarborough and York, you can look along a drainage dyke that cuts through peat beds where the lake used to shine. Beside the dyke, the soft mud is printed by the passage of archaeological pilgrims. There's no visible commemoration. Just the rectilinear fields from a later age. The site of the lake can also be reached a couple of miles away, where a narrow track slopes down from the village of Flixton to a bridge over the dyke. More fields. My favourite view of the place known in archaeological circles as Star Carr is from the bulge of chalk to the south, where a slender byroad climbs steeply above Flixton to the crest of the Yorkshire Wolds. There's a wide verge up here, and the adjacent fields are speckled with flint. Off to the right is a panel of sea covering Doggerland. And ahead, in the middle ground between the Wolds and the Moors, is the vale. Back in 9000 BC, the lake would have been mirrored in blue within its fringe of golden reeds and clearings, wisps of smoke rising from darkened patches of trampled ground. On the far side of the vale are the roofs of industrial estates and the shining pipes of Seamer Carr household waste recycling centre (it was an archaeological dig prior to its construction that led to the discovery of bones belonging to Britain's last known wild horse, radio-carbon dated to 10,007–8723 cal BC). The land may have forgotten the lake and its settlement, but Seamer remembers, the *sæ* and *mere* meaning lake or marsh pool.

The Carrs appear to have attracted an unusual density of activity. Far more typical of the time was a solitary habitation much further north, on the Firth of Forth, where a structure with a staked frame was erected on a bluff overlooking the sea and a small river. It was one of those spots that offered the bare necessities: fresh water, timber for fuel and access to a tidal shoreline for fishing and foraging. The occupants of this vantage point left chert tools, hazelnut shells and food waste. Within a decade, there was little to see of the site, the remains of the shelter recorded as twenty or so stake-holes in the acidic soil. The pioneers who came to the Forth had reached Britain's most daunting internal border. Because of the way that northern Britain rebounded upward once the burden of ice had been removed, the sea level in these latitudes was some ten metres higher than it is now. But this part of Britain was pinched on both sides by two gigantic seawater inlets: massively extended versions of the Firth of Forth on the east, and the Firth of Clyde on the west. Sea reached so far inland from both coasts that it almost met in the middle. An isthmus that may have been as little as ten miles wide connected southern to northern Britain. And that isthmus was low lying and boggy. The overland route from the south to the north of Britain was a navigationally tortuous zig-zag between mountains that stood clear of the water, and it involved multiple, hazardous river crossings. The gulf of water in front of the structure on the Forth was one of the largest sea inlets in Britain. Further progress north could only be achieved by a perilous voyage by boat or by a very long and difficult overland detour in search of the land-bridge connecting the two parts of Britain. The Forth–Clyde constriction may have impeded northward progress for some centuries, effectively forming a border between the known and the unknown.

The spread of human occupation northward and westward suggests that the thousand years following the thaw saw a progressive increase of population and pressure to try new lands. On an island of diverse habitats, inequality of resources was a fact of human life. Some groups were better off than others. Fewer resources meant covering more ground to find food and having less fireside-time. Locked into a cycle of foraging, hunting and eating, groups in areas of scarcity had little time – or will – to invest in any particular tract of land. But groups operating in areas with an abundance of edible resources were able to accumulate food and live in greater security. These better-off groups tended to be less mobile, with stronger social structures, and more developed ideas of time and space – and of territory. Territories may have had markers, but they had no artificial boundaries. Some territories overlapped; others were seen rather than trodden. A view from a bluff might 'belong' to a group who had no physical relationship with it.

Far from the Forth, an unusual range of hills provided the stage for Britain's earliest known cemetery. The Mendips were a land apart from the chalklands. Where chalk tended to produce smooth, curvaceous hills, the harder limestone of the Mendips had created an upland that – more than any other in southern Britain – appeared to have been cast by a cosmic architect. The range was roughly twenty miles long, five miles wide, and loomed more than 300 metres above the approaches to the West Country. Its upper surface was a plateau that felt completely removed from the trafficked world of the lowland. Up here, there were stands of birch and pine, and tracts of open grassland shared with bear and lynx, wolf and boar. Sometimes the lower world was obscured with blankets of white mist, islanding the plateau in clear, blue air. It was the boundary of this upland, between the hidden world of the plateau and the veiled lowlands, that held the wonders. The western edge of the plateau was pierced by the most spectacular gorge in Britain: a winding chasm walled with cliffs and anthropomorphic outcrops. It took nerve to venture along the shadowed floor of this gorge, over the tumbled boulders and toppled trees, beneath a jagged-edged ceiling. Another, smaller chasm squirmed through the northern defences of the range.

The water of the Mendips connected two worlds. In places, streams gurgled and clattered between sunlit banks like any normal watercourse, but then they disappeared through voids into the underworld. In other places, water rose to the surface, fully formed as small rivers. And this underworld was accessible. Partings in the rock provided access to cave systems which twisted and turned through darkness more intense than anything imaginable in the outside world. Some of the passages led to echoing caverns decorated with pillars and tendrils which glistened and writhed in the flame-light of torches. Where water had pooled, there were surfaces so still and deep that the living saw themselves reflected with radiance. For those who were temperamentally suited to these mysterious voids, the sense of wonder was overwhelming. And so it was for the people who trod into the dark night of the rock bearing the bodies of the departed. Between 8400 and 8200 BC, a cave on the side of the northern chasm accumulated the remains of over fifty people. The practice of interring certain individuals within caverns was a tradition that went back thousands of years, pre-dating even the glacial snap that had driven people out of Britain shortly after 11,000 BC. For some at least, caves were places associated with ancestral ritual.

Functionality was being imprinted on landscapes. Some locations were suited to hunting, others to dwelling. The dead had their places. In Britain's far west, a ridge in what is now Pembrokeshire became a production centre. Sea-level rise has turned the site into a cliff-girt headland, but back in the ninth millennium BC the people who chose this spot would have been gazing across a broad coastal plain which separated their ridge from the tideline by three or four miles. It seems they were specialists in exotic stones. The beaches hereabouts were littered with geological strangers that had been carried from the far north on long-gone glaciers. Nearly all of the stone being worked on the ridge-top site was flint that had been lugged up from the beach, but these knappers, chippers and borers had also collected bits of greensand, rhyolite and tuff. And they had visited an outcrop fifteen minutes to the east, where they'd collected pieces of blue-grey shale then worked them into small oval shapes about 2 or 3 millimetres thick. Using a spike of flint, each piece was drilled with a small hole. Hundreds, perhaps thousands, of these beads were made here and circulated throughout the region. The distribution of beads is the earliest glimpse we have of regional communications; a web of paths and waterways servicing a hub.

Northward, westward and upward they walked. High ground ceased to be a barrier. Whatever reticence the lowland migrants of Doggerland had once held for the wilder heights of Britain, had diminished by the end of the millennium. Scatters of flint were left on the Mendips, on the limestone hills north of the Carrs, on the Pennines and on the Brecon Beacons of western Britain, where a plateau some 480 metres above sea level was visited by hunters at some point between 9000 and 8000 BC. At the time, many of these places were just above the upper limit of woodland, where lines of sight were longer and movement easier. In these upland zones, lakes were a particular draw. Down south, repeat visits to many lowland sites had embedded them in ancestral itineraries. Over 800 years after the reindeer hunters on the Colne had hauled the severed legs of their prey to the edge of the floodplain, a far larger group of hunters settled on the same rib of higher ground to extract the fat, grease and marrow from at least fifteen red deer and a couple of roe deer. Further west, the habitation zone on the Kennet was still being visited over a thousand years after pioneers had lit the first fire on the valley floor. Continuity on this kind of timescale introduced an additional dimension to place-making, as if the return to sites that were 'ancient' conferred some kind of value that was historic rather than purely geographical.

Two thousand miles from the Thames, life in the Fertile Crescent was on a different trajectory. The process that had begun with a societal shift towards gatherings at cult centres and dwelling in villages had entered a new phase. Food production – the driving force behind the creation of artificial habitats – had a competitive advantage. In the long term, the availability of wild animals was declining, while climate change was making wild cereals more available. People were developing food production technologies such as the sickle, baskets and grinding slabs, all of which made cultivation even more efficient. And then there was the feedback loop linking food production and a rising population: as numbers rose, people were forced to provide more food, which was most efficiently achieved by becoming more sedentary, which reduced the time between births, which increased the number of mouths demanding food. Food production and population growth were locked into an autocatalytic spiral. Paradoxically, food producers were less well fed than people living by foraging and hunting because their populations increased faster than the availability of food. Food producers were forever committed to improving yields and finding new sources of food. Per hectare, food production yielded more edible calories than foraging and hunting, and so the solution was to look beyond the exploitation of wild cereals. By bringing edible plants and selected animals under human control, yields could be increased, and more mouths filled. It took perhaps a thousand years for the Mesopotamian inhabitants of the House of the Rivers to progress from the harvesting of wild cereals to the cultivation of domesticated crops. At some time after 8500 BC, the animals that submitted first to controlled production were sheep and goats.

The effect on the landscape was dramatic. Within a few centuries, architecture took a new turn with the appearance of rectangular buildings. And these were not loosely assembled drystone structures, but buildings raised with a variety of materials and techniques: walls were built from stones that had been mortared and filleted with smaller stones. Internally, floors and walls were sealed with smoothed plaster that had been made by incinerating lumps of limestone and mixing the powdered lime with water. Some houses were built on levelled terraces; some had two storeys. By 8000 BC, it was possible to walk along a valley past plots of wheat and grazing sheep to a small town of rectangular buildings and courtyards gathered about a circular space and a communal building – an early kind of 'power house'. It was a new world. The fabulous stone enclosures of Göbekli Tepe were abandoned and smothered by backfill of quarry rubble, grit and old bones.

Britain had trees. By the end of the millennium, elm, oak and lime were growing in the south. Hazel trees had become the most common species at the abandoned lake site of the Carrs, and birch and pine had crept beyond Loch Ness and over the fjordlands of the west coast to Britain's northern reaches. Woodland moved during the course of a human lifetime; on good soils in southern Britain, oak migrated at a rate of 350 to 500 metres a year and pine at 100 to 700 metres a year. In the far north, a variety of factors that included climate, soils and topography reduced the rate of advance but two of the trees most useful to human pioneers were widely spread by 8000 BC. Pine, which burned so well in the hearth, was common and so too was hazel, the provider of nuts which could be hoarded and stored for later use. For the first time since the ice had melted, Britain was largely tree-clad. Branches sang with finch and thrush. The eagle owl blinked and beavers colonised wooded watercourses.

People were few and far between. There was no universal pattern of occupation. Bands of perhaps twenty-five or so moved seasonally between complementary habitats: a winter base camp might be established on the coast or on a lakeshore or close to various hunting grounds which would be visited through the colder months as required. In the warmer months, some or all of the band would move to a summer hunting camp on the uplands where they could predate on grazing game. There were many variations on the basic model. Some bands may have spent most of the year within an area no more than ten miles across; others may have roamed seasonally as far as fifty miles. Woven into the foraging and hunting were other imperatives such as encounters with prospective marriage partners in neighbouring bands, and visits to sacred or ceremonial sites. These seasonal migrations along familiar routes established homelands or territories as physical areas, described on the ground by paths and hearths, stone-scatters and camps. Britain's post-glacial pioneers wafted through wood and glade like wraiths. The human imprint on the wilderness was little more than that of the dam-building, river-blocking beavers or heavy-footed elk. Left for a year or two, hearths, stone-scatters and post-holes were rubbed-out by regrowth. Reeds and trees withdrew as areas were fired or coppiced, then advanced again after people moved on. There was a disparity between the recoverable impacts caused by humans, and the wholesale transformation being wrought by earth systems.

Climate was the cosmic thunderclap. Although solar insolation had been kind, and would not peak until the period between 8500 BC and 7000 BC, it seems that there was a deleterious climatic 'event' in around 8200 BC, accompanied by a return of the cold Siberian high. The cause may have been a dump of cold, fresh water which slowed or stopped the North Atlantic thermohaline pump. Vast amounts of ice debris floated on the northern seas and temperatures in Britain may have plummeted for a while as the warm waters of the Gulf Stream ceased to brush western shores. For people dependent on the seasonal harvesting of waters, woods and glades, the effects would have been disastrous.

The ice rafting may have lasted a century or two and by the turn of the millennium the climate in Britain appears to have been back on an amenable track. In terms of human occupation, it was still a relatively empty land. Population density lay somewhere between 0.02 and 0.1 per square kilometre, with more people exploiting niches in the south and east than in the north and west, and coasts and estuaries being more populated than interior uplands. In total, Britain was probably supporting somewhere between 5,000 and 20,000 people. But a threshold was about to be crossed.

THREE

Tsunami

8000–6000 BC

The house at Howick Haven looked away from a lost land. It was built
in around 7850 BC with its door facing west. Behind it gleamed the
ocean that had taken so much of Doggerland. In front of the house,
the gentle green undulations of mid-latitude Britain rolled towards
the Cheviot Hills five miles away. The site had been carefully chosen.
Immediately to the south and west, a deeply-incised valley formed
a perimeter; to the east, the hill dropped sharply towards the beach
a few hundred metres away. Only from the north was there an easy
approach. The house was big and had been built to last. Sturdy,
35-centimetre posts set in a rough circle supported a wind- and rain-
proof conical roof that had probably been thatched. Inside, a sunken
floor stretched some six metres across – large enough to accommodate an extended family group of half a dozen or so. In the centre of
the floor was a hearth. Outside the door lay an ecotonal larder: boar,
deer and aurochs on land; pike, perch and dab in the river; herring,
salmon, sea-trout beyond the beach. Puffins, geese, duck and swan
added variety. Eggs could be collected from the cliffs; scallops, lobster,
mussels, crab, winkles, whelks and limpets from the shallows. Seals
were taken. In nearby glades grew fungi, fruit, berries and hazelnuts.
Furs could be sourced from the local population of fox and beaver,

wolf and bear, and nettles used for making cords, netting, clothing and belts. Fresh water came from the stream below the house, stone from the beach and timber from surrounding woodland. The materials left by the householders included 13,000 flints and several hundred thousand charred hazelnut shells. There were also pieces of micaceous sandstone and haematised shale – earth pigments that could have been used to create colours on a spectrum from yellow through to red-brown. Since these ocherous pigments could not be found in the local geology, they had been collected elsewhere and brought to the house. Ochre had many applications, from body art and decoration of ornaments, to healing and processing of hides. It was also a fixative in mastic and may have been used to fabricate tools or to seal the hulls of inshore boats. Land and sea fed successive generations here. The building was refurbished twice and stood for two hundred years or so. The house lasted longer than the people who first built it, so it fulfils the definition of a 'permanent structure', the first established example in Britain.

Eastern Britain had been a refuge ever since the seas rose. The lakeside settlement at the Carrs had been a tactical, upland setting beyond the reach of Doggerland's greedy tides. A thousand years later, those tides were still rising and much of Doggerland had surrendered to the expanding North Sea. Britain's east coast extended every decade. Between Doggerland's hills and eastern Britain, the great forked gulf of the Outer Silver Pit had enlarged so much that the land route from the continent had narrowed to a low-lying isthmus which rose gently to meet Britain between the Ouse and the Thames. In the sense that it offered any sense of permanence, Doggerland had ceased to be a 'land'. The house at the Haven may well have been constructed during an episode of relocation by displaced people – climate refugees from Doggerland who had trekked west to safer ground, along with their narrow-blade toolkits and distinctive traditions of house-building. Stone-scatters suggest that the house at the Haven may have been one of several sharing its coastal hill. And fifty miles further north, just beyond the mouth of the Firth of Forth, a virtually identical dwelling had been built slightly earlier, in about 8000 BC. The house at White Sands also faced away from the sea and had been sited to take advantage of a propitious ecotone of woodland and coast. With their sunken floors, cooking hearths, capacious interiors and robust construction, the post-built houses at the Haven and White Sands introduced a new phase of settlement in Britain. Their presence may have faded from sight (one of the house sites has partially toppled down a receding cliff, and the other has been overtaken by limestone quarrying for a cement works), but the principle that a strategically chosen house-site would be rewarded by the possibility of permanent settlement was established. The house at the Haven was one of the founding structures of Britain's anthropic landscape: a permanent dwelling large enough to accommodate an entire family, through four seasons, with space enough indoors for a range of tasks, including cooking.

Ahead of the house-builders, the far north of Britain was dotted with 'places'. North of the Forth, people had left flint microliths on the sheltered shores of the Firth of Tay. Small, symmetrical, pointed and tanged flints used as arrow heads told of visitors to Orkney. Travel this ambitious can only have been possible with seagoing boats, both for reaching islands and for bypassing convoluted and frequently sheer sections of mainland coastline. The simple dug-out required elementary carpentry skills to construct and was ideal for lake and river, but it was also heavy and prone to capsizing in waves. For beach launching and open-water crossings, lightweight, buoyant, high-freeboard vessels were safer and quicker, though far more complicated to build. With practice, thin, pliable saplings such as willow or hazel could be bent and lashed into a symmetrical frame and then covered with hide, which was then sealed. Propulsion would have been by paddle and navigation in these northern seas depended on eyeballs; on keeping sight of land.

During the eighth and into the seventh millennium BC, the outer ripples of hominid colonisation reached the peninsulas and islands of Britain's north-west coast. The sites chosen for their short stays or seasonal settlements were on – or close to – coastal havens like sheltered beaches and bays. Not too far from the mouth of the Clyde, deftly knapped pieces of flint were left on three remote peninsulas tenuously clinging to the mainland, now known as the islands of Jura, Islay and Tiree. North of Tiree, they reached the mountainous island of Rum and used a site by the shore at the head of the island's only sheltered loch, Loch Scresort. It's likely that Rum became a site of special interest. On a beach near the western tip of the island a very unusual rock could be found. Chalcedony came in many colours, ranging from white and blue, to green, brown and black. The Rum chalcedony was dark green, flecked with red, and it outcropped on a mountain above the beach, where millions of tides had smoothed fallen chunks into rounded pebbles. The chalcedony pebbles were taken to the Loch Scresort site where they were worked into blades and hammerstones. The people using the site also worked with flint they had found, and agate and quartz, too. The place by the loch was used on and off for the best part of 4,000 years. Eventually the soil here would be scattered with over 100,000 pieces of worked stone, along with burnt hazelnut shells and various pits and post-holes.

The dispersal of humans throughout the archipelago was a measure of Britain's graduation from post-glacial edgeland to agreeable long-term habitat. Mean July temperatures had reached 16–18°C, and mean January temperatures had risen to 5°C. The sun-filled spaces between woodland had evolved into zones rich in biodiversity. Dense stands of mature trees formed the backdrop for more open foregrounds of younger trees, saplings, shrubs and gardens of hawthorn, blackthorn and holly. Lakes, pools, marshes and outcrops of rock marked further spaces in tree cover. Herds of aurochs, red deer and other grazing herbivores helped to maintain these wood-pastures by eating the leaves at the edges of clearings. Skylarks and meadow pipits nested in grazed clearings, and the corncrake and crane were there, too. Occasionally, nature would add to the diversity of the wood-pasture by tearing apertures in the tree cover. Sudden blasts of wind – wind-throw – could take out a clump of trees in one breath. An extreme storm could flatten millions of mature trees in one night. Lightning strikes could also punch holes in the tree canopy or even start a wildfire if a flammable tree like pine was struck. Once gaps had been torn in the woodland, the spaces became weak spots, exposing trees whose root systems were unused to the physics of leverage. Clearings in the tree cover were key niches for humans; chinks in the woodland armour. And so were the great tracts of chalk upland where grass had directly succeeded the tundra and where trees were fewer.

It was on the chalk heartland of the Plain that a second type of founding structure made its earliest appearance in Britain. With its billowing grassland and scatterings of pine and hazel, the Plain was lighter and more open than much of Britain. You could see further. There were aurochs for the taking. There were springs and brooks of clear water. Strewn across the Plain were strange, gigantic slabs of grey stone which were extremely hard and bore no resemblance to the softer, white chalk. The slabs were siliceous sandstone that had formed when concentrated silica in the ground water had cemented sands overlying the chalk. When the uncemented sands were washed away, the massive chunks of hardened sandstone – sarsens – were left lying on the surface like petrified corpses. Above the Kennet, entire slopes were scattered with sarsens, some of them enormous. What role these great stones played in the cosmologies of the Plain's people can never be known but they're likely to have been woven into myths and stories and to have performed in events such as burials, sacrifices and other rites. They may have been named. Enduring relationships developed between people and place; acts of association weighted with territoriality. There were other strange forms on the Plain, too.

Up near the centre of the plateau, a couple of miles west of the River Avon, an area of grassland was patterned with long, low, parallel ridges. The ridges were a topographical memento from the Ice Age and had been caused by the successive freezing and thawing of water as it dribbled down gullies draining the plateau. Each time water in the gullies froze into ice, it expanded and prised away fragments of chalk. The gullies became progressively deeper and wider. In the millennia since the ice had melted, the gullies had become covered by a thin layer of soil and turf, but they were still visible on the land surface as ribs in the grass. And they happened to align with the solstice axis of the midsummer sunrise and midwinter sunset.

The solstitial ridges in the grass were well known to the people who had established themselves nearby. Fifteen minutes' walk to the west, in a nook of the Avon's valley, lay a home-base. For most of its course across the chalk heartland, the Avon was confined between bluffs, but at around the halfway mark, the river took a pronounced eastward meander as it was thrown off course by a promontory. There was flat land here, and a place where the river could be forded and – hard under the promontory – a cluster of freshwater springs. It was secluded, sheltered from the prevailing south-westerly winds, invisible from the open country above but easy to find for those approaching from the east because it lay beneath a long, narrow hill that pointed like a finger at the promontory and its meander. There was wood for fires and plenty of flint which could be baked in embers and then used for cooking and heating water. People based themselves here, on and off, for at least 3,000 years, from around 7500 until 4700 BC. Some places have a timeless appeal (7,000 years after aurochs were dismembered here, people built a massive, banked enclosure on a bluff above the spring). So much stone was knapped here for knives, arrows and scrapers that they left 10,000 pieces of struck flint, together with several kilograms of burnt flint and a carpet of bone.

While aurochs were being brought to ground near springs on the Avon, a group of people – perhaps the aurochs-hunters themselves – created a striking landmark close by on the Plain. Apparently drawn by the parallel ridges in the grass, they contributed their own alignment to the location by excavating several pits in the chalky soil and levering into them sizeable sections of timber. The form of these uprights is open to conjecture: they may have been upturned tree stumps, roots splayed skyward in the manner of a site that appeared later on the coast of eastern England. Or they may have been the trunks of trees that had been shorn of branches and foliage and erected as columns above the Plain. Whether they stood tall or short, the timber totems had diameters of between 0.6 and 0.8 metres. They were selected to impress. Like the ridges on the ground, the trunks guided the eye towards springtime sunrises in one direction, and towards autumn sunsets in the other. The longest and the shortest days of the year were transitions with profound meaning for those who lived beneath open skies; solstitial waymarks in the divisible cosmos. Far to the north, another artificial landmark was raised at some point between 8000 and 7500 BC, when a group of people dug a set of pits in the valley of the Dee, some fifteen miles from the sea. The seven pits were in an approximate line and at the time of their excavation, the valley was wooded with birch and hazel. The pits do not appear to be related to any practical purpose (despite twenty-first-century claims that the alignment represents the world's oldest calendar). Millennia after the sylvan totems were raised on the Plain, the location of their post-holes would be marked by white-painted, concrete discs among the bays of the Stonehenge car-park, by which time landmarks like these had acquired the generic label 'monument', derived from the Latin word for remind, advise, warn, instruct or foretell.

Cosmic architecture was of course nothing new. Sacred places of various kinds had always sustained people through the highs and lows of the seasonal round. But creating them artificially, at scale, from the ground upward, was an enterprise that demanded devotion of time to tasks that rejected the principle of least physical effort. Digging pits, felling trees and raising totems did not immediately provide shelter from the next storm or feed the family. To engage with activities not directly related to subsistence was only possible at times of surplus (or perhaps at times of extreme need, when other strategies had failed). Whatever their reason for being, monuments were the locus of belief. The many moving parts of land and sky connected destinies and the repositioning of some of those parts may have been a means of altering outcomes. Monuments would also have provided some of the social mastic that bonded communities. They were perhaps the neutral spaces where disputes could be resolved and alliances melded. Perhaps, too, they embodied the human desire to enrich the experience of existence through shared places; to commemorate or celebrate social cohesion. They may have had an astronomical function. Memorising the movements of celestial bodies could only be achieved by repeated visits to the same location and it's possible that monuments were focal points for observation, whose records would be handed down through successive generations. Whatever the rituals and practices carried out in these places of veneration, they were socially relevant and they were statements of shared ideology. Whether they were the subject of congregation or convocation or pilgrimage, there was a shared belief in the quality of the location and its suitability as a communal expression of heightened devotion. Monuments had a fixed, cosmic setting and their scale and form was related to the management of human experience. Engagement with monuments may have involved sounds, colours, fabrics and objects and their form may have changed through time as they were adapted to evolving beliefs. As physical presences, monuments were 'performers' on the landscape. They joined houses and camps and hearths and tool-working sites as real, terrestrial places which could be revisited within the infinite labyrinths of the human mind.

Two thousand years after the glaciers began to slip and liquefy, this was still a warming land. Surviving herds of reindeer drifted northward into extinction, the last-known animal dying near Cape Wrath in around 7400 BC. The couplet of cycles arising from the wobble of the earth on its axis, and from the way the shape of the orbit rotated, had a way to run in their 19,000- and 23,000-year durations and continued to increase the amount of solar warmth – insolation – reaching the upper atmosphere. Combined with the amount of incoming light being reflected by the earth surfaces, and with sources of greenhouse gases like methane and carbon dioxide, the overall result was continued global warming. But by around 7000 BC, the precession of the axis and precession of the equinoxes had reached a turning point in their cycle and solar insolation had peaked.

Climate change has a long tail. The beast that unlocked Britain from the Ice Age had knock-on effects measured in millennia. The coast was especially unpredictable. In parts of the north, the land – relieved of its burden of ice – was rebounding upward faster than sea levels were rising. The most extreme sea-level fluctuations were occurring on the coasts closest to the point where the centre of the vanished ice sheet had been at its thickest and heaviest; here, forty miles north of the Clyde, mountains now rose above tree-filled valleys. On the east coast, sea levels in the Firth of Forth had fallen an astonishing six metres in the 2,000 years since humans set foot in Britain, but in around 7500 BC, the rising volume of water in the ocean overtook the rebound of northern Britain's mountains and coasts, and by 7000 BC, relative sea level in the Forth surged upward by some four metres. Over on the west coast, sea levels rose at similar rates around Arran, Islay, Jura, Oronsay, Colonsay and Mull – all used for fishing, foraging and hunting. Established coastal sites that had been selected far above the high-tide line, were wiped out by the rising sea. Promontories became islands as sea levels isolated Shetland and Orkney and further south, the Isle of Man and the Isles of Scilly.

In the south of Britain, where seas were rising and land falling, shores retreated with every century. Bays could be stripped of sand and estuarine marshes could be engulfed in a single season. Woods died as salt water soaked root systems. Grassland was swallowed and coastal camps inundated. Sea levels had risen so far in relation to the land that rivers in the east were backing up. Valleys became clogged with peat and in some areas, dry plain was overtaken by fenland at a rate of ten metres a year. While humans felled a few trees here and there, the sea took out whole forests on the coastal plains of western Britain. All the way from the Mersey to the Solway Firth, land was disappearing. But the really huge losses of land were occurring on the other side of Britain, where the North Sea was creeping ever-southward. The plains and low hills of Doggerland had shrunk to a narrow umbilical. Glacial drift deposits that formed so much of these retreating shorelines were no defence against pounding waves and storm surges: cliffs of unconsolidated drift could be driven back at rates of between one and ten metres a year; by contrast the average annual rate of cliff recession on granite coasts was only one millimetre. Rising sea levels and collapsing cliffs removed from the record innumerable sites that bore traces of human activity. Homes were flooded by rising seawater and then eroded to oblivion. Tremors periodically shook the ground as the earth's crust continued to rebound from being compressed beneath such an immense weight of frozen water.

And then there was the rain. This was a wet land exposed to maritime weather systems. One of the many legacies of the post-glacial warm-up had been the spreading across the land of a layer of fine-textured, clay-like soils. These 'gley soils' were not very permeable and so suffered from periodic or permanent saturation by water. In mainland Britain, about 40 per cent of the soil cover was subject to seasonal or permanent waterlogging. Whenever it rained heavily or for a prolonged spell, areas with a high water table, like land bordering estuaries, or inland basins, were prone to inundation. The changing climate was recorded in the evolution of tree cover. By 7000 BC, the landscape around the Carrs had changed beyond recognition: the birch and pine that had surrounded the lake back in the early days of its settlement had been overtaken by hazel, and then by oak and elm, ash and lime. Elm had travelled to the far north of Britain, beyond the Forth and the Dee and across Loch Ness to the old tundra lands that led to the terminal cliffs in sight of Orkney.

Britain's waking was a saga of forcings and adaptations. Environmental forcing occurred at varying speeds and amplitudes. At ground level, it appeared chaotic and unpredictable, but from a cosmic viewpoint, there were patterns. The long-term rise in temperatures from 9700 BC included episodes of significant cooling. The first 'Little Ice Age' began in around 6700 BC and lasted for centuries. The most likely cause was a pulse of cold, fresh water dumping into the northern oceans. In the eastern Mediterranean region, oaks struggled and explosive storms tore away so much land that a layer of distinctive sediments was laid down on the seabed. In the already-overgrazed grasslands of the Fertile Crescent, population crashed. Icebergs sailed the North Atlantic. The effect on Britain was to shorten the growing season, which disrupted the grazing habits of herbivores and reduced the harvests of wild foods. Winters were cold and stormy. These were bleak centuries. And there was worse to come. In around 6200 BC, a gigantic lake in the north of continental America broke through its ice-dam and disgorged an enormous volume of chilled fresh water into the North Atlantic, which shut down the Gulf Stream and caused temperatures in north-west Europe to plummet. Seasons suddenly became far colder and surface air temperatures dropped by at least 5°C. The intense, dry cold may have lasted for as long as two centuries. Hazel trees, a forager's staple, were hit hard, and so were elm and alder. Britain's woods changed their hue. As whole ecosystems shut down, sea levels rose by half a metre, enough to flood entire estuary habitats and to send seawater coursing up low-lying river valleys. A global-warming aftershock thousands of miles away had changed the shape and nature of Britain in a matter of months. But that wasn't the end of it. Again, the cause was post-glacial adjustment. This time it was a tsunami.

The tsunami had been waiting to happen for thousands of years. In the millennia since the melting of the ice caps over northern Europe, a thick deposit of glacial debris had accumulated on the submerged continental shelf off Norway. When, one autumn in around 6000 BC, the earth's crust made one of its periodic shudders as it recovered from being depressed by ice, a tremor destabilised a vast tract of sloping-seabed debris, which slipped towards the lip of the continental shelf and then plunged into the submarine canyon running parallel with the Norwegian coast. Had the North Sea been calm at the time, a watcher would have seen a wave some three metres in height, moving at great speed. But this wasn't a surface wave like those generated by a storm. It was the tip of a vast column of water.

People living and working on the coastal flats of northern Britain had no warning. As the wave raced towards land, the reduction in depth forced energy upward, turning a rolling 3-metre hump on the sea's surface into a 20-metre wall of grey water. The tsunami struck the Shetland Islands first, thundering up beaches and roaring inland. The vertical run-up on Shetland was 20 to 25 metres. Entire islands were drowned. Minutes later the tsunami reached the Orkneys, then northern and eastern Britain. Anybody on a beach or estuary or coastal plain was swept away. Hundreds – perhaps thousands – of people were killed. The true numbers were never known. Although the wave had lost some of its energy by the time it reached the southern shores of the North Sea, the effect on the low lying isthmus connecting Britain to the continent was devastating. Entire communities were swept away with their dwellings and fishing camps. Shellfish beds were ripped up, fish-traps and huts smashed by the wave. Food stores were destroyed. Because Doggerland was so low lying, the wave travelled far across its surface. No sooner had the first wave receded than a second wave tore across the boiling land surface, smothering it in sand. Woodland that survived the surges died later as root systems succumbed to saturation in salt. For the people of Doggerland, the tsunami was catastrophic. Starvation and abandonment followed. With the loss of life and habitat came a recasting of the ocean's image. The values people attached to the sea and its habits were modified for generations.

The history of these early millennia is archived underground. In the Shetland Islands at a place called Maggie Kettle's Loch, an eroded bank of peat reveals a buried layer of sand some 30 centimetres thick. Mixed in with the sand are balls of peat which the tsunami ripped from the land as it tore ashore. Far to the south, a layer of sand covers a hunter's camp that was taken out by the wave (both campsite and its blanket of sand are now buried by a supermarket in Inverness). In the short term, the death and destruction caused by the tsunami made the coasts of northern and eastern Britain unattractive habitats. And in the longer term, these coasts would play a lesser role in the story of Britain because tides, currents and storms would eventually pare most of them down, along with their traces of human activity. The enduring narrative was being played out inland, beyond the reach of the erosive sea.

While earth systems could wreak cataclysmic changes, humans were only beginning to affect their habitat. The fixing of special places that attracted repeat visits, the coming of durable dwellings and construction of monuments had aggregated effects. And so too did the struggle to create living-space for a population with a long-term trend of growth. Woodland was a necessary source of food, firewood, timber for building and material for tools. But the places where things happened – where animals were easiest to kill, where foraging for fruiting shrubs was most plentiful and varied, where living room and sunlight allowed for the best home-bases – were the clearings.

Clearings were the places where groups of humans gravitated; places with meaning; places associated with territoriality. Burning sedge and clearing woodland created grassland that would attract game, sometimes in herds, which could then be hunted more safely, more quickly and in greater numbers than would be possible in thicker vegetation. Clearings that could be regularly harvested made food supplies more predictable, provided foraging groups with increased resilience to hunger, and reduced the distances they had to travel. It didn't take a huge leap of imagination – a facility peculiar to human beings – to realise that the creation of more open spaces would increase the return from hunting and fruit gathering.

Clearings could be enlarged – or created – by a variety of means. Individual trees would die if you peeled a ring of bark from the base of the trunk. With sufficient manpower, a tree could be toppled by severing its roots with stone axes and then hauling it down with ropes of honeysuckle. Where the tight, woodland canopy had once deflected gales, the gaps created by felled trees allowed the wind to take out the surrounding trees. Then there was fire. The deciduous woodland of Britain was remarkably resistant to fire; a living oak was no more likely to burn to the ground than a wet rock. But birch would burn; its bark was as thin as dried skin and was ideal for starting fires and was so oily that it would ignite when wet. And the flammable oil in needles, twigs and cones could turn an ignited pine into a column of spitting flames.

Clearing vegetation was the first major impact inflicted on land-scapes by humans. As trees and shrubs were felled, the soil that bound their roots was washed into rivers, where it accumulated as layers of alluvial silt. In the south-east, the beds of streams and rivers began to change their hue. In the River Ouse, a few miles from its outfall in the English Channel, alluvial deposits thickened the river bed as silt washed down the main stream and its tributaries. The Ouse was no giant among rivers, but its drainage basin embraced congenial and diverse landscapes ranging from the sun-warmed southern slopes of Ashdown Forest to the gentle levels of the Low Weald. Flints were readily found on the chalk of the South Downs and there were plenty of freshwater springs and hazel woods. The hackers and burners of the Ouse had no idea they were leaving their geological signature in Britain. Far to the west of the Ouse, tree-fellers were also leaving their mark. Around 6500 BC, fire was being used to clear sedge from the peat bogs of the Exe floodplain and around 300 years later, woodland on the uplands of the far west was also being cleared. The effects and the rate of wildwood clearance varied widely across Britain. In low-land areas, tree felling could be followed by regeneration, but in the thin-soiled uplands, widespread felling was followed by degeneration. In the Pennines, where oak, ash and alder had taken root on thin-ner soils, the effects of clearance were striking. As trees disappeared, animals were able to graze the green roots of regeneration and the exposed soils were rinsed of their nutrients and eroded, which led to the development of peat, whose growth was encouraged by the warm, wet climate. By 6000 BC, land modification was taking place at favoured locations the length and breadth of Britain. Many of these locations developed into geographical pressure-points as humans were drawn repeatedly back to the same 'special places'. Continuity of human occupation was the greatest lever the species could exert against the rewilding energies of woodland. Britain of 6000 BC was a place more 'known' than it had ever been, and its people were ex-erting greater-than-ever pressures on selected animal and marine resources. But after more than three thousand years of continuous occupation, the effect of the human population on the land was still immeasurably small compared to the impacts being brought to bear on this post-glacial Atlantic peninsula by atmospheric systems. It had been climate change that had eradicated the tundra and killed off Britain's horses and reindeer; it had been climate change that had clothed Britain in woodland. And it was climate change that was lifting the oceans.

Britain was drowning. Since the climax of the last ice advance in around 18,000 BC, sea levels in the southern North Sea had risen by about a hundred metres and submerged an area greater than modern Britain. On southern coasts, where sea-level rise had been greatest, people had grown used to seeing the blackened, slimy roots and branches of dead trees. Whole forests had been taken out. And this had always been prime food-gathering terrain: low-lying edgelands where hunting, fishing and foraging could be combined. Oral histories must have been saturated with flood sagas. The encroaching sea was most voracious where the coastal fringe was lowest. Where the firths of Forth and Clyde pinched inward to form a narrowing neck, rising waters virtually severed the north from the south. Between 7500 and 6000 BC, sea levels along the shores of the Forth estuary had risen eight metres, filling the valley as far as the mountains in the east, where the Clyde had also flooded, occupying Loch Lomond with sea. Whales swam where once there had been woodland. Ancient foraging grounds lay deep underwater. An isthmus no more than eight miles wide connected northern to southern Britain.

In the south, the rising waters of the North Sea and Channel were close to meeting. The plain of Doggerland, once so rich for foraging and hunting, had been reduced to a besieged land-bridge precariously connecting the ragged peninsula of Britain to the continent. At around 6000 BC, the Siberian high-pressure systems that had sent pulses of cold, stormy winters across the northern hemisphere for 700 years finally moved on, and Britain became the recipient of warmer, more humid conditions. And the rate of sea-level rise at last began to slacken. But it was too late for Doggerland. The huge tides of the

North Sea and Channel ruptured the tattered isthmus and separated Britain from the continent.

In the chaotic phase of coastal adjustment that followed the Doggerland breakthrough, the south-east acquired an estuary. The Thames and the Rhine had once disgorged together into the eastern end of the Channel, but Britain's separation from the continent split the two rivers so that they flowed separately into opposite sides of the North Sea. The new estuary opened eastward, the outrushing waters of the Thames mixing with those of the Rhine in the saltwater straits that now united the North Sea and the Channel.

The south-east was a disaster-zone. After millennia of marine invasion, the reduced plain of Doggerland finally became a physically divided land, with a stump attached to the continent and another to south-eastern Britain. The breach itself was exposed to ferocious erosion. Tidal cataracts of water sluiced to and fro through the widening gap, tearing away chunks of land and pulverising them into sand, mud and gravel that accumulated in submerged banks. And the effects of Britain's severance were felt far from Doggerland. As the waters of the two seas mixed and sought equilibrium through Doggerland's gullet, the tidal range was reduced in the Channel and in the North Sea. Before the breach, the Channel had probably enjoyed the same kind of extreme tidal range – 15 metres – as that of the Severn Estuary. Once Britain became surrounded by a free-flowing sea, the tidal range in the eastern Channel shrank to seven metres. Extensive mudflats and estuaries that had existed for millennia along low-lying sections of the Channel and North Sea coast abruptly shrank. Plankton and their dependent food chain were severely disrupted. Populations of molluscs and fish may have crashed. Critical intertidal foraging land was lost. Some of the greatest losses were experienced halfway along the Channel coast, where the tides moving down the North Sea met those coming up the Channel. This zone of relative marine stasis eventually led to Britain's smallest tidal range, of barely two metres. The creation of Europe's largest island was unlikely to have been the cause of celebration.

FOUR

Island

6000–4050 BC

The island born from the latest bout of aberrations was a continent in miniature, with a latitudinal span equivalent to one-ninth of the distance between the equator and the North Pole. Britain's head was 500 miles from the Arctic Circle and its toes 500 miles from the Mediterranean. On this tall, thin landmass were topographies ranging from barren, thousand-metre shards of rock to teeming, green wetlands. And trees. Later, this climax woodland gained the evocative title 'Wildwood'.

The naked tundra of the reindeer hunters had become incompletely clothed in an undulating patchwork of mixed deciduous woodland which stretched from sea to sea and scaled every upland except the highest mountains of the far north. Every spring, most of Britain was stippled with billions of leaves. The distribution of tree species was uneven and mobile. At its peak (say between 5000 and 4000 BC), the wildwood was a discontinuous cover of local and regional adaptations to soils, drainage, climate, weather events, disease and the attentions of unquantifiable terrestrial organisms. The widest variety of species could be found in the south-east, where lime dominated, followed by hazel, oak and elm. Local variations included pines in the eastern Fenlands, while birch occupied a niche in the wetlands of the south-west and beech on the sands, silts and clays north of the Thames. Much of western and northern Britain was dominated by oak and hazel, but elm and pine and lime grew there, too. In the far north, pines continued to cling to the Cairngorms and other mountain redoubts, while birch still performed a frontier role on the impoverished soils beyond the Great Glen. Within these regional generalities there were numerous variations: hazel and elm dominated the peninsulas of the south-west; ash had been attracted to areas of chalk and limestone, while low-lying fenlands had become colonised by alder. Of the total land area of Britain, around 60 per cent was wooded, and three-quarters of that woodland was deciduous. Around 20 per cent of Britain was grassland, and the rest was occupied by fenland, heath and moor and various shrubs, herbs and ferns.

Marooned on the new island were over 500 million mammals. The giant of the wood-pastures was still the horned, one-ton aurochs. There were over 80,000 of them grazing in lowland herds. The aurochs were vastly outnumbered by around a million boar and an even greater number of red deer. Less numerous, but more dangerous to hunt were over 10,000 brown bears and over 5,000 wolves. Damming rivers the length and breadth of Britain were nearly 100,000 beavers, whose combined effect on watercourses far exceeded any human enterprise. Woods hid 20 million moles and 42 million voles. Trees sang with birdlife. Most common – and found in all types of woodland – was the chaffinch, closely followed by the robin. The deciduous woods harboured by far the greatest number of birds, a diverse spread of species that included the wood warbler, pied flycatcher, hawfinch, song thrush, willow warbler, wren, blackcap, tree creeper, blackbird, goldcrest, chiffchaff, nuthatch and tree pipit. On damp earth near rivers, the secretive golden oriole hid its nests. Bands stalking the northern pine woods would have known the high-pitched *ti-ti-ti-ti-ti* of the shy hazel hen, a bird favoured for its sweet flesh. Patrolling the woods and open ground were a wide range of airborne predators, most commonly the tawny owl and common buzzard, which thrived on the furtive rodents that darted about the woodland floor. Familiar, too, were the sparrowhawk, honey buzzard, goshawk and kestrel. Less frequently seen were the osprey, marsh harrier, the white-tailed eagle and eagle owl. Grasslands trilled with skylarks who shared the air with crane and corncrake, grey partridge and crested lark. This was an archipelago in which humans were vastly outnumbered by other mammals. There were more elk than people.

It's unusual for cultures to be cast adrift. For thousands of years, Britain's physical connection with the continent had facilitated the diffusion of beliefs, innovations and genes. As the isthmus had thinned and as Doggerland's refugees had migrated to higher ground, perceptions had changed. While Britain had been appended – like Scandinavia – to Europe's cooler, northern rim, it had been a safe haven for a spectrum of cultural affinities built on physical connectivity. Insularity changed the nature of the place. Inundation of the land-bridge severed cultural and genetic pathways. For centuries, the new straits coursed with tidal rips so ferocious that the only safe time to tackle the crossing would have been slack water. And it's possible that the window of opportunity might have been too short to have paddled a boat with much confidence. Remnants of Doggerland survived for a while as shrinking fragments in a rising, widening sea and these residual islands would have provided stepping stones between the continent and Britain. But these were treacherous waters, with shifting sandbanks and strong currents. Time passed before a maritime culture evolved.

Britain entered an era of semi-isolation. The evolution of settlement patterns was put on hold; the big-house culture of the east – built on architectural traditions transmitted around and across the North Sea basin – does not appear to have survived severance, although continued human presence was recorded in the intermittent inscriptions left on the landscape: child-prints in the mud of the Severn Estuary; the scoops and stake-holes of occasional home-bases; the hearths and scatters of geometric microliths. In one model of human occupation, groups were less mobile, a reflection perhaps of the increasing productivity of the land, and of improved techniques for foraging and hunting. Base camps became more significant. A group would have a base camp on the coast, and a number of summer hunting camps in the most convenient, productive uplands. A variant would be a base camp on a lowland river or lakeside, with seasonal camps on an accessible stretch of coast and in uplands, too. As ever, ecotones were critical. Settlements gathered in river valleys had access to both riparian and valley-slope habitats. Other settlements were attracted to the meetings of different soil types where a greater variety of flora and fauna could be anticipated. In the Pennines, camps pitched by hunters and foragers tended to be smaller, suggesting that game was becoming less plentiful, or that more food was being procured elsewhere. A general movement beyond lighter, better-drained soils to most other soil types hinted at a rising population. It's possible that some groups had by this time abandoned the rituals of seasonal movement and become sedentary; if so, they were in the vanguard of a settlement revolution.

Four thousand years after pioneering immigrants walked west across Doggerland, Britain had little to show in terms of enduring 'places'. Home-bases and camps belonged to a pattern of way-marked mobility in which groups moved about the landscape from fishing waters and stalking glades to flint grounds, hazel woods and the myriad other habitual sources that provided the wherewithal for subsistence. People returned to the same places, generation after generation. On the north-western island of Islay, one lithic scatter eventually consisted of over a million pieces of stone. It's possible that knapping had become a conscious act of fragmentation, the working of stone into ever-smaller pieces a ritual performance associated with the conveyance away from the site of selected lithics, a practice that extended to human remains. The original discard-site became a physical place that was revisited and embellished with more 'offerings' and memories. By the sixth millennium BC (and quite possibly much earlier, too), places of remembrance included sites of human cremation. Beside a tidal inlet close to the stump of Doggerland, a one-metre-wide pit was excavated in around 5600 BC, and into it were placed struck flints and burned bones and charcoal taken from a nearby funeral pyre. The adult whose body had been cremated belonged to a community whose ancestral memories knew the fragmentation of land. They also knew enough about pyrolysis – the decomposition of substances through heating – to have created a pyre burning at around 600°C, the minimum temperature required for combustion of a corpse. Landscapes were the canonical repository for belief systems that did not require the construction of enduring monuments.

In southern Britain, rising sea levels have destroyed many of the shoreline sites from this era, but those that have survived suggest hand-to-mouth subsistence strategies with little or no motivation or time for the creation of permanent structures. Near the tip of the Lizard peninsula, a group (or several groups) used a camp close to the shoreline as a seasonal base. The scatters of narrow-blade microliths they left were characteristic of the time. Along with a range of awls, borers and scrapers were various chopping tools fashioned from beach pebbles, and bevelled pieces of slate which might have been used for chiselling limpets from rocks or for processing seal carcasses taken from the local colony. Britain was rimmed with fast-food sites, virtually all of them too ephemeral to endure as landmarks. But there were exceptions. In places, shells and bones accumulated in mounds so enormous that this constructionally hushed era could be called the Age of Middens.

At the mouth of the Taw estuary in the south-west, an ecotonal mound of boar and red deer bones also contained the shells of limpets, winkles, mussels and oysters. On the Isle of Portland, meals of limpets and winkles produced a sprawling midden covering some 650 square metres. The most extraordinary middens to have survived can still be seen on the Hebridean island of Oronsay, where five gigantic heaps accumulated at a time – or times – no more accurately estimated than between 5300 and 4300 BC. Among the fish bones being discarded were those of angel shark, thornback ray, spurdog, tope and saithe. Placed with them were the bones of great auk, tern, shag and cormorant, and the mammalian remains of red deer and roe deer, boar, otter, grey seal and common seal. On the other side of Britain, a midden close to the mouth of the Tay included food-remains from woodland, water and air: bones of aurochs, boar and hedgehog; sturgeon, cod and haddock; guillemot, gannet, puffin and thrush.

Middens could be more than trash-piles: the one on the Isle of Portland had been partially covered at some point by stone flooring for some kind of structure, while the free-standing mounds on Oronsay were conscious accumulations arising from deliberate placement of bones and shells. They were scaled to dominate the landscape of their diminutive island and under sunlight they shone with composure. Oronsay had a special meaning to the people who knew it because they lived here all year round, effectively as settlers. Among the remains in the middens were parts of human bone. At the opposite end of the midden-scale, screes of bone and shell could accumulate in the recesses of caves. Middens were not 'monumental' in the sense that they conformed to a conspicuous model, but heaps like the ones on Oronsay appear to have conveyed some kind of meaning as topographical markers – ancestral beacons, perhaps, or symbols of belonging or terrestrial reassurance on an exposed speck beyond the edge of Britain. Whatever their resonance at the time, middens were the earliest anthropic landmarks to record an emerging maritime culture.

One of the prime adaptations to the all-encompassing sea was the identification of favoured havens. From the early days of Britain's isolation, there were places on the coast of the south-east that were known for being safer landfalls. Beaches on estuaries, for example, were less likely to be barricaded by breaking waves than those on open coasts. Although the evidence is lacking, by the fifth millennium BC, hide boats must have been the offshore vessel of choice. With high freeboards and sturdy greenwood frames clad with stitched hides, they would have been able to handle surf and riptides, while their shallow draughts allowed them to slip through tidal creeks and estuaries. Four paddlers could shift cargoes of up to half a ton across open water. They were lightweight, manoeuvrable and versatile and their technical development in the south-east can only have been accelerated by the unprecedented difficulties faced by mariners following the final break-through of the Doggerland land-bridge, when tides and currents surging between the North Sea and the Channel forced seafarers to redraw their cognitive charts and very possibly to develop new techniques of seafaring, and more seaworthy vessels. Over on the west coast of Britain, where the rocky Atlantic coast had not been subject to the cataclysmic inundations that had occurred in the south-east, sea lanes had remained relatively constant over the millennia. Generations of mariners had plied the minches and lochs of the Western Isles; the bays and ragged peninsulas of Ireland. Knowledge of tides, winds and seamarks had been handed down from ancestors.

Coastal havens were key points on the pattern of sea and land routes that collectively formed a web of human movement. A beach good for boats was a place vested with mortal meaning: the first and last landfall; the meeting place; the place of transference between land and sea; between solidity and transparency. At a local, subsistence level, these routes were used for the daily round of fishing, foraging, hunting, wood-collecting and tool-making. But for groups with a greater abundance of food and better access to materials, the web could work as a network of exchange and kinship – the means by which objects, ideas and genes were transmitted between groups. The export of exotic objects or the staging of feasts could create a local or regional web of debtors, supplied by a 'creditor' who gained status. By virtue of their perishability, the furs, skins and unusual foods that passed between groups left little record on the land. But precious stones were a different matter, and in various parts of Britain, their transmission was remembered by their imperishable presence in the soil.

Britain was well-endowed with currencies. The bedrocks of this three-billion-year-old archipelago were among the most diverse in the world for a landmass of its size. Some stones were rarer than others and the imbalance created opportunities for systems of exchange – and of aggrandisement. The motives for possessing objects cover a spectrum of human conditions, from the private pleasures of association to greed and status, superstition and altruism (some objects would have been sought as gifts). Pieces of slate were carried from the windy moors of the far south-west to the valley of the Kennet and to downland camps close to the Solent. Chert from mainland Britain was taken by boat to the Isle of Wight. In the north, pitchstone from the island of Arran was distributed throughout the highlands and islands. The unusual chalcedony of Rum circulated on Skye and the mainland.

The routes tramped by the conveyors of precious stones must have criss-crossed Britain's landscapes. Paths existed as physical imprints on the ground and they were inscribed on the intricate, cognitive maps etched into memories by the repeated exchange of objects along familiar routes dotted with locations rich with associations. And beyond the immediate archipelago, a wider-reaching web connected the islands of Britain and Ireland to the continent of Europe by way of open-water sea-crossings. But these were not easy passages of water. The ancient sea lanes along the western edge of the Atlantic façade were exposed to storms, ocean swells and, in places, fierce tidal currents. It's not impossible that voyages were being undertaken between the north-western tip of continental Europe and south-western Britain, but these would have been high-risk undertakings far from sight of land. The situation in the south-east was even more fluid.

Now that sea separated Britain from the continent, diffusion of ideas and objects was less of a drip-drip-drip, than a periodic splash. Another tsunami was building, still distant, yet destined to overwhelm Britain's green, intemperate land. Its genesis was far away, beyond the eastern Mediterranean. Locked into an expanding spiral of feedback, the evolution of food production and rising populations intensified and spread beyond the mountain risings of the Euphrates and Tigris, into south-east Europe. The new ways advanced intermittently. The Balkans were reached by two routes: one by sea along the coast of the Mediterranean and into the Aegean, and the other overland through the mountains of Anatolia. The best land for food production was the alluvial soil of river valleys, and loess – fine, dark soils derived from windblown sediment distributed as the ice sheets melted. In the Danubian lowlands of the Carpathian Basin and in the valleys of the Elbe and Rhine, cultivators found lots of both. By 5500 BC, the frontiers of Europe's wilder lands were retreating at a rate of fifteen miles a generation. While the people of Britain stalked wood pastures for aurochs and foraged for nuts, berries and tubers, clearings on the continent were being cultivated with wheat, barley, peas and lentils by communities who also kept domesticated cattle and pigs and who commemorated their beliefs with raised stones, massive ritual mounds and enclosures. In places, rows of rectangular, timber-framed longhouses had gathered to form villages.

By 5000 BC, controlled food production – agriculture – had reached river valleys just a few days' walk from the Channel. But there the march of the monumentalists faltered. Climate may have been a factor. There's evidence to suggest that from around 5000 to 4550 BC the climate of north-west Europe had become wetter. High winter precipitation and low summer temperatures dramatically reduced the length of the growing season, especially in the northern, maritime fringes of the continent where rainfall, humidity and cloud cover cut the rate of evapotranspiration. Soils became more prone to waterlogging. Faced with harvest difficulties, planters and breeders working the lands of the Channel coast may have had little incentive to migrate further northward.

In Britain, evidence of human activity dissolves from around 4500 BC. It's a strange dearth. There were two more tsunami events, both in about 4350 BC, and although they were far less destructive than the tidal waves of around 6000 BC, the run-up on the Shetlands was more than ten metres and the havoc wreaked on coastal zones would have been extreme. At the time, most of Britain's communities lived close to seashores. And in around 4300 BC, a deep solar minimum turned back the glacial clock of the North Atlantic, unleashing fleets of icebergs from Greenland and reducing temperatures as far south as the Aegean. Off Ireland, sea temperatures fell and it's possible that the Gulf Stream weakened. In the Alps, glaciers advanced and in central Europe, oaks barely grew. Britain's climate switched to a drier, continental model, with warmer summers, and colder winters. The pattern lasted long enough for ecosystems to shift their ground. Seasonal rhythms and migrations would have ceased to provide expected resources. Winters of extreme cold occurred in 4120 and 4040 BC. Afflicted by earth-system spasms, from tsunamis to climate change, it seems that Britain's isolated groups of foragers dwindled in numbers. It's conceivable that by 4100 BC, the island was virtually uninhabited.

Britain was a habitat-in-waiting. As the drier, colder winters and warmer summers lifted moisture from the land, seasonally waterlogged soils lost moisture and were colonised by grasslands. Lakes shrank and perennially marshy shores were overtaken by moisture-loving grasses such as *Glyceria*, with its green and purple summer flowers and ability to grow through shallow water. Woodland – particularly pine woodland – became more flammable. Clearings were created by lightning strikes, and the incendiary woodland became more susceptible to deliberate firing. The drying climate brightened Britain's facade. As rates of evapotranspiration increased, grassland spread across coastal wetlands and estuary fringes. The edges of the archipelago became greener.

Under pressure on the far side of the Channel were several concentrations of cultural energy. Closest to Britain were people who had gathered in the hills and wooded valleys each side of the wandering Somme, where they had access to nutrient-rich, loess soil and overlying beds of flint-bearing chalk. Over the centuries, settlements with ditched enclosures had developed. The most important source of flint in the region was a complex of shafts and tunnels on the valley-sides and plateaux above a pair of rivers a couple of days' walk from the Channel coast. The mines were at the centre of a nexus of distribution; worked blades and axe-heads being exported within a radius of at least two days' walking (today the mines outside Mons, Belgium, are a World Heritage Site). Fields and pastures surrounded the mines, woodland had been cleared, and an area of 14 hectares or so had been surrounded by a double ditch and banks. Never had the lands that faced each other across the Channel looked so different; on a clear day at the Cape of the Grey Nose – Cap Gris Nez – you could peer from the brink of the continent at the shining, chalk cliffs of Britain.

The decision to embark upon a Channel crossing may have been impulsive, perhaps desperate. The people who climbed into the boats had no intention of returning. They were colonists, 'house people' whose timber halls symbolised their placing in the cosmos. And they left the continent with complete life-support systems: tools, weapons, clothing, food, animals and the young, whose coerciveness, strength, longevity and fecundity would ease the colonists through their transition in an unfamiliar land. Initial numbers may not have been great. A kin-group working as a community of founding colonists over one or two generations would have been enough to trigger the transformation that followed.

We're 6,000 years into our story, or about half of the total time that Britain has been continuously inhabited. Six millennia of global warming had converted a sub-Arctic peninsula into a temperate island clothed in twenty billion trees. If, say, there had been as many as 10,000 people living in Britain at the time, there would have been one person for every two million trees. Europe's largest island was still a lightly trodden game-park, its land surface faintly etched with foot-tracks and home-bases. The most enduring human landmarks amounted to little more than totemic timber monuments and a scattering of humped middens. The greatest anthropogenic impact had been the clearance of trees. In more intensely settled areas, felling and burning had denuded some areas of woodland, which had increased soil erosion, led to the formation of peat, and laid a layer of silt along the beds of rivers. So far, the human imprint on Britain had been slight. But all was about to change, very dramatically. The house-people intended to stay. And they brought with them a vast raft of practices, traditions and rituals that could only be exercised by altering the earth.

PART TWO

Sacred Places

FIVE

Altered Earth

4050–3700 BC

A few years ago I paddled a sea kayak around South Foreland. From wave-level the white cliffs were a blinding wall. Wrestling both wind and tide, I told myself that it would have been much easier if we'd been heading in the opposite direction, north towards the gentle beaches of Pegwell Bay. I'd grown up walking these illuminated cliffs, but down in the wave troughs of the English Channel, they were plain intimidating. There was nowhere to land; no beach. Sea and rock met at a pounding death-zone of surf.

If the house-people had crossed the Channel at its narrowest point, this is what they saw, too. And they did the sensible thing. They turned north, paddling or sailing past the truncated terminus of the North Downs until the cliffs dipped to meet the shore. From South Foreland, they were only just over ten miles from the channel that cut across the tip of Britain between Pegwell Bay and the estuary of the Thames. If they had paddled up the Wantsum Channel then turned left into the Thames and hugged its southern shore, they'd have found themselves at the mouth of the River Medway.

The earliest-known rectangular building in Britain was erected several miles up the Medway, close to a gap in the chalk wall of the North Downs. Behind the site rose the abrupt scarp of the downs, with woods of oak and ash, shading soil that was lumpy with flint; in front, the land angled gently through hazel and alder to the banks of the river, just over a mile away. The constructors felled trees and in the sheltered, sunny clearing they excavated foundation trenches and embedded several rows of posts – some of them with diameters of 30 centimetres. Inner rows of posts bore the weight of a massive roof. Between the posts, the foundation trenches were laid with shorter, horizontal timbers supporting walls of vertical planks. To people of the wildwood, the blocky form of the house would have been astonishing. It was at least three times larger than a roundhouse. The building on the Medway was 7 metres wide and 18 metres long.

Not long after the great rectangular house on the Medway was built, a rectangular stone chamber was assembled nearby from local sarsens, and inside it were placed the remains of several people. Another tomb appeared further up the Thames where the river took a long meander around a peninsula later known as the Isle of Dogs. A body thought to have been a woman's was placed in a grave cut into the sand and retained by at least one oak plank. She crouched, along with her carinated bowl and her well-used flint knife. (At the time of writing, she's the earliest-known Londoner.)

The Thames was a yawning portal. It was the closest major estuary to the continent, it was sheltered from westerlies and its long, indented shoreline offered numerous landing places. The chalk hills of the North Downs formed the backdrop to the estuary's southern shore and both shores were generously laden with vast amounts of glacial rubble which could be foraged for workable stone. Not only was there woodland and fresh water, but the tidal flats of the estuary's fringes offered a dream ecotone – a naturally cleared wetland that attracted game. Beyond the estuary, the River Thames and its convenient tides opened the way to an enormous hinterland suitable for settlement.

On the boats came exotic bowls, axes as smooth as skin and domesticated animals. The cow was far smaller than the intimidating aurochs, and much more docile. It could be bred in captivity, herded, milked and sacrificed for its meat. The pig was a calmer, biddable relative of the boar, a rapid breeder content to graze the forest floor and a life-changing source of protein, fat and leather. Sheep were even more docile and were suited to open grasslands. They could be herded in numbers, they could be milked and were good to eat. Grasses came in the boats, too. The cultivation of emmer wheat had been creeping across the continent with the new breeds of livestock. High in protein, rich in fibre, magnesium and vitamins, emmer provided stable yields on poor soils and was resistant to drought and to the kinds of fungal diseases that lurked in damp valleys and fens. It competed well against weeds and thrived on marginal land. It was an ideal crop for low-input agriculture; for people who were reluctant to forgo traditions of foraging and hunting. Emmer was also well suited to a cool, moist growing season followed by warm, dry weather for ripening. After being uprooted, the strong husks were pounded between stones to release the grains, which could be ground to flour and baked in fires. Its nutritional qualities meant that it was suitable as a staple. And the amount of land needed to grow enough wheat to form a regular, daily part of a family's needs was modest: a plot just 7 metres by 17 metres would produce 27 kilograms of wheat a year. Surplus grain could be fed to livestock. Along with emmer wheat, there were other new grasses like barley, whose nutty-tasting grains could be pot-cooked as gruel, or ground and mixed with wheat-flour for baking. Both emmer wheat and barley could be grown from seed to ripeness in less than three months. The house-people may also have brought disease. The lactose-tolerant herders, with their habit of living in confined spaces year after year with domesticated animals, could have been incubating plagues that advanced ahead of agriculture, decimating the wandering forager-hunters of Britain's coastlands with onslaughts of smallpox, influenza, measles and brucellosis.

So the initial toehold established in the south-eastern corner of Britain extended to a foothold. Initially, diffusion was slow. It took perhaps two or three generations for the core practices to travel sixty miles. But then they accelerated.

*

It's not fully light and a numbing gale is buffeting from the unseen sea. My boots slip on wet, winter-matted grass. Weathering has smoothed the pits and spoil heaps into a pattern of bowls and domes like a miniaturised version of the downland they're sculpted from. Rabbits have excavated their own caverns and tunnels, spilling smudges of chalk that gleam in the half-light. The dips are not quite deep enough to duck the wind. I linger for a few moments by a clump of ragged gorse, then try the leeward side of the down. On the crown of Harrow Hill, later settlers of prehistory have constructed a rectangular enclosure and for a moment I wonder what they made of the abandoned excavations on the eastern slope. As dawn seeps, the scoops and hummocks are illuminated in sharp relief and for the first time I can see the whole, tilted field of endeavour, spotlit by the same, waking sun that warmed this slope as the first adze cut the first turf. What rituals accompanied the incision and lifting of the earth's rough hide? Was the cutting and baring solemn or celebratory? Urgent or measured? Was the placing of the first pit significant? Did solar reasoning bring them to the morning-slope of the hill? Despite the ploughing, the Royal Commission counted 245 pits and shafts up here. It still looks like a mogul-field. Soon, sun rays lever the low overcast and spear the hill with blinding shafts. Avian chatter erupts from the nearby trees. Skylarks begin to fret. A flight of buzzards wheel above the workings. The grey plinth of the sea solidifies in the rising light and I'm amazed by how close it looks; a walk of perhaps of an hour, or two hours, say, to allow for the tangled woodland that would have filled the hinterland between shore and down. In fair weather, the travelling time from the labyrinths of Spiennes would have been as little as three days. These are the first large-scale, permanent earthworks to modify Britain.

Continuity was provided by chalk, the bedrock they knew from their own lands. On the continent, it tracked all the way from the Seine to the lowlands of the Somme and beyond to the rise and fall of downland behind the narrowest part of the Channel. On the British side of the Channel, the chalk presented itself as a herd of reflective cliffs cantering above the blue water from North Foreland to beyond the Isle of Wight. Perhaps they had noticed the thin, horizontal black lines drawn as if by charcoal point across the white face of the cliffs; if so, they'd have known roughly how far beneath the surface each bed of flint lay. They chose to dig in sight of the sea on the ten-mile block of chalk downland between the rivers Arun and Adur. Over time, extraction sites developed on four adjacent hills: Blackpatch Hill was pierced by over a hundred cavities; the next hill to the east had forty or so; four miles to the east of Harrow Hill, a 600-foot hill accumulated more than 270 dark voids. West of the Arun, more holes pocked Long Down and Stoke Down. Nothing like it had been seen in Britain before. Above the swelling wildwood, downland domes were ripped and riven, their entrails spilled. Whatever language the excavators used when they communed with their landscape, it was utterly unlike that of the people who had lived in Britain since the ice melted.

Upper seams could be reached by excavating an open pit, but deeper flint required the boring of a shaft. It was the shafts that spoke most of alien ways. Pit-digging had been part of the forager-hunter repertoire for millennia, and so had the practice of entering caves. But the creation of artificial, subterranean systems was new to Britain. Below ground, in dark constrictions, hunched figures wielded picks and shovels fashioned from the antlers and shoulder blades of red deer. The deeper voids extended downward for ten or more metres with side galleries and tunnels radiating outward to follow the flint beds – a technique already practised in the chalklands on the continent. One of the shafts on Cissbury Hill dropped more than 13 metres. Tracking seams of flint, the side galleries and tunnels – with roofs as low as one metre – grew longer and longer, sometimes meeting those of adjacent shafts so that they evolved into three-dimensional labyrinths. At least one of the horizontal tendrils grew to a length of nine metres. By punching parallel lines of holes into the soft rock with the pointed tines of antler picks, entire blocks of chalk could be prised clear. Rock falls were common. In places, columns of bedrock were left intact to support the roof. The further removed from daylight, the harder the chalk and the better the flint. Far below the sunlit woods and grassland, all senses were deadened: no colours; no breeze; no sounds. What fears tripped the heartbeats of those who dared to descend where all was black and damp and clamped by stone? To economise on effort, galleries were backfilled with debris, but vast amounts of loose chalk and stone had to be hauled to the surface with ropes and baskets and then tipped on growing mounds. Much of the hard-won stone would eventually be discarded during the tool-making process since flaws in the flint were often revealed at a late stage of processing. One cubic metre of excavated debris might release sufficient flint to make only four or five axe-heads.

They didn't need to dig for flint; ample supplies of raw nodules could be picked from the surface. There was more to these holes than stone. The locations most favoured for sinking shafts were often far from sources of drinking water and shelter, on the upper curves of the chalklands. Remoteness seems to have been a factor. And superstition. Sometimes the bones of humans and animals were left underground. So were strange chalk objects, and pictures and patterns: a deer's head, and a depiction of what might be an aurochs. At the entrance to one of the galleries on Harrow Hill, a burrower etched into the chalk a pattern like a lattice, or a grid. It's hard to avoid the conclusion that the whole, fantastically laborious and hazardous exercise was in some way connected with transition to another world – a netherworld. The act of passing through the surface would have been well suited to rituals that connected the living with ancestors; that sought bonds with ancestral roots and spirits. Departing the bright, aromatic, busy crest of the Downs for a descent into chilled darkness was an emotive experience that lent itself to transformative states of mind. And the activities up on the Downs – both ritualistic and extractive – would have involved groups of people. Rituals required numbers and so did the excavation of pits, shafts, galleries and tunnels. All were communal enterprises. A subterranean complex of say 300 to 500 cubic metres represented 600 to 1,000 worker-days of excavation, an undertaking that might have been executed by a group of four or five people over half a year. More probably, these hidden places were carved out episodically, over decades, or longer. They were modifications to the land on an unprecedented scale and were reflections of a profound commitment of skills, organisation, labour, time and shared belief. The digging, hacking and hauling may have been a part of the ritual itself: an act of creation; remembrance; communication. The nodule released from the clasp of deep chalk then raised into the light and crafted into a ground axe-head, was transformed into an object of reverence. Among all the axe-heads that were carried from their chalkland sources to other locations in south-eastern Britain, those from this nest of excavations on the South Downs constituted the single largest group.

After the last rubble basket was set aside, each excavated complex became a fossilised belief; a monument sculpted within the billions of calcareous shells that formed the Downs. With their central shaft and radiating branches, the larger labyrinths resembled inverted trees. The forces brought to bear on Britain's landscapes in around 4000 BC were unprecedented. Nothing is known of the beliefs that drew people beneath the downland but fundamental to their way of life was a will to construct landscapes; to model Earth to their own ends. The underground places of the South Downs were Britain's largest artificial structures.

Above ground, the house-people migrated through Britain, raising more of their tall, slab-sided, timber halls. The rectilinear method of construction was based on a simple post-and-lintel frame that could be repeated as required to create structures of many shapes and sizes. The principle in house-building of resting one horizontal element on top of two vertical elements was not new; it's likely for example to have been used in the construction of the roundhouse at the Haven, 4,000 years earlier, and was probably the standard method of supporting the roof above a round-house doorway. But long before the great transformations of 4000 BC, people on the continent had multiplied the binary post-and-lintel to create structures assembled from repeated rectangles: the rectangular, timber-framed longhouse. At some point in this process, they developed a joint to connect post and lintel – and indeed to connect any two pieces of timber meeting at a right-angle. The basic concept was an evolution of the socket-and-haft joint that had been used for millennia in the assembly of tools like mattocks. In its latest, constructional application, the 'mortise' was the cavity cut in one end of a section of timber, to take the projecting 'tenon' on the end of a second section. If the tenon was long enough to protrude right through a mortise, it could be pierced and fitted with a wedge that locked the joint together; on the continent, such a method was being used to build the rigid, rectangular timber linings of wells by around 5000 BC. Applied to buildings, the locked rectangle could be used for doors, windows and entire walls. The mortise-and-tenon changed the face of the land. The masters of the binary cosmos understood the tensile properties of timber; they knew how to take accurate measurements and they were skilled at carpentry. Their rectangular houses were more versatile than roundhouses and could be built to far larger dimensions. Unlike circular or curvilinear structures, rectangular buildings were straightforward to divide and expand, and also offered the possibility of upward extension through the raising of a second storey. Most significantly, the only limit to their size was human will and the availability of level land and materials. Versatility and scale were synonymous with prestige and power.

On the continent, rectangular buildings had become a virtually universal standard for agricultural communities, reflecting a need perhaps to define more precisely – and elaborately – the idea of place and to construct more protective settlements. But the rectangular buildings that began appearing in Britain were relatively isolated. Typically, they were raised on low-lying sites, close to a river, with convenient access to higher land. In those respects, they conformed to the old ideals of the ecotone and of the river as a valued axis of resources and communication. Some of the buildings gathered accumulations of pottery and grain; some may have been used both by the living and the dead. Massive and immovable, they were expressions of axe power, technical superiority and permanence. They were inviolable statements of presence; territorial claims in a land of wandering peoples. They were not in use for more than two or three generations, and none was rebuilt or gentrified. Perhaps they should be seen as frontier structures; pioneer homesteads with attitude. Once the extended families who built them were established, these timber block-houses were abandoned as kith and kin dispersed to settle in smaller, less enduring structures. The 18-metre Medway building was massive, but others provided the kind of floor area that conventionally would have suited a single family. A stone's-throw from the River Colne (not far from the reindeer hunters' camp of Chapter 1), at least four rectangular houses of the early fourth millennium BC were built on the valley's level, well-drained sands and gravels. One of the houses measured 9.8 by 6.5 metres, comparable in floor area to a traditional round-house. A couple of the others were slightly smaller, but a fourth was some 15 metres by 7.5 metres. Internal walls divided the interiors into two and a couple of the houses had deeper foundation-slots that may have taken the heavier timbers needed to support the weight of an internal mezzanine floor. (The sands and gravels beneath the houses are now being extracted to provide aggregate for a concrete plant capable of producing 120 cubic metres per hour of ready-mixed concrete.)

Along with the halls, and tombs and mines came a third terraforming impulse. Agriculture required clearings. There were cattle and sheep to graze, crops to plant. Felling was hard work. Using the latest ground-stone axe-heads, it could take a month to clear two hectares of woodland. The wildwood couldn't last. Agriculturalists didn't belong to the wilderness, but took the view that it should be controlled through the breeding and consumption of domesticated animals and cultivated grasses. To agriculturalists the land was the quarry; a resource to be exploited. Plants and animals were there to be bred and harvested; clay was there for pottery; timber for burning and building; stone for tools. Where foragers had harvested their habitat, agriculturalists mined, quarried and clear-felled. Withies and reeds, mud, timber and stone were there for building huts and houses; fibres from plants and animals could be spun into yarns for making cloth.

Where flint and chert had sufficed for the forager, the agriculturalist turned to deeper geologies, searching for stones that could be worked into sickles and querns, buffed axes and adzes. The energy driving this attitudinal revolution was derived from a diet of novel, nutritious foods that could be produced faster, in greater volumes and more reliably than foragers and hunters could ever hope to extract from fixed locations. Foraging and hunting had been a culture that stretched back to the beginning of time. The new ways were a culture, too, but a culture that closed histories. Agriculture had been advancing across Europe for over a thousand years and Britain was one of the last untapped reserves.

The people of Britain were a relic race. They were among the last of their kind in Europe. Indigenous peoples retain their identity – and territory – through isolation. This was the end-time. First contact between hunter-foragers and incomers from more developed worlds is usually a collision of bewilderment, followed by assimilation – or annihilation – of the indigenes. The encounters between the sea-people and Britain's wanderers of wood and glade conformed to history. Mutual incomprehension must have been accentuated by the long history of the Channel as a barrier. On the continent, the last generations of forager-hunters had been wiped out or converted to agriculture in time beyond memory. But the width of the Channel had meant that Britain had lived on in primitive isolation. By the time the boats touched shore, the cultural gulf between the continent and Britain had never been so great.

It's not hard to imagine Britain's dispersed communities – if indeed there were many – being unable to resist such a contagious basket of innovations. Those not wiped out by disease or killed in clashes were 'converted' through cultural osmosis. The speed and geographical track of transmission – slow and then accelerating ever faster from the south-east – suggests that the new ways were introduced by a small number of colonists whose ideas were adopted and transmitted by successive generations; one extended family of, say ten people, with six of them able to work the land, will have expanded to around a hundred land-workers within three generations. There were three broad phases: initially, a group of foragers would have access to agriculture, without taking advantage of it. This was followed by a rapid transition phase, during which agriculture provided up to half of their diet. During the third, longest phase of consolidation, the contribution of agriculture to the diet became increasingly important. It appears that many of Britain's early agriculturalists were peripatetic, continuing the tradition of seasonal migration, moving on with their livestock from pasture to pasture, foraging and hunting, but returning perhaps to fixed clearings or home-bases where they could plant or harvest cultivated foods. Mobility only increased the rate of transmission. Agriculture was rampantly reproductive. It's possible that climate change had reduced Britain's foragers and hunters to such a basic level of subsistence that the adoption of domesticated animals and cultivation was snatched as if it were famine relief. The new ways offered increased food security and status. Animals and plants were sources of prestige and power and so were the houses and ancestral sepulchres. A monument was an effective means of controlling space and local resources. Home became a location monumentalised with specialised, enduring structures. The old idea that natural features should form the focus for human endeavour was substituted by a belief that places were capable of greater amplification if they were constructed by humans.

By 3950 BC, the new landforms and agricultural 'things' had crept far inland to the Cotswold Hills. At some point between 4185 and 3990 BC, a local community on or close to Crickley Hill shifted a huge amount of soil into a low mound some 10 metres long and 4 metres wide. Not far away, and a little later, a mound with a trapezoidal ground-plan was raised. It had a forecourt, with an entrance leading to internal, transepted chambers. Long barrows like this performed a similar role to the enormous timber halls: both were 'houses'. Both acted as core structures that bonded social groups in life and death. The new ways were not just spreading; they were becoming more elaborate. By 3900 BC, the ways of the 'house people' had been adopted wholly or in part across south-eastern Britain to a line roughly from the Wash to Lyme Bay.

Britain's grassland and wood was replete with ready-made niches for domesticated species. To spread, an agricultural economy needed a suitable climate, fertile soil, ample water and space. The pattern of warmer summers and cooler winters that began in around 4100 BC did not peak until around 3800 BC, when conditions reached their driest. So for three hundred or so years, the growing season had been getting gradually longer and lands suitable for grazing domesticated animals and cultivating crops had extended steadily northward. Britain's population was so sparse that the land appeared to be empty. Old ways were subsumed.

The dam burst. Around 3800 BC, the rate of transmission turned from a trickle to a tsunami and within a period of time that may have been as little as a couple of generations, new landforms flooded west and north: beyond the Tamar; beyond the Forth and Clyde. In the distant north, where the River Dee rushed below the Grampian Mountains, massive rectangular buildings rose from the valley floor about fifteen miles from the sea. These were halls perhaps, rather than houses, conceived on a scale even more ambitious than those built by the early-generation settlers of the south-east. On the north bank of the Dee, oaks were felled and raised as posts to form the walls of a gigantic building 20 metres long and 9 metres wide. Seven massive purlin beams ran the length of the building and helped to support a gable roof. Internally, partitions were installed to create separate rooms. On the other side of the river, less than a mile away, an even more imposing hall was built. At 24 by 12 metres, it must have been among the largest buildings in Britain. One of the timber halls on the Dee stood beside pastures for grazing animals and plots of arable land being used to cultivate emmer wheat, barley, bread wheat and linseed. Stands of hazel fringing the plots may have been used for coppicing.

At the other end of Britain, pioneers worked to tame the largest wetland in the south-west. Beyond the Mendip Hills and the caves that concealed the bones of ancestral Britons, glittered a vast inland waterland of meres, rivers and marsh divided by low ribs of dry land. This was traditional territory for hunters, fishers and gatherers and it had much in common with the lost world of Doggerland and with the watery Carrs way up beyond the Wolds. Pools and channels between islands and marshes drained idly towards the great estuary of the Severn. Out on the islands, 400-year-old oaks towered alongside lime and ash. There were lofty elms too, and smaller stands of hazel and holly, and willow, alder, birch and poplar, apple, dogwood and writhing tendrils of ivy. Above the wetland, one of the ribbons of dry land, the Polden Hills, had already been partially cleared. Here the oaks were younger and smaller; regrowth after felling that had occurred a century or so earlier. During the winter of 3807–3806 BC, the people of the Poldens relieved the woods of 4,000 metres of planks, 2,000 metres of heavy rails and 6,000 pegs. They supplemented the fresh timber with pieces recycled from an old trackway. Working together, they laid a planked causeway over a mile long, all the way from the Poldens, across the wetland to the island of Westhay. For strong labourers, it wasn't a particularly onerous exercise; ten people working flat out could have built the causeway in one long day. It was barely wide enough for two people to pass. Walking its planks was to walk on water. During its short life, the causeway was used as a medium for interacting with the world below. Among the objects dropped into the waters beside the timber pilings were a polished flint axe that had been carried from the workings on the South Downs, and a jadeite axe that had come from the foothills of the Alps.

How much the struggle to tame wetland and wildwood motivated the urge to monumentalise landscapes is hard to say. Time was unkind to felled timber and the wooden structures built by the house-people had limited lifespans. The pair of stone axeheads submerged in a West Country swamp were rare symbols of longevity. Both of the halls on the Dee went up in flames. Many other halls would also be burned and their physical contribution to future landscapes would be no more than stains in the soil and beds of flint chips, bone, seed and pottery.

The hall by the Medway disappeared; within centuries, wood-rotting fungi and termites had returned its timbers to the earth and the building's post-holes now lie beneath the Channel Tunnel Rail Link. The wilderness bit back. Britain's temperate climate and fertile soils unleashed vegetative counter-attacks. Untended clearings were overtaken within a couple of seasons by recolonising growth. Untrodden routeways would be enveloped. Within ten years of its completion, the raised timber trackway across the wetland below the Polden Hills was unusable. Eventually it would be swallowed by encroaching peat. Wood was part of the living world. Trees grew from seed to sapling to maturity, old age and finally death, dust and rebirth. There were parallels with the human condition. The reminders of impermanence were incessant. A receptive climate is not the same as reliable weather. Tracks flooded. Crops failed. Animals died. Kids went hungry. While all else rotted, the visible reminders of ancestral endeavour would be works of earth and stone containing ancestral remains; sacred places.

The surge of activity that propelled timber halls and agriculture to the far north and west was accompanied by waves of funereal house-building. The all-inclusive cosmos was overridden by a world view in which values and beliefs were guided by ancestors, spirits and related beings. Their presence was real enough for them to be accommodated by an evolving family of burial chambers and mounds and long barrows. When constructed, they had a visual equivalence on the landscape with timber halls. After burning, halls may well have lived on in the landscape as memories, but physically, they no longer existed. Burial mounds and barrows had a permanent physicality. Long barrows and long cairns were ancestral sepulchres, scaled up. But they were neither inert not set aside from the affairs of the living. Unlike the timber halls, barrows were permanent fixtures, to be remodelled – and in some cases completely rebuilt – by future generations. Memory was a place.

Sepulchres, like agriculture, were a communal enterprise. Both required many hands. Looking after flocks and herds and cultivation plots was labour intensive, and often had to be shared beyond immediate family, a social contract that required food to be divided within a community – or even between communities – in times of need. It was sound insurance to construct enduring links with neighbours near and far. The piling of thousands of rocks and the shifting of soil required a consensus that the location and form of undertaking was socially appropriate. The act of construction symbolised the shared values and beliefs of that community. These were managed undertakings, and they reflected the role that designated places played in fixing religious and cosmological space. The long barrows and mounds were societal mastic, binding communities to constructed identities, their solidity a stand against the chaotic habits of the cosmos.

The brawn for building monuments came from the archipelago's expanding population; the time came from intensive food production. Foraging and hunting were wide-ranging strategies that had to be pursued all the year round; agriculture was more localised and seasonal. People working with domesticated animals and cultivated crops had busy seasons and quieter seasons. During fallow periods, they were released from the repetitive grind of rearing and digging, and could be coerced into construction projects. New-build became the architecture of the modern mind. The idea that a weathered tor, or a stream junction, or an ancient oak could be the sole marker of human affairs became an old superstition. Natural landmarks still played a role where they appeared to be invested with spiritual force, but they had become signposts to the main event – rituals that revolved around artificial structures. 'Look-at-me' sepulchres became an expression of existence. The impermanent skim of huts and muddy tracks was pinned to the land by ancestral structures. Moving the earth made memory imperishable. And it also empowered. Conspicuous construction at death was used by leaders to legitimise their superiority. An infinite history of mobility and a proven ability to adapt to a wide range of terrains and climates, was evolving into a tethered bond to a specific spot, expressed monumentally. The notion of 'place' was becoming grander and more complex. Days and weeks consumed by construction projects were dynastic investments: laborious, communal commitments that would give future generations a place to cherish and defend. They would be viewed as fixed moments in time, points on the human chronology of settlement. For a community intent on establishing territory, the erection of each monument was a marker in time and space, anchoring their narrative to their landscape and opening lines of communication with the cosmos.

The first, sketchy lines of a new map began to solidify on Britain's rumpled landscape, a map that recorded human geographies. As various groups in the archipelago began to express their social ideals through the form of their sepulchres, new localities and regions took shape. In the west, where stone was more plentiful, a combination of tradition and geology led to a variety of tomb designs. Some were assembled from rocks into closed polygonal chambers; others incorporated a stone-walled passage linking the inner chamber to the outer world. By 3800 BC, passage tombs and sealed, stone chambers could be seen in four clusters along the coasts of the Irish Sea: on the promontories that formed the southern end of Cardigan Bay, on the island of Anglesey, on sea lochs north of the Clyde and on the facing coast of northern Ireland. A fifth cluster appeared on the west coast of Ireland. Another early tomb appeared on a low-lying isthmus of promising soil at the mouth of Loch Etive on the coast of north-west Britain. Here, a group of people levered enormous stones into the form of a polygonal chamber and covered it with a rounded cairn. Later, the tomb was modified by the addition of a passage.

The western part of the map was given more weight by the distribution of a kind of tomb that featured a huge – usually flat – capstone placed on top of tall stones set on their ends in such a way that they formed a portal at the tomb's entrance. Seen from the front, they could resemble a rectilinear post-and-lintel shape. Over 400 of these portal tombs eventually appeared across Ireland, around Cardigan Bay and in the south-western peninsula of Britain. The most amazing feature of these tombs was the size of some of the capstones. Construction demanded a superhuman battle with gravity. And the capstone setting appears to have been carefully undertaken, for most tilt slightly towards the back. It's possible that the act of raising the capstone was in itself the purpose of the enterprise; that the elevation of a terrestrial slab towards the sky symbolised a transition between two worlds. Perhaps the levitated capstone was a reference to a mythology in which sky and earth were one. Portal tombs were related to rivers, the life-force that made Britain habitable to humans; most seem to be aligned with a sheltered valley, facing the source of a stream, or a confluence or a distinct change of course. Exposed, high ground was generally avoided, and locations tended to be discreet rather than dominant and related to the motion of the streams they observed. Like the polygonal chambers and passage tombs, the megalithic nature of the portal tombs meant that their cores were permanent landmarks, virtually immune to weathering.

In the east, generally less enduring building traditions evolved. The relative scarcity of stone was a factor, but there may also have been cultural reasons for using timber. Along the coasts of the Channel and North Sea, human remains were often placed in timber-built mortuary houses. Forms could differ but often involved the splitting in half of a tree trunk, often oak, which was then set vertically as two D-shaped uprights. Paired, oak uprights were also raised as founding posts of timber halls. Their placing was fundamental to the rituals of hall and mortuary house. Both could also include timber chambers. It was the practice of entombing timber mortuary houses in soil or stone that made them permanent landmarks.

At around this time, the wildwood suffered a species crash. The disease – like the agriculturalists – had arrived from the continent. The likely culprit was *Scolytus scolytus*, a beetle that bred in galleries occupied by the sticky spores of a fungus, *Ophiostoma ulmi*, which then contaminated newly hatched adult beetles, who then flew to healthy elms, where they contaminated another tree with the pathogen. The fungus blocked the tree's system of water transport, which cut off nourishment to buds and new shoots. The fallen elms created openings in the wildwood that allowed the growth of grass, bushes and herbs; ideal grazing for domestic cattle. Various practices of the time may have encouraged the spread of the disease. Woodland margins that had remained fairly constant for centuries were more resilient to disease than margins created suddenly by felling, and they acted as peripheral walls of protection for mature elms thriving within the wood. Once the woodland margins were felled, the mature elms were exposed to invasion by beetles. The practice of pollarding also accelerated elm decline because the trimming of trees encouraged them to grow faster, and rapid growth made the trees more prone to disease. The disease eventually took out half of Europe's elm population.

Elm disease spared Britain's tree-clearers millions of hours of back-twisting labour by increasing the space available for pasture and planting. But there was a cost. Along with yew, the pliable elm had produced the best poles for making bows. And it had been a prized raw material for timber devices subject to multi-directional forces because its interlocking grain prevented it from splitting. Elm had also been a source of food: the seeds were rich in protein; the bark could be cut into strips, boiled and eaten; and the leaves were a steady source of fodder for domesticated animals. Suited to pollarding, it could be harvested, and each time an elm was felled, new suckers produced a clutch of successors. Elm was an agriculturalist's tree: simple to propagate, fast-growing, long-lived, and able to thrive on a variety of soils and locations. It didn't even mind salt, so thrived around the archipelago's coasts, too. Everyone knew the elm. Elm – with hazel – was the most common tree in most of Ireland and western Britain, and elsewhere it was in the top five. Elm was ubiquitous, indestructible, a part of everyday life. And then it died. The fallen elm was symptomatic of wider-ranging instabilities. The warmer, drier centuries that had given agriculture a toe-hold on the archipelago, were followed after 3800 BC by less reliable seasons. In the north, the frequency of storms increased. Temperatures began to fall.

SIX

Temenos

3700–3000 BC

We took a walk on Windmill Hill. There was no one else there and the blown grass was speckled with purple-blue clusters of bellflowers. Their crimson stem is said to be attracted to the remains of the dead and in Cambridgeshire they're known as 'Dane's blood'. But that's a later story. In the valley below, we could see the roofs of Avebury. The hill swells to 195 metres above sea level and is wrapped on three sides by gentle valleys that carry the trickles of the infant Kennet towards the Thames. The watershed between the basins of the Thames and Avon – between the North Sea and Atlantic – can't be more than twenty minutes' walk away. Could they have known, I wonder, that this hill marks the great divide between east and west? The Pole of Liminality? On Windmill Hill you're on the dome of the world, touching the over-arching concavity of the sky. Sections of ditch and bank can still be traced around the upper slopes. Close-up, they look like curves in the turf but seen from the air they're recognisable as three circuits of nested roundels. There is purpose in the symmetry, beauty in the place and mystery in the meaning. All that can be said for certain is that Windmill Hill was one of eighty or so similar sites that expressed a common belief that people had reason to gather together in considerable numbers. They were the first artificially enclosed, open-air spaces devoted to large-scale human interaction. Wembley Stadium, Glastonbury, Aintree and the open-air theatre in Regent's Park began as roundels in the turf.

The earthworks on Windmill Hill were probably excavated over a period of 20 to 55 years, so it's possible that a single individual – or just a couple of generations of the same family – may have been present for the construction of the entire monument. They would have seen the episodic creation of a landmark that dwarfed all others in the region. The first enclosure was some 90 metres across and defined by a modest, segmented ditch. This was followed by a much larger enclosure of 360 metres which enveloped the original within longer sections of ditch backed by earthen banks and possibly timber. The final circuit was some 220 metres across, and was constructed between the other two, with both ditches and banks. It's as if the community of constructors were so impressed with their initial circuit that they built a far larger one around it, but when they decided on a third circuit, the huge size of the second, 360-metre enclosure put further enlargement beyond their reach, and so they compromised with a final, medium-sized circuit between the circumferences of the first two. Perhaps the trio of circuits should be seen as three separate monuments, each an expression of collective endeavour at a particular time.

What were they for? Their size is significant. In the fourth millennium BC, these enclosures were the largest artificial landforms in Britain. The outer ditch on Windmill Hill enclosed an area of nearly nine hectares. For the football-minded, that's larger than ten international pitches. Given the date of construction, the scale is astonishing. The other eighty-odd comparable enclosures known in Britain range in size from 0.4 to 10 hectares. They were not intended to be defensive. Set at intervals around the encircling ditches were 'causeways' or gaps where the original land surface had been left intact to facilitate passage between the exterior and interior. The appearance and size of the circuits was certainly monumental, and construction would have required organisation of labour on a communal basis.

Windmill Hill probably consumed around 62,000 worker-hours. Teams of fifty people, labouring ten hours a day, would have taken perhaps a month to construct each of the two smaller circuits; two months for the largest. A human commitment on this scale would only have been possible if food surpluses were available. The size of the circuits and points of access suggest that they were used by the wider community for the kinds of practices and rituals that reinforced social bonds. In that respect they can be seen as 'social enclosures' – gathering-places constructed specifically for large groups of people who shared a common interest in cohesiveness. The enclosures may have been stages for activities such as worship, feasting, remembrance of ancestors or for settling issues related to territory, grazing, water or crime. In their final form, social enclosures like Windmill Hill that developed into multiple circuits of ditch-and-bank were complex spaces that would have suited processional rituals. It's also possible that the internal spaces were conceived as 'non-places' separated from the outside world by their encircling earthworks – neutral or liminal voids or sanctuaries set aside for deities or ruling elites.

The idea spread fast. Social enclosures – like the mines, the timber halls and the barrows – were an imported form, appearing first on coasts closest to the continent. The earliest social enclosure in Britain was probably built on the Thames estuary shortly before 3700 BC. By the middle of the thirty-seventh century BC, they had reached the chalk heartland. Within seventy-five years, open-air spaces had been enclosed in a contiguous region that extended from Lyme Bay on the Channel to the Severn and the Wash – although several more appeared in isolated spots as far north as the Firth of Tay. The greatest number were clustered in the south-east, either on the chalk grasslands or a couple of days' walk from them. They proved to be an enduring land-form and were used for at least 400 years.

Build a massive landmark and it attracts a host of lesser landmarks. When the social enclosure was constructed on Windmill Hill, it was probably the first monument to appear in the district. But during the thirty or so years that it took to construct the three circuits of ditches and banks, another monument was raised above the headwaters of the Kennet. On the far side of the valley south of Windmill Hill, a long barrow was built on a spur of the downs. It was huge. Seen from above, this sensational chalk-work took the form of a tapered, stone axe-head. The eastern end of the barrow was dominated by massive sarsens arranged to form a courtyard, behind which a stone-lined tunnel pierced the blackness. Openings on the side of the passage led to four stone chambers – two on each side. And at the far end of the passage, a larger chamber terminated the journey into this artificial underworld. Eventually, the mound stretched to over one hundred metres and topped a height of three metres. In terms of soil-shifting, it represented around 16,000 worker-hours, equivalent perhaps to the effort required to construct one of the smaller circuits just across the valley on Windmill Hill. Poised upon the rolling, green grassland, it was a luminous, chalk-white monument. The social enclosure on Windmill Hill and the long barrow on the facing valley-side were in sight of each other and separated – or connected – by the Kennet. The dialogue between builder and built was written on the land: the people whose remains were interred in the barrow probably would have been living during the construction of the third and final circuit on Windmill Hill, and conceivably been among the excavators of the first two circuits.

Britain became a barrow-land. The spread of social enclosures was accompanied by a magnification of long barrows as timber structures were replaced by mounds of stone or soil. Scale mattered. Some of these mounds reached widths of 25 metres and lengths of 50 metres. One end was usually higher than the other, and ditches extended along each side. Long barrows were physically enduring and varied in form according to the local availability of stone, or soil type or belief system. Many were modified through time by successive generations of believers. Large numbers of long barrows sprang from the grass-lands of the chalk country, clustered according to local congregations. A group of barrows accumulated on the downs inland from the Isle of Portland; another cluster developed on Cranborne Chase, a third cluster on the Plain and a fourth on the watershed where tributaries of the two Avons and Thames rose from chalk springs around Windmill Hill. Individually, barrows varied widely and reflected both the passage of time and regional variations in rites.

To anyone taking a mid-millennium walk across the chalklands of southern Britain, the bone-white walls of enclosures and barrows were more than eye-catching. These monumental, timeless structures were nodes around which human life revolved. But most of the structures raised during this architectonic revolution would never be remembered by the land. Timber didn't endure and the earthen banks, ditches and mounds of entire localities would be erased by later communities and by the erosive forces of nature. Out in the clearings, the ephemeral landmarks that recorded the everyday comings and goings of Britain's increasingly tethered population often failed to outlast the generation who made them. Unless they were in regular use, trackways, droves, pastures, huts and homes were reclaimed by regenerative tendrils of vegetation in a matter of years. Most people lived in isolated houses or hamlets that would disappear without trace from the surface of the land. The record being accumulated by the land was partial. Of the many structures created by the people of Britain during this episode of readjustment, only the long barrows would survive as substantial landmarks.

By mid-millennium, Britain had become a monumentalised island of varied traditions. Many factors directed the evolution of regional style, but underpinning them all was geology: the incredibly varied range of rocks and soils that floored the human habitat of Britain. Across much of the south-east, vast swathes of land were deeply laid with sands and soft clays, flaky mudstones and friable sandstones. The 'rock' that formed the skeleton of the south and east – chalk – wasn't really a rock at all. Its optical brilliance and role as the mother lode of flint were not matched by its qualities as a building stone; most chalk was too crumbly to bear weight and it was susceptible to weathering. In odd locations there might be a scatter of sarsens, or an outcrop of sandstone, or a stray chunk of stone imported by long-gone glaciers from western or northern regions, but in general, the builders in this part of Britain were dependent upon timber and soil. A monument required work-squads wielding antler picks, mattocks and axes to cut ditches and raise banks, fell trees and trim timber for palisades, construct revetments and related buildings. Further north and west, stone and the physics of leverage made it possible to create architectures that were not only more diverse in terms of texture and form, but more importantly, enduring.

Insularity – and connectivity – also played roles in the development of regional traditions. It was during this critical millennium that the largest insular block of land in the wider archipelago, Ireland, diverged from Britain. As early as 4100 BC or thereabouts, there had been a foretelling that Ireland would embark upon its own cultural trajectory, when a domesticated cow was brought ashore long before cattle came to Britain. By around 3500 BC, communities in Ireland were fitting their megalithic tombs with 'wings' or 'arms' that extended from the front of the main structure. The effect was to create a courtyard or portal that led towards the tomb chamber. The basic design allowed for much local variation. Tomb chambers could be constructed in pairs, facing each other across a central courtyard, or in line with each other, opening onto a peripheral courtyard. At least 400 of these tombs were raised in Ireland. Unique to Ireland also was an extensive pattern of stone walling that developed on the west coast. Between the edge of a sheer cliff and the lower levels of a mountain, people devoted hundreds of hours to clearing loose stone from the fertile grassland along this coastal strip, piling slabs of sandstone into a rectilinear grid of small enclosures. The walls incorporated several court tombs and covered an area greater than 12 square kilometres. Some of the walls were over a mile in length, and most were roughly parallel and up to a couple of hundred metres apart from each other. Shorter, cross-walls created further sub-divisions. Here and there were stones too large to shift, like the one-metre lump of granite that had been dumped millennia earlier by a glacier. Compared to the plots on the Dee in northern Britain, these western cliffs were exposed, but this was an era of mellow weather and winds, and these slopes were ideally suited to grazing. In so much back-bending, repetitive stone-humping there must have been motives beyond the penning of livestock, for herds do not require containment such as this. Perhaps the walls were a sprawling, communal signature along the peripheral lands of the Atlantic seaboard; a physical claim to space; a pattern of ownership. Maybe the walls had more to do with tethering people than with containing cattle. They were certainly monumental. In some of these small stone boxes, wheat and barley were cultivated. At least one of the cultivators had experimented with a new method of sod-busting. By attaching a pointed stone to a length of timber hitched to oxen – bullocks that had been castrated to make them more docile – ground could be broken with less effort than could be achieved with hand-tools. These scratch ploughs, or ards, cut through the topsoil and created a furrow into which seed could be dropped.

Landscapes are better at recording ideas than their means of transmission. No trails connect the walling of western Ireland to comparable patterns of stone in other regions. That was not true for the courtyard tombs of northern Ireland, which had a structural affinity with tombs on the British mainland around the Firth of Clyde. Less than twenty miles of sea separated the coastal havens of northern Ireland from those of the Kintyre peninsula, a distance that could have been paddled in a day. At various coastal sites on the mainland of Britain, and on the Isle of Arran, too, communities fitted tombs with crescent-shaped forecourts similar to those in Ireland. Further evidence of communication across the North Channel was left by axe-heads of Irish porcellanite that were distributed across northern Britain, and pitchstone from Arran that was carried by boat across to Ireland. Axe-heads left numerous trails of transmission. Painstakingly formed by chipping, abrasion and polishing, the axes were very beautiful – and impractical – objects whose colours and patterns reflected the original geology of the stone. Sources were typically remote, as if the axe itself brought with it the spirit of 'other'. Besides the Isle of Arran, axe-stone was quarried from the fringes of Snowdonia, from the Preseli Mountains and from the coasts of the far south-west, where at least a couple of locations could only be approached by sea. The extraction sites were specialised landscapes and access to them may have been restricted to 'keepers' trusted with the topographical knowledge without which the seams could not be found. One of the most extreme sources developed the longest lines of transmission.

High in the mountains of the Lake District, a prized band of stone linked a family of peaks. The stone was epidotised tuff: volcanic ash that had solidified into hard, dark rock similar to flint. An eagle's eye view of the tuff would have seen it as a dark, intermittent procession of outcrops that stretched for some 12 miles at a height of 500 to 900 metres above sea level, west from Langdale Pikes to Scafell Pike and then back east to Glaramara. Among the outcrops, the most inaccess-ible appear to have been valued the highest. On the barbed ridge of the Langdale Pikes, stone-extractors balanced on narrow ledges as they battered and prised blocks of tuff from the cliffs high above the valley. So much rock was removed from one of the peaks, Pike o' Stickle, that it gradually acquired a stepped profile. Several artificial 'caves' were also created, and a scree of nearly one million man-made flakes. The dale below the Pikes must also have been a 'special place'. One of the most spectacular valleys in mid-latitude Britain, its only means of easy access was from the east, where a foot traveller had to pass through a constriction that formed a natural gate in the dale. Beyond the narrows, the inner dale revealed itself as a long, level-floored sanctuary stretching for some five miles between crags until the Pikes eventually appeared high above. A source of axe-stone even more extreme than Pike o' Stickle could be reached – with difficulty – three miles west of the Langdale Pikes, where the tuff resurfaced on the highest peak in southern Britain – Scafell Pike. Just below the 3,206-foot summit, pits were dug to reach the tuff and a natural blockfield broken up for axe material. I went there while writing this book, and was wondering how the 'keepers' ever found their way down from Scafell Pike in mist. Then I came across the tilted cross-hatched slab that marks the route through Dropping Crag to the twists and turns of the Corridor Route. Polished by millions of feet, this patterned slab is like a signpost to safety. Three miles further along the band of tuff, a spot near the summit of 2,560-foot Glaramara was exploited, too.

The axe quarries of the Lake District had much in common with the flint mines of the South Downs, which had fallen out of use by the middle of the fourth millennium. In both cases, the stone was being extracted for reasons beyond utility, since raw material for tools and weapons was available at far more accessible locations. They were, in some way, sacred sources that were known – or imagined – by people living far away. Lake District axes were spread in great numbers across Britain, from the Channel coast to the Firth of Forth and beyond, with the east attracting the greatest density. High-value stone from other sources travelled long distances, too. Gabbro axes from the Lizard were carried beyond the Thames to the flatlands east of the Wash and to the Wolds north of the Humber. Porcellanite axes from Tievebulliagh, the most important source in Ireland, were far rarer, but ended up at points ranging from the Channel coast of southern Britain to the Outer Hebrides.

Axe-heads did not travel alone. They were units of exchange imbued with economic value. And with them came unrecorded beliefs, ideas, innovations and people. Each piece of stone passed through many hands and lands, providing its bearers with reasons to exchange information, genes and goods. The human interactions that travelled with the axes are lost, but their numbers and distribution make clear that Britain had become a well-connected place and that some people must have developed a spatial awareness which reached far into the island's most remote recesses. Their mental maps spanned thousands of square miles.

Something happened. The flood of new agricultural ideas and structures that had remodelled British landscapes had come from the south. But during the mid-centuries of the fourth millennium, innovation appears to have spread from the north. There had been disruptions from the sky. From 4100 BC, the climate had been kind to

breeders and growers, with the driest decades clustered around 3800
BC. Thereafter, conditions became less favourable as solar intensity
dropped and sank to a deep minimum by around 3500 BC, which can
only have reduced the resilience of communities exposed to colder,
stormier winters. Britain's maritime climate was fickle at the best of
times, but any amplification of extremes could lead to crises on the
ground. With so much soil prone to waterlogging, cultivation was
particularly vulnerable to episodes of wet weather. Slight variations
in the seasons could be disastrous for groups that depended on crop
yields to get them through the winter. Failure of a harvest could
lead to food shortages, violent conflicts, political disruption and
the dislocation of exchange networks. The economy that had spread
across Britain with the adoption of more intensive food production
was dependent on full bellies. Food shortage provoked a reversion
to a traditional default: mobility. From around 3650–3600 BC, cereal
cultivation began to decline and the number of settlements fell, too.
People continued to forage for wild foods, and to keep domesticated
animals, but they became less dependent on permanently occupied
sites, turning instead to the more mobile lifestyle of their ancestors. It
was as if the agricultural experiment was faltering.

Just as a warming climate had created circumstances conducive to
agriculture rooting earlier in the south, a cooling climate forced early
adaptation from the north, where the effects of climate change would
have triggered earlier tipping-points. Climate tends to play a shadowy
role in the cultural affairs of the deep past, and this particular solar
minimum is no exception. But for whatever reason, at around this
time, northern communities developed an affinity for what can only
be described as 'elongation'.

Enclosures became longer. Raised banks of unprecedented length
were constructed. Some increased in length episodically. On the
boundary between the cold highlands and more liveable lowlands
of the Forth, a community pushed out a mound that eventually
measured 322 metres. Forty miles away, another burial mound was
lengthened incrementally until it reached over 1,800 metres. Both

of these elongated features were close to timber halls. Many other elongated mounds appeared. The act of lengthening may have been some kind of priapic appeal to declining productivity; it may have been related to emerging rituals of procession; it may have been a sign intended to be read from the sky.

The idea of an elongated sacred space – let's call it a *temenos* – gained currency and retained a connection to other 'sacred' landmarks. Temenoi were constructed in many variants. Some were as short as 100 metres. Widths could range from 150 metres to a narrow 20 metres. Some had rounded corners, others right-angled; some occurred singly; others in pairs. Ten miles to the south of the Carrs, the great vale marking the northern terminus of Britain's chalk country, at least four temenoi were raised at a bend in the largest and broadest valley cutting through the chalk hills of the Wolds. One of these temenoi took an ingenious, tilted, dogleg route that offered a constant visual relationship with long barrows on the horizon to the west. It was an integrated landmark, placed to relate with both natural and artificial features.

When communities in the south of Britain took to elongation from around 3650 BC, they did so on an even greater scale. One of the chosen locations was a strip of land on the chalk heartland. The part of the Plain they selected had been a 'special place' for at least 3,000 years; the place where foragers had erected their totems of pine beside the strange stripes on the ground. Some 700 metres north of the long-gone totems, work teams marked out a temenos some three kilometres long, and varying in width between 100 and 150 metres. Like the totems, the temenos was aligned ENE–WSW. In one direction, it pointed towards sunrise in spring, and in the other, towards sunset in autumn. Warmth and cold; the rise and fall of nature; birth and death were etched into the turf by this gigantic, solar earthwork on the Plain. Sixteen miles away, an even longer temenos developed on the corridor of chalk country running from the Plain towards the south coast. Its creators aligned their monument with existing long barrows and with the midwinter sunset – a course that connected ancestors with cosmology, humans with nature. It was one of the largest earthworks in Britain. Initially five kilometres long, its length was doubled in a second phase of building. The outline of this colossal temenos was formed by a ditch and a revetted bank two metres high. Excavating the ditch, raising the banks and building the revetments consumed around half a million worker-hours.

The temenos was by far the largest type of monument ever constructed in Britain; it was more methodically laid out and it demanded more man-hours to construct than anything previously. Temenos builders were not interested in physical bulk, or height, but in abstract notions of space and time; the temenos can be seen as a shape rather than a structure. That shape was an elongated expression of a rectangular house-plan, with parallel sides and one at least of the ends being closed. The outline was defined by ditches or posts or pits. For a while at least, the rectilinear form, stretched for emphasis, was imbued with tremendous power. Giganticism was intrinsic to the monument, reducing its human adherents to specks. Viewed from one end, the temenos appeared to taper to infinity, like a sun ray. Most temenoi were constructed on level terraces of gravel, with the effect that reference points were more likely to be found in the sky than on the surrounding land surface. They were perhaps a form of communication with the gods. Temenoi were 'clean'. Unlike the social enclosures, the internal space of a temenos did not accumulate rugs of crap left by feasting and gathering. Inside a temenos, you'd look in vain for hearths and rotting aurochs or devotional gifts. The act of construction itself was a memorable social event requiring the gathering together of many people to create a new and vast landform that could transform entire landscapes. These sacred enclosures did not just unify local communities, but came to represent a belief system that claimed to be universal. The belief that drove their form and construction was so powerful that it spread throughout the island, subsuming regional architectures. It was a form of expression that meant as much to a community in the uplands of the far north as it did to those of the far south. Their appearance marked a turning point in the evolution of society. The temenos was the line in the turf that separated Britain from its past. And separated Britain from the continent. It marked the beginning of a more insular tradition of monument building. There was something else, too.

Temenoi contributed to a process of localised aggregation. A recurring theme in the long evolution of Britain is the way certain sites were re-used repeatedly through time, as if the peculiarities of their location held a common attraction, whatever the age. Many temenoi were connected through orientation, line of sight or proximity to other, artificial landmarks. Increasingly, certain locations were being developed as ritual landscapes – socially acknowledged zones of ritual activity whose scale could place enormous demands on the human workforce. Ritual landscapes were fundamental to finding an accommodation with the cosmos. In many cases, the temenos appears to have been the feature that sought to find common ground among dispersed monuments. It was the message, the river, through which truths flowed. If climate change had been the factor in stretching the all-British temenos, it followed a certain, geographical logic: as an exposed, northerly island kept habitable by a passing ocean current, Britain would have been one of the first – if not *the* first – European land masses forced to respond adaptively to a cooling climate.

Around 3350 BC the general deterioration in the seasons coincided with population collapse. As conditions across Britain and Ireland grew cooler and wetter, plots were retaken by weeds and scrub; saplings crept across pastures; tombs were sealed with rubble. From around 3300 BC, long barrows ceased to be built. In the four centuries or so since they had appeared on British landscapes, over one thousand had been constructed. Gatherings at social enclosures came to an end. No more temenoi were built or used. On that west-coast site in Ireland, peat bogs crept down the slope like a black, seeping fungus and smothered the rectilinear grid of stone walls. A forest of pines rooted on the peat. By 3200 BC, the population of Britain had fallen back to a level barely above its pre-agricultural maximum a thousand years earlier, a thinning of density that spoke of faltering life-support systems. In this leaner Britain, the land took a breather. In some regions of southern Britain, cereal cultivation ceased altogether. Only in the far north, and in the islands off the north-west coast, did cultivation and settlement continue at a meaningful level.

The mid-millennium solar minimum was just the start. Things got worse. A lot worse. The fourth millennium BC turned into the most climatically chaotic episode since Britain's convulsive emergence from ice and tundra in the tenth millennium. By 3200 BC the world was heading into another solar minimum that didn't bottom out till 3000 BC. In the most serious failure of the thermohaline system since Britain had been glaciated, the warm waters of the Gulf Stream faltered. For the second time since the glaciers left Britain, the archipelago was gripped by a Little Ice Age. The glacial disruptions of the seventh millennium had been bad, but back then Britain had been a lightly populated land whose foragers and hunters retained the precaution of mobility. When the world's climate reorganised itself in the fourth millennium, Britain's population was far larger and many of its multifarious communities had put down roots in fixed, monumentalised locations. As the planet's climate systems became drier and cooler, the grasslands of the Sahara reverted to desert and chill clamped the Aegean. Icebergs returned to the North Atlantic and Britain's winters were racked by storms and cold.

The social backdrop to this era was one of upheaval. Conflict escalated into warfare. Not the sporadic, dynamic encounters of old, between edgy, nomadic bands, but full-on, static confrontations. Many burial chambers now contained the corpses of men and women slain in interpersonal violence. Some were interred with flint arrowheads embedded in their bodies. Enclosures had been attacked, their defenders falling before the withering fire of archers. A two-metre bow of yew was a lethal, long-range weapon of mass destruction. Strengthened with sinew and bound with a criss-cross of webbing that acted as a shock absorber, the weapon was silent, rapidly reloaded and accurate to 90 metres. An adversary could be dispatched in just over one second. Ritual earthworks found a new role as defensive structures.

One thousand years of agriculture had felled more trees than it had raised walls. The elms had gone. Swathes of wildwood had been cleared. Here and there, the archipelago was pimpled with grassy mounds that covered ossuaries of ancestral bones. The slumped banks and silted ditches of enclosures could still be seen. In places where people had settled, there were patchworks of pastures, the odd plot under cereals and webs of muddy foot-routes used for collecting firewood and water, for managing woodland, for herding, hunting and for collecting wild foods. Houses were so slight that they left little in the way of record in the soil. Britain's early agriculturalists were archaeological phantoms shackled to seasonal cycles of subsistence on a frequently intemperate island.

Two thousand miles away, a city had risen from the Fertile Crescent.

SEVEN

Stone Eternity

3000–2400 BC

The gradient between the undeveloped and developed had never been as extreme. The part of the Fertile Crescent that flowered in such a spectacular fashion was Lower Mesopotamia, the level alluvial lowlands of the Tigris and Euphrates downstream of modern Baghdad, where the rivers slithered in tepid meanders towards the Persian Gulf. The mosaic of complementary ecological niches included lagoons and marshes, well-drained plains ideal for cereal cultivation and riparian zones suitable for water-intensive crops. Slightly raised 'turtlebacks' on the plain provided ideal locations for settlements. Had the climate not changed, perhaps this would have remained a region of small, shifting settlements, but in the second half of the fourth millennium BC, conditions became much drier. As the annual floods reduced in frequency and volume, the rivers silted and the plain spread. It was the reverse of the Doggerland experience. The Sumer gained fertile land that was also less prone to flooding and therefore suitable for longer-term settlements. But to provide enough food in the increasingly arid climate, they had to construct artificial irrigation systems, which required immense pools of labour and organisation to maintain. Water and work increased the plain's capacity to feed an expanding population. Locked into an ascending spiral, population growth and agricultural productivity created the world's first cities.

While the people of Britain slept in isolated clusters of houses constructed from timber, reeds, turf and in places, stone, the citizens of Lower Mesopotamia lived in urban centres like Uruk, mass-producing pots on wheels, building with bricks, working with metal tools, shifting lumber on wagons, speeding in chariots and keeping records with writing and mathematics. Cities were seedbeds of innovation. By the end of the fourth millennium BC, Uruk had a population nearing 40,000 and covered 250 or so hectares. Internally, the city was segmented into a religious quarter, zones for administration and production, and a residential district where houses were typically composed of rooms each side of a courtyard. At the core of the city were two clusters of monumental buildings which would have been beyond the comprehension of anyone in Europe. One of these buildings had recess-buttressed, plastered walls that rose to a height of six metres and surrounded a central hall with a podium in one corner for a statue. The temple's construction would have taken 1,500 workers about five years, labouring for ten hours a day. And the White Temple was not the largest building in Uruk. At the foot of the temple's terrace was a partially subterranean building whose bitumen-mortared limestone walls covered a labyrinthine ground plan measuring some 30 by 25 metres. The roofing for just one of Uruk's temples – and there were several – would have required between 3,000 and 6,000 linear metres of timber, probably pines hauled or floated from the Taurus or Zagros mountains. Fundamental to the success of Uruk was its ability to control its own cosmos, in particular water. Out on the plain, the city's vast resources of labour had used sweat, baked bricks and bitumen to build far-reaching systems of ditches, channels and canals, taming the nearby Euphrates in much the same way that their ancestors had tamed wild animals. The city lived by water, feeding from it and using its rivers and canals as transport routes. And Uruk got even bigger. At its height, the city may have been a regional centre for as many as 80,000 or 90,000 people. Other, smaller cities dotted the plain. Irrigation agriculture and population density, buttressed by security, stability and social differentiation, had created a new kind of power-base. The polities of Lower Mesopotamia were so socially complex and physically huge that they outpaced their competitors in south-west Asia, ultimately leading to the world's first example of colonial expansion. Uruk 'city-states' broke out of their alluvial heartland and occupied adjacent lands where they established trading diasporas, controlling resources and overland routes. Some of their colonialist enclaves were well planned, massive and speedily built. Sumerian colonial expansion was launched by city-centred, state-backed enterprise.

Cities were the ultimate human aggregation: the developmental apex of a process that began with a cluster of social, political or economic activities and culminated as a commanding centre of human networks. They were defined by their extreme size, by their economic complexity and by roles as the seats of secular and religious authorities. Cities also had a mutually supportive relationship with agricultural land; they were too large to look after themselves, while the surrounding agricultural communities depended upon them to provide markets and food-processing technology. Cities tended not to develop in isolation, but connected to the networks of other cities. How much the people in Britain knew about the cities of the Tigris and Euphrates is impossible to say. But these hot, exotic conurbations were on routes of exchange that connected them to the Mediterranean, and so was Britain. Whispers of enormous walled settlements on unimaginably wide rivers must surely have reached the wet and windy northern archipelago of streams and trees, where the age of stone had half a millennium to run.

Per hectare, Britain's soils were not producing sufficient calories to support dense nucleations of people. Social groups, be they extended families, clans, cohesive bands or far larger regional 'peoples', had to range widely to survive. The bonds of loyalty between people may have been close, but physically they were often dispersed. The long barrows, the enclosures and the temenoi had played a critical role as social tethers, periodically drawing people together and reaffirming connections. And by 3000 BC, monuments were more important than ever. After centuries of miscellaneous vicissitudes, the archipelago that emerged from the fourth millennium BC had been taken over by chiefdoms whose constituencies were dependent on herds and flocks. And upon centres of congregation.

Ritual landscapes required time, labour, food and organisation, but most of all, grass. Rich pastures bred and fattened the livestock that laid the ground for long-term, ambitious, interconnected building projects. Ireland had a lot of grass and at the time, may have had a larger population than Britain. Passage graves had grown progressively more elaborate and massive as competitive aggrandisement reached new heights. Belief systems and styles of construction imported from the continent were being superseded by local inventiveness. One of the most striking sites could be found midway along the coast of eastern Ireland, where the sheltered estuary of the Boyne opened the door to a navigable river that meandered inland through brown-earth country. Four miles from the coast, the river took an idle southerly loop, and within the loop a ridge rose to 60 metres or so. The lower slopes of the ridge were planted with various grains, while the upper slopes were used for both cultivation and grazing. Little tree cover remained. Partway down the forward slope of the ridge, overlooking the river, a spectacular tomb rooted ancestors to this part of Ireland and linked the living with the cosmos. Inside the tomb, a stone passage 19 metres long led from the outer world to the inner chamber, which had three recesses. Great care had been taken to angle the passage so that at sunrise in midsummer, rays lanced its length to illuminate a stone carved with a triple spiral. The passage was covered with a mound of earth 80 metres in diameter and kerbed with stones. The combined weight of material moved was something like 200,000 tons. Along the south side of the mound, dwarfing the small black void of the passage entrance, a massive revetment formed a blinding frontal facade of white quartz that had been carried from the Wicklow Mountains several days to the south. The rounded granite boulders that studded the facade had been sourced in the Mourne and Carlingford areas.

A mile from the great tomb, on the highpoint at the western end of the ridge, was another gigantic mound, also decorated with imported quartz and granite. Inside were two tombs, one with a 40-metre passage and the other with a 34-metre passage. Eventually, the western tombs would be surrounded by no fewer than eighteen smaller mounds. Stones at both tombs had been inscribed with chevrons, concentric circles, lozenges, spirals and cupmarks. So many tombs came to occupy the ridge of land above the Boyne that workers in the plots were always in sight of at least one. Natural features and man-made monuments – both old and new – were connected across many square miles to form an extended ritual landscape that functioned as a centre of power for the surrounding region.

Connectivity to the continent wasn't a factor in the development of spectacular ritual landscapes. Wealth and well-being accrued from the immediate neighbourhood. The Boyne was at the heart of its own geographical cosmos, centred on the Irish Sea. The same could be said of a second, even more elaborate ritual landscape. In sight of the British mainland, relatively level, brushed by the warming Gulf Stream, dotted with freshwater lochs and carpeted in luxuriant grass, the Orkneys offered some of the best grazing in the entire archipelago.

The Orkneys shared with Doggerland a history of rising waters, with the crucial difference that they had started higher and were rimmed with resilient rocks rather than erosive glacial junk. Nevertheless, the generations of people who had lived on this sea-girt northern archipelago had seen the ocean rise century by century. Since the ice had withdrawn from northern Britain, the seas around Orkney had lifted by close to 30 metres, taking out much of the prime land that had been used by pioneering foragers. Where there had once been a single, large island, by 3000 BC there were several smaller islands. On the largest island that remained, the two biggest freshwater lakes happened to have developed alongside each other, separated by a long, thin isthmus of dry land that ran like a wide causeway between the waters. The two lakes were rich in fish, fringed by reeds and the grassland was dotted with cattle, sheep and goats. Most of the birch and hazel woodland had disappeared over a thousand years earlier, so Orcadians had little need to devote time to tree felling. As a settlement zone it was a near-perfect mosaic of ecosystems: the sea was a short walk away for marine foods; the lakes and streams provided drinking water and fish; there was building stone in abundance; domesticated flocks and herds, and 200 square miles of lush grazing. The people here were familiar with abundance. They were rich.

Ritual landscapes existed in space and time. They were accumulator sites. A founding monument would attract a growing family of attendant monuments that were modified in meaning and form as they crept across the immediate land over hundreds of years. Ancestral structures were no less important than new-build. Even the houses on Orkney were monumental. On a well-drained site at the southern end of the isthmus between the two lakes, skilled wallers built a group of freestanding, roughly circular, stone buildings clad with turf. The interior plan of a couple of the houses echoed the rectilinear, cruciform shape of chambered tombs. The Barnhouse dwellings had a square, kerbed hearth in the centre, with stone furniture and recessed box-beds. Two miles from the other end of the lakes, another group of houses developed. These, too were built with a single rectangular living space within a curved outer shell of drystone walling. The homes at Skara Brae were fitted out with stone beds and dressers. With soft furnishings of cloth and skins, and painted interiors, these were snug, resilient homes. The half-dozen houses evolved into a single, contiguous development with party walls and communal passages linking entrances. Paved areas and drains improved cleanliness. Both settlements were in use for hundreds of years, probably from around 3100 BC.

The people of Barnhouse and Skara Brae shared a remarkable, ritual landscape. Only a mile or so from Barnhouse, a passage grave like no other was raised on a gentle slope at the end of the eastern lake. At its core was a chamber 5 metres wide and 3.8 metres high, with a corbelled roof of overlapping stones which rose upward to form a drystone dome. The skill of the stonemason, the intuitive grasp of gravity and weighting, the dexterity of assembly, made this one of the most remarkable structures in Britain. Some of the stones shifted to create the Maes Howe tomb weighed three tons. The internal masonry of the walls was shaped and smoothed so that the chamber was an acoustic cell. Thin, incised lines on the stone may have delimited painted decoration. At each corner of the chamber rose a large, upright monolith, while square apertures opened onto three smaller side cells. A low-ceilinged seven-metre passage connected the confined internal splendour of the chamber with the bright, expansive outside world. The passage had been precisely aligned and sloped, so that the light of the sun would shine along it for the three weeks before and after the shortest day of the year. The entire, stone edifice was smothered with a symmetrical mound of earth, circular in plan.

Circularity was a recurring form in the evolving landscape of the Orkney isthmus. The simple purity of a circle had appealing connotations. It was a form repeated endlessly through the known world, familiar in eyes, the full moon, the sun, in tree-sections, water ripples, nests and beaver lodges. It represented the cycle of life through days and months and years. The circle had no end. It was the cosmos. Every point on the circle was equidistant from the centre, and the circle divided the plane upon which it rested into two areas: the interior and the exterior. Circles could be created on the ground with a rope and a peg, and then defined with a ditch and a bank. It's possible that the great tomb hidden inside the circular mound of Maes Howe evolved from a circle of stones that originally occupied the site.

A few hundred metres from Maes Howe and the Barnhouse dwellings, a circle was laid out on the southern end of the isthmus. With a diameter of 61 metres, the ditch and bank surrounded a circle or ellipse of twelve standing stones. The entrance to the circle faced Barnhouse. Construction of the monument – the Stones of Stenness – can only have been a communal effort, and it consumed over 12,000 man-hours. These were big stones, weighing at least nine tons. The ditch was four metres wide and had been cut a metre or so into solid rock. Filled with water, it formed a reflective moat around the standing stones, separating two domains in much the same way that the isthmus did. With the addition of a second, much larger circle further along the isthmus, the ritual landscape of the lakes became even more elaborate. Brodgar's rock-cut ditch measured 123 metres in diameter and encircled a megalithic ring of sixty stones. Over 14,000 cubic metres of rock had to be hacked, prised and hauled in a construction project that took the best part of 80,000 worker-hours. The Ring of Brodgar covered an area eleven times greater than Stenness and was a measure of the rising numbers of people being drawn to the isthmus. Other chambered tombs and standing stones congregated around the twin lakes, but the greatest monument of all was set upon the spine of the isthmus itself.

Over a period that may have spanned as much as 700 years, a magnificent enclosure – a temple complex – took shape. Behind a massive drystone wall that was roughly rectilinear in plan stood more than fifty buildings, the largest of which was 25 metres long and 18 metres wide, with walls more than 5 metres thick. At the time, it may have been the largest roofed structure in northern Europe. And this roof wasn't lumpen with hides or clods of turf, but laid with thin slabs of overlapping stone that had been trimmed into rectangles. Paved walkways threaded the complex and much of the stonework was carved and coloured. Orcadians had graduated far beyond subsistence. The exterior wall of the complex was of a magnitude that expressed power and cultural security: it was no less than 6 metres thick and 3 metres high, opening at one end to a mountainous midden; at over 5 metres in height, it was probably the largest ritual rubbish heap in the entire British archipelago. The two wetlands and their dividing isthmus, the pair of circles and the ceremonial centre between them, the drystone settlements and fabulous tomb had become an integrated ritual landscape that embraced the natural and the manmade, the imagined and the physical, the ancestral and the modern. The ritual landscape at the meeting of the lakes on Orkney eventually covered a larger area than many of the urban centres of the Euphrates.

As if to counter the magnetism of the great ritual landscapes of the Boyne and Orkney, construction at a centre on mainland Britain accelerated at about the same time. The southern part of the Plain, where the Avon snaked between curves of chalk, had been a 'special place' over 4,000 years earlier, when people had been drawn here by the clear waters of the river and by the herds of aurochs grazing amid the scattered sarsen stones. It was the parallel ridges aligned with the axis of the solstice that had motivated the raising of pine totems here. Much later, around 3500 BC, monumentalists had constructed their great temenos some 500 metres north of the ridges. But the building works that transformed this tract of grassland for all time, began at the turn of the millennium. This was where, in around 3000 BC, an elite dynasty or community made their mark on the land in such a way that this place would eventually draw more than one million visitors a year. They integrated the strange, linear ridges with a new, circular cemetery.

Markers were placed and trees felled; pits were hacked with antler picks; tons of soil shifted. A circular enclosure some 110 metres in diameter took shape, defined by a bank and outside it, a ditch. The main access to the enclosure was a gap in the north-east quadrant, which opened directly onto the avenue formed by the two, parallel ridges in the Plain. Ancestral tokens were placed in the bottom of the ditch and fifty-six holes were excavated in a ring around the inner edge of the bank. The holes, and various other pits dug into the bank and ditch, were used for interring the cremated remains of people associated with the site. At this stage, the enclosure was not so dissimilar to others in Britain. But the circle-makers on the Plain had connections to a range of weathered hills on the far western coast of Britain. In the interconnected Britain of 3000 BC this was not unusual: the island was criss-crossed with routes used for conveying stone axe-heads from the remote 'holy mountains' of Langdale and from other high-value western outcrops. Cattle and pottery were moved from region to region. But the stones quarried and levered from the Preseli Hills were of a new magnitude: they were up to two metres long and weighed four tons. Eventually, eighty of them would be hauled and heaved from the Preselis to the Plain, a straight-line distance of 140 miles; some 250 miles if the strain had been shared by boats from Milford Haven to the inner reaches of the Severn Estuary. At the time, it was the single greatest consignment of building materials ever shifted in Britain. Fifty-nine Preseli stones were raised in a circle within the enclosure on the Plain.

The circular enclosures of Orkney and the Plain were central to a new belief system that spread quickly throughout Britain. Size varied, but the form was fairly constant. The basic 'henge' was usually circular in plan, with a gap – or gaps – to permit passage, and it was constructed by excavating a ditch and piling the material outside the ditch to form a bank. Many henges incorporated megalithic rings, although there were plenty of examples where circuits of stones were raised without the ditch and bank of a henge. Megalithic rings proliferated through Britain, but were raised mainly in the west where geology more frequently presented bedrock as an available local source. Eventually there would be over a thousand of them, enclosing areas that ranged from a couple of hectares to as much as 12 hectares. The basic ditch-and-bank henges were not dependent on local stone and had a wider distribution than megalithic rings. In some places, henges developed in clusters. Up on the crest of the misty Mendips, an arrangement of three circles was excavated, equidistant at 59.4 metres, in a line, with a fourth circle a little further away. On Dartmoor, a pair of circles at Grey Wethers were built six metres apart. Further north, in the lee of the Pennines on the gravel floodplain of the River Ure, three huge henges were erected in what was almost a straight line. The likelihood is that these henge-groups were not constructed as complete ensembles, but developed with time as extra circles were added to accommodate a greater range or scale of ceremonies.

Many of these henge-sites appear larger than could have been justified by the local population. On the Isle of Lewis in the Outer Hebrides, a ritual landscape developed on a promontory above the most pronounced sea-loch on the exposed western coast. Local cliffs were relieved of thirteen stones that were erected in a 12-metre ring whose centre-point was an upright nearly 6 metres high. Converging on the circle was a cross-shaped arrangement of tall, thin pillars, longest to the north, where the megalithic figures strode along the promontory for more than 80 metres. Shortly after the stones were raised, a chambered cairn was fitted into the circle, its entrance facing the rising sun. Several more stone circles rose from other points along the shore of the same, sheltered sea-loch. Lewis cannot have supported a large population. Perhaps Loch Roag was the centre for a wide circle of dispersed people who journeyed to this haven as pilgrims at certain times in the calendar.

Common to all ritual landscapes was a concentration of people with a shared belief. That belief seems to have been shared throughout Britain's disparate regions. The henges of the Hebrides and Orkneys, the Plain, the Pennines and Mendips were built by local populations who all saw the circular bank and ditch, with or without pillars of stone, as a focal point for ritual. The regional variations arose from geology, tradition and geography. By virtue of their proximity to the continent and role as a bridgehead, the south-east and Thames emerged as a leading entrepôt for imported ideas; the rock-rich coasts and islands of the Atlantic facade produced the archipelago's most astounding megalithic craftsmanship; the flint-rich, chalk heartland of the Plain had a history of nurturing insular innovation. Physically cut off from the continent, home-grown development was becoming increasingly significant. Britain had no conventional villages, let alone cities. Ritual landscapes, with their episodic human congregations and enduring structures, were the closest the island came to experiencing urban settlement. But complex, built landscapes were multiplying.

Less than a day's walk north of the stone henge by the Avon, the ritual landscape that had developed around the social enclosure of Windmill Hill and in about 2800 BC the long barrow of West Kennet gained its own megalithic ring. (Much later, the village of Avebury rooted within the ring.) The old long barrows and the wandering banks of the ancient enclosure up on the hill recollected ancestors, but this watershed landscape spoke of youth: of streams and springs. In a world where the direction of a river mattered as much as the volume of its flow, anyone walking west from the estuary of the Thames would ignore the various streams joining from the north or south, and be led, hour by hour, upward towards the chalklands that formed the Plain and its associated ribs of downland. Trending west, and confined increasingly between steep, wooded folds, the river eventually turned abruptly northward to enter a short, broad vale of extraordinary fertility. The vale was almost level, and in the moist, warm climate of 2800 BC, the high water table fed its pools and stream with a constant supply of crystal water. It was a place where trees and grasses thrived, and people too. Maybe this vale marked the rising of Britain's greatest river. A mile or so up the vale, a few hundred metres from the dip where the river rose from the earth, a huge megalithic ring of thirty or so sarsens was raised upon a low knoll. It was 104 metres in diameter and one of the largest circles in the archipelago.

There was nothing static about these ritual landscapes. They were long-term development zones in which monuments were dynamic installations. The ancient tomb or earthwork was a statement of ownership that linked land rights to ancestors. Time itself buttressed a claim. Building monuments to ancestors connected the variable – and often unpredictable – spans of human existence to the eternal rhythms of earth and sky. A rejuvenated tomb or cemetery was the means of controlling time and space. Over the next 1,500 years, many henges would undergo serial modifications as successive generations altered their form to suit current beliefs. How much the evolution of these belief systems was tripped by 'events' is impossible to say. Sequences of extreme weather, comets, disease and the fallout from distant volcanic eruptions may have played a part in prompting modifications of ritual landscapes. The emergence in Europe of a common language – Celtic – lubricated the exchange of ideas. Some ages, some congregations, were more energetic than others. But the stones and the earth would be bequeathed to the next generation. They were expressions of continuity in a timeless world that gave so few years to the ant-like routines of humankind.

The Avebury henge was modified in about 2600 BC by the addition of an enormous defensive bank and ditch. Roughly D-shaped, it had an internal diameter of nearly 350 metres and enclosed 11.5 hectares. The ditch was 7–10 metres deep and, at its top, some 23 metres wide. The bank created by spoil rose like a cliff, 17 metres from the square-bottomed base of the ditch. The hacking, digging and hauling would have taken around 156,000 man-hours, equivalent to a team of 100 people labouring for at least three years. As an exercise in excavation, it was unprecedented. But that wasn't the end of it. When sarsens were set round the inner edge of the ditch, the monumentalists of the Kennet had succeeded in creating the largest megalithic ring in Britain. It's hard to escape the conclusion that these periodic upgradings were symptoms of competition with the neighbouring super-site, eighteen miles to the south.

The stone henge near the Avon was redeveloped along ever more spectacular lines. Over the four hundred years since its initial construction, various timber structures came and went, and a second circle of twenty-five or so Preseli stones was raised just over a mile away, on the west bank of the river, although this one was much smaller, measuring just 10 metres in diameter. Then, during the mid-centuries of the millennium, the henge underwent its most dramatic transformation. By 2500 BC, it had grown to become the tallest artificial stone structure in Britain. From the core of the monument, there now rose five towering trilithons arranged in a horseshoe which was aligned with the midwinter sunset and midsummer sunrise. Each trilithon had two uprights capped by a gigantic stone lintel – an innovation that re-emphasised the uniqueness of the site. To secure the lintels, mortise and tenon joints had been chipped into the top of the uprights and underside of the lintels. Just getting the stone for the trilithons to the site had been a staggering undertaking, for they had been hauled overland for twenty miles or so from the spreads of siliceous sandstone above the River Kennet, beyond the northern edge of the Plain. These were far larger pieces of rock than the stones from the Preseli Mountains. Some of the sarsens weighed 25 tons. Dragging them through the woods and brooks was hard enough, but the route taken could not avoid the long haul from river level to the circle site on the Plain. Like the earlier stones, the new sarsens were painstakingly dressed with muscle-power and hammerstones, smoothing their surfaces and eroding all affiliation with their origins. Outside the five trilithons, and on the same axis, a double circle of Preseli stones was raised, and outside that circle, a much taller circle of thirty sarsens was erected, capped with thirty lintels, to form a continuous, elevated ring. Various other stones and structures occupied spaces within the circles. In one, relatively short phase of modification, the stone henge had been upgraded into the most spectacular megalithic ritual arena in Europe.

Stonehenge was part of a complex that incorporated lodgings for the workers who built the latest phase of the henge, and for the thousands who streamed on foot each winter from all quarters of Britain. A close-packed settlement of at least a hundred wattle-and-daub houses sprang up two miles north-east of the henge, on the banks of the Avon. In the weeks approaching the midwinter solstice, this place thronged with upward of 4,000 people. At capacity, this was one of the largest congregations in northern Europe. Some came from as far north as the lands beyond the Tweed. And the pilgrims flocking to the Avon drove cattle and pigs. Winter nights crackled with flame and burning fat as the butchered beasts were roasted on open fires. The feasting was prodigious. There was no place like it in Britain.

In its vast, interconnected entirety, the complex on the Avon was unique in its scale, spectacle and in the human resources consumed in its construction. Its initial form and successive remodelling over the centuries were British events, unifying enterprises that reflected the vision of a single island culture. For several centuries, this was more than just a 'place', it was *the* place, its megalithic silhouettes touchstones in the impermanent world of the pastoralists. Some thousand years or so after the first cities developed on the alluvial flats of the River Euphrates, the people of Britain were beginning – with their crowd of houses on the banks of the Avon – to experience high-density living. Customs were changing, too. People were being buried in single graves. Round barrows were becoming more common.

In around 2470 BC the thronged settlement on the Avon was itself monumentalised by the excavation of a gigantic henge ditch that cradled the houses like eggs in a nest. It was a massive, communal exercise. The ditch was some 5.5 metres deep and the bank of spoil outside was some 3 metres high. The finished henge was the biggest in Britain and encompassed 17 hectares, an area nearly twice as large as the social enclosure on Windmill Hill. The people who built what became known as 'Durrington Walls' left a tantalising trace of a ma-terial new to Britain, and one that had already revolutionised cultures on the continent. Among the cuts by stone tools in the chalk blocks used for the henge were a couple that had been incised by a thinner blade. The same kind of thin, sharp blade was also used to shape the timbers used to form a wooden henge on the east coast. Copper was softer than flint but lighter to wield and quicker to use. Where a stone blade was most effectively struck at an oblique angle to a tree's grain, gradually slicing deeper and deeper into the timber from opposite directions, a copper blade, being thinner, could strike the trunk at a more perpendicular angle. At around the time that copper first appeared in Britain, the size of trees being felled suddenly increased from trunk diameters of 30 centimetres to a metre. Metal was a material of mass transformation.

EIGHT

The Great Polyhedron

2400–1500 BC

The most spectacular, final flourish from the age of giganticism can still be seen on the valley floor of the Kennet a mile or so from Britain's largest henge. It's among the most enigmatic monuments I know. Archaeologists who've burrowed like moles into the chalky core of the structure have been able to compile the slender thread of a chronology.

Work began in around 2400 BC when a number of small, artificial mounds were piled close to the riverside. Two of the mounds have been found. They were less than half-a-metre high and the insects and seeds within them indicate that the Kennet valley was open grassland, well grazed by animals – perhaps cattle and sheep. In one of the mounds were bramble seeds, sloe stones, yew berries and fragments of hazelnut shell, so there must have been trees or scrub nearby. Cereal chaff suggests that there may also have been cultivation plots in the vicinity. In this productive setting, space was cleared for a far larger mound. Stray sarsens may have been heaved to one side. A fire was lit for feasting (or at least a hearty meal) and the ground seared by a hearth that was left scattered with charcoal, the charred shells of hazelnuts and two pigs' teeth. The ground was prepared, turf lifted and the topsoil of silty clay scraped away to expose the subsoil of clay, embedded with flint. On this 'cleaned' surface, a low, circular heap of honey-coloured gravel was piled, about 10 metres in diameter and just under a metre high. The gravel had been sifted from the subsoil, or it had been carried from the Kennet; either way, the composition of this starter-mound required effort that was symbolically rather than structurally motivated. The small, golden stones had meaning. At some point later, a 16-metre perimeter was marked out around the gravel mound using posts or stakes. Ceremonies may have occurred around this pale mound before it was smothered by a mixture of topsoil, subsoil and turf – possibly the spoil that had originally been cleared from the site. By now the mound had risen to waist height and been modified by the excavation of pits a metre deep into its upper surface and sides. More elaborate ceremonies may now have occurred before construction resumed as baskets or hides loaded with topsoil, turf, gravel, clay, chalk and rounded lumps of sarsen were lugged to the site. The various materials, contrasting both in colour and texture, created a layered effect on the mound, while the sarsens were me-thodically spaced like seeds within the mix. The mound had swollen to a diameter of over 30 metres and a height of over 5 metres. At this point, there seems to have been an acceleration of ambition as the monument evolved into a far larger, henge-like form. A peripheral ditch and bank were added. Unlike a true henge, however, the bank was raised *inside* the circuit of the ditch – which may not have been continuous, but segmented. The scale was impressive: part of the ditch was 6.5 metres deep and 6 metres wide and it had a diameter of over 100 metres. Meanwhile, the mound had been enlarged again and changed colour as bank after bank of chalk was piled against its slope. The rising monument on the Kennet was already a marvel, and almost as large as the circuit of sarsens a mile up the valley at Avebury.

Construction continued. It was as if the people of the vale had to reaffirm their beliefs through serial enlargement. The ditch was backfilled, and another excavated at a greater radius. Then another. And another. The mound grew higher and higher. A fourth circuit of ditches was cut, and a fifth. Subsidence was prevented by the raising

Rannoch Moor, where the ice was thickest 12,000 years ago.

Cheddar Gorge in the Mendip Hills, a 'sacred place' by 8200 BC.

The rump of Doggerland. Sandbanks off the coast of north Norfolk.

The chalk cliffs of a new island. Swyre Head, Dorset.

Top An impression of 'peak wildwood' (5000 to 4000 BC) at Rothiemurchus in Cairngorms National Park, where remnants of native pinewood cling to thin, mineral soils.

Above Pioneering flint mines from around 4000 BC and a later enclosure on Harrow Hill, South Downs.

Above right The portal dolmen or tomb of Pentre Ifan, Pembrokeshire, Wales.

Right Long barrows became prevalent after around 3800 BC. This is West Kennet, near Avebury, Wiltshire.

Henges, defined by an earthen enclosure, are found only in Britain and began to appear from around 3000 BC. This one is on Salisbury Plain.

Houses at Skara Brae, Orkney, were occupied between around 3100 to 2500 BC.

Work on the largest man-made mound in Europe began around 2400 BC. Silbury Hill is in sight of West Kennet long barrow.

Round barrows became common from around 2200 BC. These are on Normanton Down, Wiltshire.

of successive rings of chalk rubble which acted as internal revetments. Within the chalk were embedded clusters of sarsen, these geological strangers of the Plain having lost little of their wonder. The mound finally stopped rising when its summit stood 31 metres above the vale and the final diameter had stretched to 160 metres. Observed from the surrounding hills it took the form of a cone, but a cosmic viewpoint from above would have revealed that the base of the mound seemed to have nine sides and that the summit was rectangular. It appears to have been a polyhedron. By the time the last antler pick was laid to rest, half a million tons of material had been shifted by hand and the great polyhedron had become the largest artificial mound in the world.

Today the ritual vale is cauterised with tarmac and sprayed with the engine vapours of howling steel. The Great Polyhedron has been demoted to a 'hill': Silbury Hill. You have to walk the paths and flex the retinas in order to reconstitute this blunted, grassy cone as a bright, tapered spectacle of improbable scale. To the people who gazed up from the meadows beside the clear-watered stream, this must have been a source of amazement: a cosmic beacon compiled from the valley's constituent parts: turf and gravel, chalk, soil and sarsens. The earth had moved. You could see the Polyhedron from the top of Windmill Hill, from the long barrow at West Kennet, from the henge at Avebury. It was the all-seeing white eye at the focal point of the vale. Construction had probably continued for three or four generations, perhaps with stops and starts. Conceivably it was built much quicker, in a concentrated series of bursts. The multi-generational commitment to a project beyond the needs of food and shelter speaks of a belief so powerful that it could be inherited and sustained from birth till death. Everybody who had been a part of this four-million-worker-hour enterprise was part of its story. The hill embodied an ancestral narrative.

Who built it? And why? The Great Polyhedron wasn't a burial mound; despite the efforts of investigative burrowers, no human remains have been found within its layered interior. Rather, it seems to have been a viewing platform that allowed a small group of people to see things – activities and places – denied to ground-level mortals. The Polyhedron's tip provided a panorama that included the interior of the palisaded enclosure at West Kennet and most of the surrounding ritual landscape. Perhaps the Great Polyhedron was part of a wider transformation taking place at the time. As the mound rose from its meadows – possibly slightly before – two twin-rows of sarsens were erected, one leading west from the Avebury henge across the Kennet, and the other heading south-east, down the valley to a group of circular posts and stone monoliths on a spur of Overton Hill. They may have been processional avenues between ceremonial sites or have symbolised ancestral or supernatural pathways. It's likely they followed existing paths or routeways. At one point on the line of the western stones, an existing enclosure had to be demolished to make way for the new 'avenue'. Whatever their purpose, the avenues created enduring physical connections between nodes of human meaning. Movement was being monumentalised.

Seventeen miles away, on the southern reaches of the Plain, the monuments of the Avon also underwent a series of modifications. It seems that the lower, riverside circle was stripped of its Preseli stones, which were hauled 2,000 metres up the slope to Stonehenge and repositioned as an inner ring within the five great trilithons. Where the stone circle at the riverside had stood, a 35-metre henge with a ditch and bank were built. The ditch of Stonehenge's earthwork enclosure was recut, too. Some time later, a 2,500-metre avenue or processional way was laid out to link Stonehenge with the new henge down by the Avon. The width of the avenue – 22 metres at its upper end and broadening to 30 metres near the river – hints at the importance of the ceremonies taking place, and perhaps the large numbers too.

The Great Polyhedron, and the latest round of modifications at Avebury and Stonehenge, tell of changing rituals. These were strenuous, collective acts of imposition on existing ritual landscapes – and perhaps on populations, too. They may have been built by indigenous Britons to mark the adoption of new beliefs. Or could the Great

Polyhedron have been the final defiant act of a culture that was confronting its own demise? Whatever its motivation, the Kennet's gigantic, stratified mound was unprecedented in conception and scale, and can only have arisen from exceptional circumstances. It stands to symbolise the end of an era in which all-powerful chiefs could coerce the masses to move unimaginable quantities of earth or staggeringly huge stones. Stone-centred cultures dissolved in the crucible of the metal age, traditions and practices followed for millennia became obsolete, new power-bases sprang up, old hierarchies collapsed. The adaptive jolts can be read in the modifications made to long-term ritual landscapes. The forces exerted on the land surged in intensity from around 2400 BC.

Britain was an island in flux. Population had risen to levels greater than ever before and in the south and east, densities may have been heading for 10 people per 100 square kilometres. More of Britain had been cleared than at any time in the past and land was being worked more intensively. Irregular patchworks of fields could be seen the length and breadth of the island. Pioneering settlements used ards, mattocks, spades and hoes to break untaken land. Fields coalesced into extended areas of managed ground, the balance between arable and stock-keeping tending to depend on soils and the local climate. Fields were typically square and usually covered five hectares or less. In arable areas, their size may have represented the amount of ground that could be broken with an ard-plough in one day, their symmetrical shape arising because they were 'cross-ploughed' to increase the fertility of the soil. By ploughing a plot a second time, at right angles to the original direction, more nutrients were brought to the surface. Typically, a homestead would be attached to a cluster of fields that had grown organically as needs increased.

The 'farm' had arrived. The word would soon be seeded in the European lexicon by people working the plains of Latium, who used the adjective 'firmus' for strong, stable, powerful, constant, sure, true, *firm*. Farms were the new ritual landscape. Fields were places of communion; sites of daily devotion. It was to these angular plots that people were called to reaffirm their beliefs that soil and beast would provide. In an unpredictable world, these places of labour had their own reassuring, fixed parameters. Thousands of scattered communities developed local networks of fields and enclosures, tracks, bridges and causeways that were incised into – and raised above – the land surface. Many of these artificial features would endure for millennia, inherited, repaired and embellished, generation after generation.

In this increasingly busy land, contact with the continent was also running at unprecedented levels. Incomers were bringing new ideas and products, while people from Britain were travelling overseas. It had been metal-working immigrants who had opened the mine in Ireland which was producing most of the early copper in circulation. From around 2400 BC, well-armed warriors were buried in single graves on the chalklands of southern Britain along with deposits of barbed-and-tanged arrowheads and the polished stone wrist-guards they wore to protect their forearms from the painful whack of a re-leased bowstring. In the burials too were decorated pottery beakers which may have been used as ceremonial drinking vessels. A small number of them were immigrants who had probably landed in family groups over a period of time at various havens along the southern and south-east coast of Britain. These people-of-the-bow also brought delicate spatulas of bone and antler and ornaments of copper and gold – status symbols that could be used to showcase power, gender, social differences. The glossy, crafted objects they coveted had an im-mediate appeal in relatively undeveloped Britain.

One of these well-tooled, wealthy immigrants – or one who had adopted their ways – was buried in the upper fill of the recut ditch at Stonehenge. An adult male, he'd been placed in his grave along with a stone wrist-guard. His body had been pierced by three barbed-and-tanged, stone-tipped arrows. The dead bowman must have been a figure of some significance for his body to have been laid to rest in Stonehenge at a time when this was one of the most important ritual centres in Britain. Only three miles away, on the other side of the Avon, another body was placed in the downland soil at some point between 2470 and 2280 BC. He, too, was an archer, and he'd lived for between thirty-five and forty-five years. With him were fifteen arrows tipped with barbed-and-tanged arrowheads of flint, two polished wrist-guards of stone, flint scrapers and knives, boars tusks that may have been used as leather-working awls, a bone clothing-pin, a belt ring of shale and three decorated pots. Those grave-goods would not have looked too out-of-place in contemporary Britain, but the archer also had three copper-bladed knives, a pair of patterned gold body ornaments and a 'cushion stone', a small, portable anvil that could be used for metalworking. Close to the archer lay the body of a younger man, aged between twenty-five and thirty, also with gold ornaments. The foot-bones of both men shared the same unusual structure. The two were related, the older man having grown up close to the Alps and the younger man having spent his teens in central or northern Britain. They had both travelled.

As traditional beliefs waned, so did the motivation to raise spectacular ritual centres. There was one final bout of significant remodelling of Stonehenge as eighty or so Preseli stones were uprooted and moved into fresh configurations. Inside the horseshoe of trilithons, the central circle of Preseli stones was repositioned as an oval. And a new setting of Preseli stones was raised as a single circle between the trilithons and outer sarsen circle. This final arrangement would remain – albeit degenerated by theft, vandalism and decay – for the next 4,000 years. The modifications to the stone circles on the Avon and Kennet had petered out by 1900 BC.

The slowdown in megalithic monument-building was part of a general weathering of interest in stone. Post-glacial Britain had been founded with flint, and it had been the flint mines of the South Downs that had helped to raise the timber halls that had arrived with agriculturalists. Flint was still the stone-of-choice when Silbury Hill was raised, although by then the centre of extraction had shifted from southern downland to eastern Breckland. You could walk to Breckland from Avebury's monuments in three or four days by following the chalk belt that rolled north-eastwards towards the stump of the old continental land-bridge. The longest, broadest and most varied of Britain's chalk belts underwent a series of character transformations between the Plain and the North Sea: from sprawling, lofty downland above the Kennet and Thames, it mutated into the narrower band of the Chiltern Hills and then became a wider, softer range as it continued north-eastwards beside the flooded, peaty Fens, where it underpinned the low 400-square-mile plateau of Breckland. The chalk here was overlaid with layers of glacial gravel and sands that supported deciduous forest dotted with clearings. In one of these clearings, where the land sloped gently to a winding stream, flint miners had established a settlement.

I went back to 'Grimes Graves' while I was writing this chapter. When I was a child, visitors were allowed to bring their own torches and crawl unaccompanied through the mines. In the spectral shadows cast by the yellow glow of our bicycle lamps we met the miners, crawling on whitened knees, picks in hand. In places there were chambers large enough for all of us to wedge together. Today only one mine is open. You descend with a hard hat and a guide. After a safety briefing. It's still worth the visit. From around 2600 BC, this clearing in the Breckland forest was one of the largest flint mines in Britain. The stone the miners were after lay deep in black seams separated by layers of chalk. In places, that meant digging downward for 12 metres and then burrowing laterally. Subterranean networks of tunnels developed. Ritual deposits of pottery were left below ground by the miners. Spoil and flint from the shafts and tunnels were hauled by rope to the surface and spilled in circular ridges around the shafts. It was up here that the flint was worked into axes and discoidal knives. Over a period of about 400 years, successive generations of miners toiled at the site, extracting flint from one or two pits or shafts each year before moving on to a new plot. Today, there are 433 shafts and pits in this Breckland clearing, a silent, butterfly-flecked greensward pocked with craters and rimmed by black trees.

The megalithic zenith had passed. It was as if all vitality had fled the rough-hewn stones that had held generations in thrall. Across Britain, the old communal tombs had their passages blocked for the last time. In the pine forests of Breckland, scrub spread across a wasteland of spoil heaps, pits, quarries and hundreds of abandoned flint mines. Stone tools had been used on earth for over two million years. Britain's topsoil concealed handaxe flakes going back over 500,000 years. Working stone into tools was almost as old as language. The knock-tinkle of knapping was as natural as birdsong. But as the third millennium BC crept to a close, the rhythmic percussion of stones being worked slowly died and Britain's topsoil accumulated abandoned tools. Flint-mines, and the chalk lands, were competing with other geologies. Flint would continue to be used alongside metal for tools and weapons for at least another 2,000 years, but its role as a key raw material had come to an end. As the light waned from stone and monumental earthworks were reduced by erosion, millions of man-hours of labour were shifted, grain by grain, by rain and wind. Ditches acquired beds of silt; banks slumped. One of the more poignant sights as dusk closed over the age of the megalithic henge was the burial by wind-blown sand of the neat little stone houses close to the western shore of Orkney. The dressers filled with sand, the stone drains clogged, roofs collapsed and pastures disappeared beneath barren dunes. The entrance to the long barrow at West Kennet was sealed with huge sarsen slabs.

The old order, in which people congregated to fix their place in the cosmos with enormous works of stone and soil, was history. For 2,000 years, earth and stone had been worked into increasingly spectacular architectural forms. Ritual hotspots from Orkney to the Avon had been embellished with tombs, avenues, circles, pillars, altars and enclosed spaces of various sizes and shapes. Already, many of these landmarks had become historical monuments; by the time the trilithons were raised on the Plain, a thousand years had elapsed since the gigantic social enclosures had begun to appear on the chalklands above the Kennet and Avon. The extravagant scale of the earthworks and the resistance of stone to weathering won both sites a permanent place on Britain's land surface. Timber circles and buildings would crumble and rot – or burn – but much of the manipulated stone and earth remained in situ to form the visible prelude to our island story. The various families of megalithic structures and earthen forms that scattered the archipelago were the monumental foundations of Britain's anthropic landscape. And the greatest and latest of them all, mysterious Silbury Hill, is an impregnable rebuke to the armies of theorists who have sought to decode the beliefs of Britain's monument builders. We can never know why they were built, nor what happened at them. The bone and stone, bank and ditch, are inert leftovers whose only means of communication are location and form. All reason is mute. Many of these monuments would have been mysterious in their own time, and this perhaps is their preferred legacy: in the absence of knowledge, they're best seen as gymnasia for the human imagination.

Technology ages landscapes. In post-megalithic Britain, ancestors would be recalled with memories rather than monuments and labour would be devoted to the living rather than the dead. Success could be measured through acquisition: of trinkets of gold, tools and weapons in bronze, lusty bows and fancy beakers. And the land was a great provider. Britain's three-billion-year geology included every kind of rock useful to an early prospector, including accessible copper ore. Extracting the ore was a messy business, and in the long term made flint-mining look like a few pin-pricks on Britain's topography. Where they outcropped on the surface, copper veins could be opened by hacking at the exposed rock and by excavating trenches. They could be exposed by setting fires and allowing thermal expansion to weaken the rock. Veins could also be followed underground where they occurred in fissured bedrocks like limestone. Once extracted, the rock was pulverised with hammerstones and then the ore was smelted in pit furnaces. Substantial quantities of firewood were needed.

Over a period of five hundred years, the mines on the lake island in Ireland produced 25,000 to 30,000 tons of metal. And from at least 2100 BC, various sites were being exploited in western Britain.

There were veins on the curve of Cardigan Bay, on the island of Anglesey and on the limestone promontory of Great Orme, where the ore occurred in a series of stepped scars and could be extracted by open-cast mining and trenching. The Orme became so important that miners hacked and burrowed between three and six miles of underground tunnels and fissures that dropped an astounding 70 metres and extended laterally for over 240 metres. Orme's mine broke all records in Britain for an artificial subterranean void. Inland, the scarp of Alderley Edge hid copper ore, and so did the Manifold valley in the southern Pennines. Open-cast mining was used at Copa Hill too, where a copper lode outcropped 420 metres above sea level in the Ystwyth valley. At its peak, Copa Hill was a smoky, noisome gulch where grimy, sweaty labourers hacked at veins with antler picks, chucking the rock back to others who smashed it with hammerstones. Close by, trees were felled and split for the fires set against the cliff and to feed the furnaces that produced the glowing puddles of copper. The works at Copa Hill changed the geochemistry of the surrounding peat – the first recorded evidence in Britain of atmospheric pollution.

Metal brought to Britain a more intrusive form of production centre. For every ore quarry, there had to be woodland for fuel, a site for smelting and a source of human labour. But this was the new currency. Copper axes, daggers and trinkets that shone like the sun were exchanged throughout Britain and Ireland. Gold started to circulate, too. Unlike the stone treasures of old, metal artefacts could be melted and reworked into entirely new objects. It wasn't long before prospectors found a new ore. The magma that had welled through the earth's crust millions of years earlier to cool into the monumental granite tors of the south-west, contained fissures packed with a valuable mineral. Cassiterite was rare in continental Europe; Britain had lots. Nearly as hard as quartz, cassiterite crystals were easiest to recover from the beds of streams where they had been washed by erosion from local bedrock. Heating the ore released trickles of silver metal – tin – that could be mixed with copper to produce a new alloy that was harder than both: bronze. Experimental alloys of bronze had already been cast in the Aegean and eastern Europe, but two or more centuries of copper-smelting had allowed the metal-workers of Ireland and Britain to develop a flair for innovation. With the discovery that tin could be used as a toughening agent for copper, the archipelago's sweaty smelters found themselves at the forefront of a technological revolution. Not only that, but they had access to plentiful supplies of both copper and tin. The great leap forward occurred between 2200 and 2000 BC. After seven thousand years of continuous inhabitation, Britain not only caught up with the continent, but overtook. It would be the best part of two centuries before metalworkers of western and central Europe regained parity with Britain and Ireland. The bronze breakthrough was a triumph of ingenuity over insularity. Copper, a 'native element', could be separated from its ore simply by heating; bronze, an alloy, required the mixing of two metals. Technologically, it was a huge leap into the future.

Bronze facilitated mass destruction. Looking back, the flint axe had been a clunky, friable tool for reducing woodland. Striking an unnoticed knot, or swinging too hard or at the wrong angle, would break or shear the stone. Large trees were not worth the effort. Copper-headed axes made it possible to tackle greater girths, but the soft metal quickly blunted or bent. It was possible to hammer copper blades straight again and then resharpen them. But it was still – compared to bronze – a very slow process. Bronze was hard, resilient, and it took a sharp, fine edge. A team wielding bronze axes could bring down a tree in a fraction of the time that used to be taken with flint. When a large oak was required for a ritual burial on the east coast, no fewer than fifty different bronze axes were used in its felling. The stump was upturned, its roots clawing the sky at the centre of a timber wall set into the edgeland between woodland and sea.

Other great changes were in play, too. It was during this metallurgical dawn that Britain imported from the continent the domesticated horse. The harbinger of hypermobility and a counter-force in a world that was becoming more sedentary. The horse was a symbol of power and prestige and its appearance in Britain enhanced the supremacy of bronze-wielding elites. And at last, the wheel came to Britain, too, the forming of its components owing everything to the availability of new bronze carpentry tools. The first wheels were solid timber discs fabricated from planks of alder, braced with dovetailed strips of oak and pegged with ash dowels. When attached to an axle mounted beneath a timber platform, the first carts and wagons began to evolve. The wheel had countless applications, but the most significant in shaping the land was its role in short-haul shifting of heavy freight. The wider communications network, evolved over millennia by feet and hooves, and threading for the most part through wooded lands, was neither wide nor smooth nor clear enough for wheeled traffic.

Bronze tools also revolutionised boat design, making it possible to hew and carve oak planks with far greater precision and speed than was previously possible. Planks with matching bevelled edges could be fitted together so closely that, with the aid of caulking, the joins were relatively watertight. A new kind of boat hull evolved, made by sewing together with yew fibre carefully trimmed timbers and planks. Sewn plank boats were far larger than log-boats, with a higher freeboard and greater stability. Up to 18 metres in length, they could carry five or six tons of cargo, or several animals, or up to twenty people. Where log-boats were suitable for rivers, lakes, estuaries and sheltered coastal waters, the new sewn-plank boats were capable of coast-hugging voyages, and perhaps open-water crossings, too. They evolved over the centuries, becoming broader and more rigid, but the basic concept was so successful that they remained in production for a thousand years and led to new kinds of 'boat places': boat-building yards and trading stations.

Those who lived into the new millennium have been amazed by the pace of change. It must have seemed to many that the cosmos was spinning off its axis. Gold, copper, bronze, horses and wheels were just part of the story. Britain's diverse geology continued to release a stream of rare and wonderful stones, among them jet and amber from the east coast, lignite from the far north and shale from the Channel coast. Amber was either imported from the far side of the North Sea, or picked from the beaches of eastern Britain. Pioneering work on the mixing of metals to create bronze laid the ground for other innovative processes. Making a faience bead required the correct proportions of crushed quartz and lime, followed by moulding, glazing and firing. Weaving and dyeing produced brightly coloured linens, while more traditional skills turned out intricately crafted carpentry and leather-work. The modern world was weaponised with dagger-knives and battle-axes. A combination of resources and home-grown technical ingenuity was creating a new generation of production centres.

A more resilient Britain was emerging, capable of developing new technologies. Original thinking had left its mark on the land back in the fourth millennium BC with the development of the strange temenoi enclosures that could be found nowhere on the continent. The ritual landscapes of Orkney and Stonehenge were unparalleled. Beating Europe at bronze by a couple of centuries can only have empowered indigenous innovators. In the new dawn of the second millennium there was a sense that the seas around Europe's largest island were functioning as filters, permitting the passage of the best ideas and repelling the worst. The encircling waters of Britain sep-arated a terrestrial sanctum from the greater world in the same way that the liquid circle at Stenness placed the stones in a world apart from the isthmus. It was Britain's geographically removed location, maritime and northerly, that spared it continental disasters. While Britain had been feeling its way through the introduction of bronze, the entire civilisations of Akkadia, Egypt and Mesopotamia all collapsed as climate change seared the lands beyond the eastern Med-iterranean with terminal drought. In the Indus Valley, the monsoon ceased for two centuries. Cities were abandoned and their sanitation systems clogged with dust. Britain, meanwhile, enjoyed a stable and relatively mild climate. Lands were being transformed, but by whirl-winds wrought with technology and geopolitics.

The scale of communal belief and enterprise necessary to construct shared tombs, massive monuments and extended ritual landscapes was swept away by the rise of hierarchical elites. Burial and memory were enshrined in single tombs. From around 2200 BC, the land surface became progressively dimpled with hundreds, and then thousands, of circular mounds covering single burials. Size and shape varied. Some round barrows were as small as 5 metres in diameter while others swelled to over 40 metres. Frequently sited as eye-catching domes on the skyline, these reproducing hummocks were often aligned to form a row that dominated the view. The uniformity spoke of a common belief, expressed locally. Round barrows in the chalklands could look particularly impressive. With the urn containing the remains of the cremated body placed inside a grave within concentric rings of timber fences, turfs cut from around the grave were piled into a mound of say 30 metres in diameter and then covered with chalk carried from a surrounding, protective ditch. Under the bright light of a clear sky, these gleaming mounds shone from the grassland. The continuing pre-eminence of the chalk heartland was reflected by an accumulating rash of barrows that studded the gentle grassland domes like rivets on a bronze shield. From around 1900 BC, the interment of fabulous grave-goods alongside the departed told of a rise of new elites and of the central chalklands as a continuing focus of power. Eventually, some thousand or so round barrows were gathered on the banks of the clear-watered Avon, centred on the traditional gathering ground at Durrington Walls (although the barrow-builders appeared to have been repelled by Stonehenge itself, and – with a few exceptions – left a green belt of open land around the towering grey trilithons). Barrows signposted other centres of power, too. In the western lowlands not far from the River Dee, a woman was buried in a stone-lined grave along with a beautiful gold cape, a necklace of amber beads and a bronze knife. She died between 1900 and 1600 BC, and her grave too was marked by a heaped barrow. Round barrows became the most common prehistoric monument in Britain, with as many as 30,000 protruding from the island.

Where monumental stones and earthworks would endure, round-houses and other ephemeral structures would disappear from the surface of the land. Perhaps the greatest surprise is that 2,500 years of farming had not imprinted the land with fields on a greater scale. Agriculture had been a start-stop process in temperate Britain and although people had been cultivating domesticated crops since at least 4000 BC, arable subsistence had failed to take root as a widespread strategy. The landscapes that had evolved reflected a culture that depended more upon keeping animals, and on a continuing habit of foraging for wild foods, and on fishing and hunting. The corridors and clearings through the wildwood had evolved through time into the trails and grazings of herders and shepherds. Woods were fired or felled. Open areas frequented by pastoralists had waxed and waned with changes of climate and population; old pastures abandoned for new. Pastoralism had a fluid relationship with the land, following the rivers and pools of grassland; it had no fixed abode. These tethered pastoralists had lived with a relatively light footprint, their most enduring landmarks being the scattering of funerary monuments and ritual landscapes. But in the middle part of the second millennium BC, there was a seismic shift in the way landscapes were used. Imposing monuments ceased to be built. Gods and ancestors were usurped by the plough.

PART THREE

Field & Fort

NINE

Field Pattern

1500–800 BC

I'm lying on one of the tors that poke from the acidic domes of Dartmoor. The granite is warm and rough and the summer sky is snagged with sheep-wool clouds. The tor is roughly 20 feet high, and the elevated plinth of its topmost block makes me feel as if I'm floating above the moor. Far below is the deep cleft of the river. What did they use to call it, back in the second millennium BC? The roots of the Dart's name are a little later – the first millennium BC – and are taken to mean 'the river where oak-trees grow', a special place perhaps in what had become deforested upland. This is still the great watercourse of the Moor, rising at over 1,500 feet and tumbling downward as the West and East Dart, twinned streams that combine their flows at the confluence known as Dartmeet. I can't see Dartmeet. It's only a mile or so to the north, its waters screened by those oaks. Eastwards, downstream, the convex slope is daubed with variegated greens of grass and bracken. As I stare, faint shapes begin to appear, linking to form a pattern that covers the entire valley-side. Spread before me is the imprint of a prehistoric field system.

Half an hour later, we're wading the bracken. In winter, under a low light and a dusting of snow, the field edges stand out as crisp, rectilinear etchings. Even in summer, the banks of turf and stone are clear to see. Parting the rampant ferns are the remnants of fields and houses; lines and circles of forgotten geometries. On these abandoned heights, it's a challenge to imagine the ground beneath our boots as possessed and tended. Back then, the houses stood tall and were thatched or turfed. The field-banks may have been topped by hedges. There were ditches too, now filled with silt. People lived and worked, laughed and bickered among these patterns of piled stone. The rectilinear field-banks of Dartmoor have come to be known as 'reaves', from an Old English word, *raew*, meaning 'a row'. There was good soil between the reaves, but it was littered with angular granite boulders – locally-known as 'clitter' – formed by frost action during the Ice Age. Dartmoor's heights were better suited to pasture than cultivation. The angle of reaves relative to gradients may have been intended to create tripartite units, each with a section of valley, slope and upland. The open upland was probably grazed as a common, available to the whole community. These vestigial reaves are the fossilised conductors of an intensifying agricultural economy that spread like capillaries of molten metal across southern Britain from the middle centuries of the second millennium BC.

Britain was filling up and the only means of feeding additional mouths was by taking more food from the land. Existing plots were worked more intensively, and marginal lands were taken for crops and pastures. Changes in the climate did not help. For the third time since the glaciers had pulled back in the tenth millennium BC, Britain began to be gripped by an extended cold interval; a 'Little Ice Age'. From around the fifteenth century BC, Earth received less heat from the sun. The dip in solar input lasted until the eighth century BC, long enough for the complex, interrelated heat-exchange systems of the planet to alter their intensities and impacts. The edge of Arctic sea ice shifted south, and so did the winter westerlies, which slid away from Britain and struck the western Mediterranean instead. Winters in northern Europe turned from mild and wet to cold and dry. Within this prolonged, chillier era of reduced solar input, there were three especially deep troughs, the first occurring at around 1450 BC. The climatic shift altered yields and forced agriculture to adapt. The shift to cooler, wetter conditions was accompanied by a rise in the water tables of bogs that lasted for 200 or so years.

Woven with human and atmospheric forces was the perennial, energetic glow from the source of power and innovation that had always invigorated Britain's destiny: Europe. As ever, this chilly, northern archipelago was lower down the settlement hierarchy than the continent. At around the time that farmers were gazing from Dartmoor's windy tors at the frontier of their advancing field systems, the Mediterranean island of Crete was dotted with towns, palaces and villas, complete with storerooms, cisterns and bathing pools. Continental Europeans had emerged from the woods into the brash consumerism of a bronzed, trans-continental age. New technology and ideas, people and wealth spread on an expanding network of communications. On the limits of the landmass, Britain was too remote to be arterially connected yet close enough for partial inclusion. Southern Britain – and southern Scandinavia – were drawn progressively into the fringes of the European exchange network. Scandinavians were rich in dried fish, cattle, sheep, furs, seal-skins and seal oil; Britain could offer tin, copper, gold, hides, cloth, hunting dogs and slaves. From the continent came new crops like the bean, the pea and spelt, and innovative strategies for preserving harvests. Lined pits began to be used to store feed through the winter, and an ingenious granary appeared, raised on four posts to allow the air to circulate and to protect grain and animal feed from rodents. Wool and looms spread and the numbers of querns increased.

In the expanding European exchange-zone, it was the southern sea-havens and rivers that became the main conduits of material wealth. Participation in this glossy world of acquisition and alliance came with two strings – or gigantic hemp ropes – attached: membership was not available to subsistence economies or to remote communities. To join the exchange-zone, you needed surplus products and established communications with corresponding economic zones on the continent. Best-placed to achieve that end were fertile lowlands close to south-eastern coasts. Primary entrepôts developed on sheltered Channel inlets and along the estuary and river banks of the Thames. Secondary networks reached along coasts and inland. The effect in Britain (and Scandinavia, too) was to further segregate the north. Producing surpluses that opened the way to exchange was achieved by maximising productive capacity: by intensive agriculture. The middle centuries of the second millennium BC saw the landscapes of southern Britain reorganised in what became the island's greatest communal effort to date. Thousands of square miles of informally worked lowlands were overtaken by field systems and linear land divisions.

The crucible of the new economy was the farm. Concentrations of fields, enclosures, trackways, droves and homesteads grew to be as complex and as integrated as the ritual landscapes of earlier times. Just as the henge, temenos, stone circle, tomb and avenue had been connected through time and space to form extended physical areas, farms and their attendant structures were woven into far-reaching economic landscapes. Agriculture anchored the activities of farmers to the field-walls and ditches, banks and hedges that defined their claim on the land. These were the new, familial monuments. The collective efforts of communities were directed to utilitarian ditching and hedging, track-laying and the raising of multiple, low-level boundaries. Burial mounds rose no more; the dead were cremated and their remains placed in urns which were buried in 'urnfields'. The resting place of the dead now resembled the workplaces of the living.

The farm reformation stretched over hundreds of years, long enough for there to have been many phases of change and evolutions of landform. But the guiding principle throughout was integration of food production. Key to this was the development of efficient field systems together with an infrastructure of supporting features. Field systems of varying degrees of uniformity came to cover a greater land area in Britain than any other artificial landform. These were far more than plots for grazing and growing; they were badges of prowess and the people who controlled them were the new masters of the landscape. Size and form mattered. The most mechanistic type of field system was a rectilinear grid, laid across the land with little or no regard for topography. The neat, angular geometry of these coaxial field systems must have been magnificent to behold, not least because they were so unlike the curvilinear monuments of old. Many were laid out fairly rapidly in what appears to have been episodes of controlled construction. Slight anomalies in alignment led to untidy 'gang junctions', where sections of a field boundary being built by separate work-gangs met. Some of the longer boundaries ran for several miles. Coaxial field systems could be huge. The grid above Dartmeet covered 3,000 hectares, an area 250 times larger than the ditched and banked enclosure at Avebury. On Rippon Tor, a grid of coaxial fields also covered 3,000 hectares. Far to the east, on the lowlands by the Thames, a variety of coaxial field systems covered areas ranging from 5 to 15,000 hectares. On the chalk downlands of the Plain, different coaxial field systems followed similar orientations, trending north-east to south-west and completing grids that varied in size from one to nine square miles. Some of their main axial boundaries ran for two or three miles and their uniformity suggests that they may have been created in a single, enormous undertaking. The construction of coaxial field systems was a communal enterprise far beyond the scale of anything previously seen in Britain. Collective wills were at work, either in the form of committed families or as organised groups of agriculturalists. One of the reaves on Dartmoor deviated no more than 3 metres over a distance of 1,400 metres, a feat of alignment that can have been achieved only through accurate surveying (most likely with poles) and a quest for rectilinear precision. So immutable was the rationale driving coaxial systems that they frequently overlaid existing land divisions and monuments. On the gravel terraces of the Colne, early field boundaries had apparently respected an ancient temenos by skirting its edge, but by 1500 BC, the pressure to increase productivity at this site had become so intense that ditches and banks sliced across the monument. (Maintaining the tradition of serial desecration, the site is now smothered by Heathrow's Terminal 5.) Over time, field layouts expanded and contracted. The system was adaptable. By 1000 BC, rectilinear field systems had become the dominant type of land-use in Britain south of a line from the Severn Estuary to the Wash.

Not all field systems were perfect grids. Bounded plots could aggregate over time in a piecemeal fashion, taking their cues from contours, soil fertility and the individual whim of the wall-builder. Sometimes there were a number of construction phases on varying axes. On the stony uplands of the far west, less-ordered field systems may have arisen because control of layout diminished on marginal lands. An example of these irregular field systems evolved on the clitter-strewn slopes of Bodmin Moor, twenty miles west of the Dart. This too was an ancient land, dotted with cairns, circles and tombs that had been standing for thousands of years. Up here, to the west of the River Tamar where knobbly tors gazed down from bulges of granite a thousand feet above sea level, the soil was good and the airs generally gentle, but walls were seldom straight and fields accreted in higgeldy-piggeldy growths alongside their attendant roundhouses. Farmland fitted into suitable spaces, occupying the soils that were most fertile and best drained. The land between the rivers Lynher and Fowey was not untypical. The two rivers ran roughly parallel, four or five miles apart, tumbling seaward through steep-sided valleys. On a pass linking the two rivers stood a trio of ancient stone circles, arranged in a line with touching circumferences. A ten-minute walk north of the circles, a far earlier tor enclosure gazed over the grassland from a ridge. Across the hill from the tor and the circles, several clusters of fields developed on south- and west-facing slopes. In places, the fields appeared to be coaxial, running in parallel from a long boundary wall, but generally these were a ragged patchwork of fairly random shapes that appeared to have accumulated over time. At least six field-groups developed here, together with attendant clusters of roundhouses. There were more fields seven miles to the north-west, on the edge of the uplands, where Rough Tor commanded huge views over the coastal lowlands. Along the skirts of the tor were roundhouses and stone cairns which had grown in size as farmers rolled boulders to clear grazing land. The farms on these western uplands can only have been worth the effort if more accessible, low-lying land had already been taken. Southern Britain was growing short of wild land. Far to the east, on the chalklands of the Plain, these 'aggregate' fields seem to have developed later, filling spaces between existing coaxial layouts, their slighter boundaries and informal layouts hinting that organised planning had been overtaken by a guerrilla-approach to field-making.

Two-and-a-half thousand years had elapsed from the arrival of the first 'farmers' in Britain to the evolution of the first field systems. It had not been an overnight revolution. The intermittent surges of agriculture, from pioneering plots tended by first-generation colonists, to a widespread, systematised economy, had taken a very long time indeed. But by the late second millennium BC, many southern farms were matured models. Everything had found its place. Integrated into field systems was the suite of agricultural 'furniture' that made it possible for farmers to optimise the productivity of their land. The livestock conduits were droveways, together with a variety of associated cattle and sheep runs, compounds, stockyards and ramped watering holes. Within these task zones, farmers could undertake the routines of animal management: checking for disease and injury, separating categories of stock and relieving their herds of milk.

At the heart of every farm was the roundhouse, a landform that had already pocked Britain's surface thousands of times over with its characteristic, irregular circle of post holes. House size could vary but in southern Britain, the ring of structural posts that supported the roof was typically 7 to 9 metres in diameter, so the full span of the house might have been as much as 10 or 11 metres – large enough to accommodate an entire family and to provide work-space for domestic tasks. Doorways tended to face southwards to maximise infiltration of daylight (there were probably no windows) and porches deflected wind and rain, and provided a symbolic transition between exterior and interior worlds. Heat for cooking and warmth came from internal hearths, which must have filled the upper levels of the house with smoke (apertures to release smoke would have caused sparks to fly and therefore increased the risk of roof-fires).

Joining farms and roundhouses in this increasingly integrated, and predominantly southern, web were production centres ranging from seasonal salt pans to boat-builders, carpenters and textile weavers. And holding the whole fabric of production and exchange together like the frame of a loom was an infrastructure covering thousands of square miles: tracks and droves, bridges, causeways, jetties, staithes, rivers and sea lanes connected regions to each other and to northern Europe. Long-distance trackways rose and fell along the Chilterns and chalklands linking the Wash with Salisbury Plain. Another long-distance ridgeway may have run along the crest of the Cotswolds. Timber tracks were laid in the huge West Country wetland west of the Mendips. Distances shrank and so did exchange-times between one region and the next. The Thames was bridged in its lower reaches. Where the river ran through a series of braided channels (underneath what is now Vauxhall Bridge) a series of tidal islets offered footings. The place chosen for the bridge was close to a perfect settlement site, where a tributary ran into the Thames. This was no mere footbridge, but a massive structure at least five metres wide and resting on oak posts 40 centimetres or more in diameter. Two carts could have passed each other safely in midstream. These were capital projects, where 'capital' was measured in labouring-days made available by increased agricultural productivity. Several miles upriver, and a little later, another bridge was built across the Thames. Other rivers gained bridges too. Further west, close to the main estuary on the Solent, a solidly built bridge was constructed to cross the River Test. Two parallel rows of uprights were driven into the river bed, and then planked to form a crossing nearly 2 metres wide and over 20 metres long. Local woods had been sourced for oak, alder, ash, hazel and willow, and the various timber components had been formed with bronze tools. The use of mortise and tenon joints, pegs, notches and bevels was indicative both of carpentry know-how and an understanding of structural forces.

Climate and European connections continued to split Britain. The impacts of colder, wetter weather increased with latitude and altitude. Intensive farming was a southern, lowland phenomenon. The north lay on the other side of the fence; societies of surplus and of subsistence cohabited the same island. As vales, coastal strips and plains accrued fields and pastures, mountainous regions began to look 'other'. A new physical map took shape, in which two Britains were depicted: the intensively farmed and the lightly farmed. The single greatest area of 'other' lay north of the Solway Firth, where the vast tract of virtually contiguous upland constituted almost half the length of Britain. In these chillier latitudes, there was a widespread expansion of farming into marginal areas as agriculture crept further uphill and inland, but the extensive coaxial field systems of the south were rare. Patterns of walls and hedges tended to accrete in an opportunistic fashion, taking their cues from existing field remains and from the current needs of the farm. In places, cereals may have been cultivated on open ground rather than within the confines of walls or hedges. Cairns of stones accumulated where fields and pastures had been cleared and woodsmoke oozed from circular farmhouses. In some locations, groups of houses clustered into small, nucleated settlements.

The widening north–south divide brought with it a divergence of ingenuity. The more testing climate and topography of the north pro-voked architectures unseen in the south. Across the hilly borderlands between the latitudes of the Solway and Clyde, the lack of level land made it necessary for house-builders on sloping sites to engage in geo-engineering. Platforms were hacked into slopes, the debris quarried from the uphill sides being piled on the forward edge of the platform to create level areas large enough for the floor of a roundhouse. In places, the platforms tracked along hillsides sharing the same contour to create linear settlements. Occasionally, natural platforms could be utilised to save effort. Some of these engineered sites were as high as 300 metres above sea level, but these were not impoverished huts. Houses could measure as much as 10 metres in diameter and were built with cavity walling then finished with cobbled paving and in-terior walls of plastered wattling: on a north-facing slope high in the hills of the upper Clyde valley, thirty or so platforms accumulated over time. Some of the houses were rebuilt at least twice. They were probably not all occupied simultaneously, and at least one was used for animal byres or pens. A couple of the houses were built with twin rings of wattle-and-daub panels, the intervening void seemingly stacked with turf for insulation. Aside from its chilly, northerly aspect (which can only have increased the benefits of cavity walling), this settlement was well sited on the main overland route between southern and northern Britain, up the Annan valley from the Solway, and down the Clyde to the fertile lowlands between the Firths of Clyde and Forth. (A location of timeless appeal, the site of these platformed, insulated homes now stands beside the A74(M) a few minutes south of Abington service station.)

North of the Forth, areas of farmland dotted with clearance cairns and lined with trackways and field banks also had double-walled houses at altitudes as high as 450 metres. These too were comfortably large, with diameters reaching 11 metres. Out in the far west, on the islands of the Inner Hebrides, were other distinctive styles. On Jura, at least one house had a massive, two-metre-thick wall lined on the inside with a wattle screen. On Islay, a house was built with slab revetments. On the Outer Hebrides, an innovative settlement took the form of three aligned roundhouses and an extremely unusual, multi-cellular house. Out at the furthest limits of European reach, the architectural imagination that had produced the wonders of Orkney endured as a tradition.

Back in the south, intensive farming permanently marked the landscape as natural non-conformities were replaced by the rationalised boundaries of agriculture. Time itself reinforced the artificiality of farmland. Hedges thickened into impenetrable barriers. Walls grew ever-sturdier as each plough-turned stone was humped to the edge of the field. Season by season, ditches were cleared and recut; trackways hardened beneath the feet and hooves of each generation. Farmed hill-slopes generated their own, strange signature: repeated ploughing could modify contours and turn a tilted field into a defined, physical landform – a plot where so much soil had shifted downhill that endeavour was monumentalised. The dragging of ards across steeper gradients caused soil to slither downward until it reached more solid, unploughed ground at the lower edge of the field. Over decades and then centuries of sloped ploughing, topsoil migrated to the bottom of each field, helped on its way by the conveying effects of rainfall. Over time, these hillsides developed characteristic steps or 'lynchets' (a word derived later from the Saxon 'hlinc', or ridge). On extended slopes, well-developed lynchets looked like flights of terraces climbing from valley to skyline. Many of these levellings were probably used to graze livestock. Their gradual evolution on the land surface may have been caused by the laws of physics but their appearance reinforced the sense that farmers were able to direct their own destiny. Some of the lynchets on the Plain still rise to a height of six metres. Intensive farming was the single most transformative land-shaping strategy inflicted by Britain's people on their habitat. A threshold had been irreversibly crossed. Animal 'husbandry' was a marriage of species; a domesticated bonding of human and beast. The ditches, fences, walls and hedges that contained and controlled animals were culminating symbols of subjugation: imposed habitats that visibly removed animals from the wild beyond.

Farming killed the wildwood. Clearance had been habitual for thousands of years, but the agricultural campaigns of the second millennium BC were of unprecedented scale and permanence, especially south of the Severn–Wash line. Among the building materials needed for a single, large roundhouse were up to 200 oaks aged between 50 and 90 years. Fencing stakes to provide for a year's worth of farming would have required as many as 300 ash trees and a hectare of hazel for wattles. A community of thirty people needed managed woodland of some four square miles. Ash and hazel were cut to stumps to provoke growths of long, straight, regenerative shoots which could be harvested – coppiced – for poles and food for cattle. Ring-barked trees sprouted growths of low-level saplings which could be grazed by cattle. By rooting for mast and acorns, foraging pigs hindered the regeneration of beech and oak. The coppicing and harvesting was an ancient practice, but now it was endemic. Local woodland was as 'possessed' as enclosed farmland. In these managed woodlands, the mature, close-packed, totem trunks of the wildwood were replaced by misshapen torsos and gaps which were exploited by successional trees like birch and colonised by grasses and bracken. Thinned woodland trembled each spring with carpets of small blue flowers. The lime, so common in the early days, had virtually disappeared. In the south of Britain, stands of untouched wildwood were increasingly rare. River valleys had been cleared. Great tracts of downland and other uplands of the south were largely treeless. And so was the main part of the Thames Basin. In many areas, soils were becoming more acid as they were rinsed by rainfall and denied nutrient input from trees. Soil migrated towards valley floors, clogging drainage and forming flood-meadows. Erosion unleashed by tree-felling was accelerated by tilling and completed by wind and rain. In delicate upland habitats, tree clearance had increased soil saturation and tipped huge tracts of land into peat formation. Further north, the bronze-axe massacre was also well advanced. The lower slopes of the Pennines had been largely cleared, and so had the lowlands of the Lake District. Woods of oak, hazel and birch shrank as individual trees died and failed to be replaced. Some of the damage was caused by low-intensity grazing, some by the spread of peat bog and the deterioration of soils. By the early part of the first millennium BC, blanket bog was creeping across abandoned fields. Only in the far north were extensive forests still intact. The land of glens and beinns was a thousand or so years behind the south, and spring winds still whispered through the needles of swaying pines.

With the wildwood went the wildlife. The populations of bear, wild boar and beaver crashed. The pelican flew no more above fenland. But the most conspicuous casualty of the farming onslaught was the magnificent aurochs. These powerful beasts had always favoured low-lying, level, fertile land, but so did the farmers. Aurochs took to grazing in surviving tracts of marginal wetland, but eventually they lost this last-chance reserve, too. By around 1350 BC, Britain's largest mammal had been driven to extinction.

During this transformative era, cause and effect are the tangled shrubbery betwixt field and wildwood. The solar minimum of 1400 BC was followed by two more, in 1000 and 800 BC. As the mini-Ice Age tightened its grip, a massive high pressure system developed over Siberia, sending freezing pulses across Europe. Oscillations in the pattern caused catastrophic droughts and famines in the eastern Mediterranean, accompanied from 1225 until 1175 BC by an onslaught of earthquakes. The effects of climate and quake were recorded in many lands. From around 1208 BC, Egypt found itself under pressure from invading Libyans and 'Sea Peoples'. The palace systems of Mycenae collapsed in about 1190 BC. The empire of the Hittites fell at about the same time. And in the Fertile Crescent, the central powers in Assyria and Babylon lost their hinterlands as nomadic pastoralists poured in from the east. Among the knock-on effects was the erosion of societal stability and the quenching of royal bronze foundries that had fed long-distance webs of exchange through Europe. Ancient centres were destroyed by fire and the countryside deserted.

In Britain, there was no 'civilisation' to extinguish and evidence of human adaptations is thin on the ground. Colder, wetter conditions would have made life difficult for marginal farmers. Climate change was probably related to the eventual abandonment of the coaxial field systems on Dartmoor, for at some time before 1000 BC, the painstakingly walled reaves on these south-western uplands ceased to be used and the slopes above the Dart and its adjacent river basins reverted to wild pasture. If farmers were driven from Dartmoor by climate change, it's likely that other marginal lands above the 300-metre contour were being abandoned, too. It appears that many of Britain's uplands had by 1200 BC been overtaken by blanket bogs. The farms on the chalk downlands above the River Cuckmere were left during the prolonged cool-down after 1400 BC. Destructive farming practices may have been a factor. Besides the multi-century climate shifts, there were much shorter-lived catastrophes that left traces on the land. In 1159 BC, a volcano 500 miles from Britain blew its top. There had been plenty of eruptions already on Iceland, but this was a big one. The blast hurled 5.6 billion cubic metres of cinders, ash and pumice into the atmosphere. Much of this volcanic shrapnel – or tephra – drifted across the Atlantic and fell on northern Britain. For weeks the skies were dark. Animals and crops died. Trees stopped growing.

Come the turn of the millennium, a new landform had emerged from the interwoven influences of climate, population and European variables. The practice of enclosure was hardly new. Britain's social enclosures, temenoi and henges were as spectacular as anything constructed on the continent. This time, however, the enclosures were smaller and more common. Settlement enclosures drew new, physical lines on the land, separating the farm from the unenclosed tracts of 'wild' that still constituted the greater part of Britain. It was a process rather than an act. Many of the roundhouses around which farm-life centred were removed from the surrounding landscape by various combinations of ditches, banks, or palisades. In the past, the zone occupied by a house or group of houses had been imperma-nently defined on the ground. Settlements were fairly diffuse entities and there was little impulse to construct barriers between home and the world of wood and glade. But the boundary-led fields revolution turned 'home' into a territory. The basic architecture of roundhouses may not have evolved much, but their place within their landscape had changed. There were practical reasons for enclosing a settlement. Britain was still patrolled by predators like bears and wolves, and compared to the subsistence homes of old, farmhouses had become vaults of valuable assets. Quern-stones, sickles, looms and bronze heirlooms were worth stealing. And there was another security issue particular to intensive farming. With beasts like wild boar still on the loose, there was a risk of domestic animals inter-breeding with their wild relatives. A farmer's most valuable asset of all was breeding stock. From their protected roundhouses, families sallied forth to their fields, withdrawing at dusk to the sanctuary of their corral. The more successful people were, the more they had to lose. In winning against the wilderness, the architects of artificial landscapes looked increasingly defensive.

Beyond the basic roundhouse enclosure came a more elaborate expression of place. In the hotter economy of southern Britain, property was a convenient expression of success. For the upper tiers of society, a humdrum roundhouse would not do. Aggrandisement took a variety of forms. There were rectilinear enclosures that surrounded a centrally placed roundhouse, such as the one built on low-lying land beside the Blackwater Estuary, a premium location with excellent communications and access to grazing on the salt-marshes and adjacent gravel terrace. Other forms of premium property included D-shaped enclosures and palisaded riverside settlements. Common to most aggrandisers was a desire to be seen. Their properties were frequently in locations chosen for their visibility. Routeway sites in particular guaranteed the notice of passing traffic, and also provided immediate access to exchange networks. When it came to showing off, conspicuity and location were everything. A particular ostentation among eastern magnates was a circular, massive enclosure. Most frequently found close to river systems, it was defined by a bank and an enormous ditch – in one case nearly four metres deep. Some of these circular 'ringworks' had two ditches. Banks could be as high as three metres and topped by timberwork. Internally they could range in size from 40 to over 120 metres in diameter. The ditches were not designed for defence; their multiple entrances and fastidious concentricity were architectures of the private realm commanding respect and admiration. Inside were small circular structures, and often a large, central house whose entrance always faced the main approach through the enclosure's bank. By the time they were abandoned, the interiors of many of these circular enclosures would be strewn with fine pottery and discarded bronze.

One other kind of contemporary enclosure merits a mention, not so much because of its scale or frequency, but because it appears to have developed as a communal focus for the gathering and exchange of goods. One of these communal meeting places was built a mile or so to the east of the highest point on the chalky whaleback of the downland ridge separating the vale of the Thames from the Kennet. These days, nothing rises from the ground, but under the right conditions from the air, the dark cropmark tracing the timber-laced earthwork can still be seen. There were several entrances and the site seems to have been chosen for its aspect overlooking a variety of different landscapes. The wide range of pottery that passed into the enclosure recalls its role as a focal point for producer goods. The site is inconspicuous to the brink of invisibility, but its role as a hub for attracting a variety of products speaks of the emergence of a new kind of central place whose role included the exchange of goods. Ornaments and weapons left in the ground at other, similar sites suggest that they too were located at the meeting point of community territories, and were being used as centres for distribution of valued objects. By 900 BC, these communal meeting enclosures appear to have been replaced by sprawling, unenclosed feasting centres.

Britain of 800 BC was a land under unprecedented stress. Population growth, climate change, depleted woodland and diminishing wildlife were in tension with agricultural productivity. A tipping-point was reached. Bad seasons could cause human dislocation. On a local scale, a run of several bad harvests might be sufficient to tip a marginal farm beyond the brink, provoking abandonment and migration to less unreliable ground. On a greater scale, communities shift, conflict flares, disease erupts. Britain had been visited by several climatic crises during the latter part of the second millennium, but the nadir came with the solar minimum of 800 BC. With a population greater than at any time in its past, the potential for disruption in Britain was enormous. During the next era, Britain's landscape was reinvented.

TEN

Fearful Height

800–400 BC

The horse gallops the Ridgeway towards the sinking sun, riderless and as white as the clouds that sail the downland skies. It is midstride, with quick knees and flying fetlock, its tail, croup and crest a single sweep of momentum. It measures 110 metres from nose to tail. It's a memory-horse, a fleeting image. Look at it for more than a moment and it deconstructs. Two of the legs are detached. The flank and shoulder have no volume. It's a graphic abbreviation; a defleshed skeleton laid on the tilted screen of the grassland to be viewed from the vale below. The horse's creation may have been a ritual. It certainly demanded concerted, organised planning and effort. The outline would have been marked out before teams of pick-wielders and adze-heavers removed the turf and subsoil to expose the chalk bedrock. Since its birth more than two thousand years ago, it's been well groomed, a process that involved filling the indented outline with fresh chalk that had been 'puddled' – crushed and mixed with water to form a paste which set hard as it dried. Tradition sustained the Ridgeway steed. Where and how this equine monument fitted into the social narrative of its creators is impossible to say. It might be in flight, or beating the bounds. It might be a territorial marker. It could be a symbol of reassurance, appeasement or apprehension. Its meaning may have changed through time, reinvented by successive generations who tramped to the site equipped with scouring tools. The White Horse lived through the transition from a Britain of fields and farms, to a Britain of forts and boundaries. Optically stimulated luminescence dating suggests that it was created between 1380 and 550 BC.

It was the third trough in solar input – the one of around 800 BC – that was probably the most disruptive, especially on the continent. While insolation dropped, the enduring influence of weak winter pressure systems in the North Atlantic continued to drive winter westerlies – and stronger winter precipitation – through the Mediterranean to Asia. Winter snows covered more of the land, and took longer to thaw. Average temperatures dropped by 1 to 2 degrees Centigrade. Fed by summer rain and snowmelt, rivers swelled and accelerated. Lakes expanded. Flooding increased. Settlement zones along lakesides and rivers became untenable. As the decades passed, Alpine glaciers extended and tree-lines slid 300 or 400 metres down mountainsides. People switched from cultivating cereals to planting hardier oats and rye. Livestock became more important. Winds became damper and less reliable as the means of drying food. In the eastern Alps, salt mines became cultural hubs as brine was taken up as a preservative. In the plains of northern Europe, water tables rose, feeding bogs and fens. Droughts and famines returned to the eastern Mediterranean.

Climate change is never fair. In Britain, the unholy alliance of pressure systems and reduced solar input produced the most intense climate shift since the sudden change of temperature that had unshackled ecosystems – and people – nearly 9,000 years earlier. This time it was a shift in the opposite direction; a reversion to a less hospitable climate. But the nature of Britain's elongated geography meant that some regions were more exposed to extremes than others. Colder, wetter conditions hit hardest in the uplands of the west and north, where fields were abandoned and the upper limits of cultivation slipped downhill from 400 metres to 250 metres. Settlement was dislocated. Bogs became much wetter, and bigger. In the huge, West Country fenland between the Mendips and Quantocks, local communities had to construct raised trackways of timber and brushwood to prevent settlements becoming isolated by rising waters. Cereals became a less-reliable source of food. Unpredictable harvests increased the dependence on livestock. In the east and south, the effects of climate change were less severe, but overall, it's possible that the growing season shrank by as much as five weeks. Intensive farming seems to have become unsustainable. It's also likely that wresting maximum productivity from the land for so long had increased the acidity of soils and reduced yields.

The crises were recorded on the land, and may well have begun considerably earlier than 800 BC. Across the south of Britain, swathes of field systems were abandoned. From the chalklands of the Channel to the Wolds of mid-Britain, rectilinear grids of fields were obliterated by radical new boundaries: linear earthworks that sliced through field systems as if they no longer mattered. Typically, the new earthworks took the form of a ditch bounded by banks, but they could run for miles; compared to conventional field ditches, these were excavations on a massive scale with a purpose that reached far beyond livestock control. On the Plain, over forty miles of them would still be visible two thousand years later. Most were V-shaped in profile, bounded on both sides by banks of earth and varying in width from two to eight metres. They ran for at least 500 metres and the longest ones stretched more than nine miles. Some of them followed ridge lines along the downs, with additional banks branching off to form large blocks of physically demarcated land. Many appear to have been aligned to optimise their visibility from afar, and also to accommodate existing settlements and burial mounds. A suggestion of coordinated division can be seen on Sidbury Hill, where no less than six linear earthworks converged at a single point. There were variations on the basic form. In at least a couple of locations, two parallel ditches were excavated and the soil piled between them and possibly topped with a hedge-row. In other cases, timber palisades were erected beside the ditches.

For whatever reason, the patchwork patterns of field systems that had accreted across so much of southern Britain over hundreds of years, had become obsolete. Intricate field systems had been replaced by enormous blocks of land. And the new divisions were long-term. These were not one-off constructions but living boundaries which were regularly cleared of silt and debris. On the Plain, linear earthworks began to appear between 1200 and 1000 BC, but they were still being refurbished and added to through to 600 BC.

With the linear earthworks came a new form of enclosure. This, too, was constructed on the ditch-and-bank principle, but it could be found on hilltops. Common to these lofty enclosures was an absence of buildings, although some did have circular and four-post structures – possibly huts and fodder-ricks. By the standards of the day, these hilltop enclosures were large; some were enormous. Many covered 15 or more hectares. On the highest point in south-east Britain, straddling one of the chalkland's principal ridgeways three miles south of the River Kennet, the Walbury enclosure covered 35 hectares, an area more than three times larger than the 'social enclosures' of the fourth millennium BC. One of these hilltop enclosures appeared briefly in the early pages of Chapter 5 while I'd been waiting for the sun to rise on the abandoned flint mines on Harrow Hill. The wind was insane, and while searching in vain for shelter to take notes, I stumbled – literally – on a more recent earthwork. It was a banked, roughly square enclosure. Each side was 60 or 70 metres in length. Midway along the western bank was a gap for an entrance, and in the north-eastern corner another gap. I was amazed by the scale of the bank. Even though it had been constructed nearly 3,000 years ago, it still reared at least five feet above the land surface. When the Harrow Hill enclosure was excavated back in the 1920s, archaeologists discovered a large quantity of ox skulls – so many, in fact, that if the number of bones found in the trenches was an indication of the enclosure as a whole, there must have been the remains there of at least a thousand oxen.

A variation of these ditched-and-banked hilltop enclosures was built west of the chalklands, high above the Avon, where the river ran through a deep defile on its way to the Severn Estuary. This was the single largest construction project this district had ever known, enclosing 32 hectares of the plateau with a wall over one mile long and some three metres thick, revetted with timber and stone. It was an artificial space on an unprecedented scale. Within the enclosure on Bathampton Down accumulated the bones of oxen and pigs.

Construction of the linear earthworks and hilltop enclosures must have been a physically demanding, communal effort. They were too big and too coordinated to have been the outcome of piecemeal, individual enterprise. The sheer reach of the linear earthworks and the scale of the enclosures are suggestive of sweeping new land claims. And these territorial statements were meant to be seen from afar; they were highly conspicuous proclamations. The vast quantities of animal bones that accumulated within some of the enclosures told of their function. They were primarily stock enclosures, but also the locations of episodic feasting. The likelihood is that they were used at specific times of the year for the various tasks associated with husbandry, like castration, redistribution and culling, and that the linear ditches associated with the enclosures were there to corral and drive the animals. The enclosures seem to have been part of a dual farming system, in which smaller, lower-lying farms concentrated on cultivation and the hilltops were used for stock control.

One of the more remarkable sites from this era developed on a prominent spur not far from the rising of the Avon on the northern part of the Plain, where several linear earthworks converged on an enormous mound of debris. Even today, the East Chisenbury mound is some 180 metres in diameter and 2.5 metres high. Inside it is a mass of animal bones and miscellaneous cast-downs ranging from broken pottery to spindle whorls. The bones have been estimated to represent the consumption of 600 cattle and 3,800 sheep, annually, for a century. That is an awful lot of grub, and far more than could be consumed by the occupants of a single farm. Another slightly smaller mound covers an area of 3.5 hectares on a greensand ridge three miles from the northern slopes of the Plain. As much as 50,000 cubic metres of debris accumulated in a heap a couple of metres deep on what is now the edge of Potterne village. Analysis suggests that it was used by a cattle-rearing society who visited regularly with their dogs. Dung in the heap had been burned and dumped rather than used for manuring fields. The mega-middens of East Chisenbury and Potterne may have been central gathering-places, rather in the manner of long-gone Göbekli Tepe on the Euphrates. At pre-ordained times of year, people from surrounding districts would have streamed over the hills and up the vales to congregate for the butchery of livestock, for feasts and ritual celebrations. The concept of the 'central place' has been an intermittent visitor to this story ever since the construction of 'social enclosures' back in around 3600 BC. It recurred with the huge ritual landscapes of Orkney and Stonehenge but its reappearance after 800 BC was suggestive of a central place that had no dependence upon pre-existing, formalised, enduring structures.

It wasn't only livestock that headed uphill. Select settlements in southern, central Britain showed an upwardly mobile trend, too. Sizeable clusters of roundhouses continued to exist in traditional, low-land, open locations, but certain groups – very likely elites – began to take to the high ground, building fortified walls around roundhouses on hilltops. These 'elite-forts' were built on a massive scale and their construction required the cooperation – or coercion – of more than a few families. Typically covering one to three hectares, they were sited on hill summits or at the ends of spurs. Delimiting each fort was a massive wall. Storage pits within the walls provided some level of resilience against food shortages.

There was no overriding template for elite-forts; the location, size and shape of each was a response to local topography, manpower and beliefs. The locations chosen for these menacing enclosures were often related to earlier ritual landscapes, as if the fort-builders were drawn towards the silhouettes of ancestral endeavour. A 1.6-hectare fort was built on the prow of the westernmost extremity of the Marlborough Downs. At 195 metres above sea-level this lofty spot had views over the lowlands leading to the Severn Estuary. The fort-builders followed the outline of the prow, creating a triangular enclosure. But the western apex of the prow was already occupied by a pair of bowl barrows. Rather than obliterate them with their own earthworks, they deviated the line of their ditch from the inner rampart so that the burial mounds became an integral part of the fort; more than that, being at the apex of the fort, the barrows appear to be the focal point of the settlement. I once spent a night in this fort, with its constructors for company and the stars for a ceiling. Twice in the night a helicopter touched down by the rampart; an unexpected reminder that such places are invested with varied and timeless attachments.

Up on the Ridgeway, the White Horse attracted its own corral. A roughly D-shaped enclosure, some three hectares in area, was built on the crest of the down some 200 metres ahead of the galloping beast. A lofty spot with avian views over the vale below, it had been selected by earlier generations for the site of a long mound and a round barrow. A more recent linear earthwork crossed the down, and the enclosure appears to have been built astride it, apparently acting as a portal between two neighbouring blocks of land. A trackway passed from the downland, through the enclosure and out of the other side. The enclosure was defined by a single timber-walled rampart, with entrances through the western and eastern walls, and outside it, an enormous U-shaped ditch, 757 metres long, 7 metres wide and 3.5 metres deep. The original height of the rampart is not known, but had it been two metres, around 700 trees would have been needed. Excavating the ditch would have required teams consisting of someone with a pick, a shoveller and a number of carriers to shift the debris to the rampart. Ten teams, working eight hours a day, seven days a week, should have been able to complete the task in four months. The whole construction exercise would in itself have been a socially reinforcing enterprise, locking its workers into a calendar and probably punctuated by various rituals and ceremonies. The possible scope of these rituals is hinted at by the horse carved into the chalk downland beside the fort. The two features are roughly contemporary; conceivably, horse and fort were constructed by the same people.

Tramp Britain today, and land-stress can be read in banks, ditches and stones from the Lizard to Cape Wrath. On an archipelago so latitudinally elongated and so topographically and geologically diverse, the effects of climate change and cultural shifts had never been uniform. Developments in the north and south, west and east, were seldom synchronised or similar. In western and northern Britain, the centuries of cold, wet and uncertainty that intensified in 800 BC, saw a progressive solidifying of anxieties. Between the Tees and the Forth, protective palisades were erected around many settlements. Some had double circuits of palisades. In the far north-west – the rocky scatter of the Shetlands, Orkneys and Hebrides, together with the mountainous mainland they faced – the stone that formed the structural core of the roundhouses meant that they attained a monumental longevity, enduring for generation after generation. Typically, they had a circular stone wall and internal divisions separating dwelling spaces from storerooms, a work area and a central hearth. Walls could be an impregnable three to five metres thick. Floors were sometimes flagged with stone. At a place on Orkney now called Bu (a Norse word meaning 'estate'), a house with an internal area of some 70 square metres was protected from the elements by a massive, roughly circular wall five metres thick. Inside, local Orcadian flagstones had been set on edge to form head-height partitions which divided the interior into a number of discrete spaces. The largest space was paved, and perhaps used as a byre. At the core of the building was a room a couple of metres across containing a hearth, a built-in cupboard and a cooking tank. A howling northern gale would have been all but imperceptible to anyone warming themselves at the fireside.

The cold wet phase began to ameliorate at some point between 650 and 550 BC. But climate trauma has consequences and the cultural shift that had been imposed on Britain's communities by the most severe aberration since the glaciers had melted was irreversible. Insecurity and instability were the new normal. Across the centre of southern Britain, communities withdrew into fortified hilltop settlements. Defence against marauders was more critical than immediate access to fresh water, timber and communications. The architecture of paranoia was closed and fearful: typically, these 'community forts' covered around five hectares, with perimeters that took advantage of contours and gradients. A single external ditch provided soil which could be heaped into a rampart faced with timber or stone – and in places, both. Two formidable gates, usually on opposite sides of the fortifications, provided options for controlled access. Cowering inside the walls was a huddle of huts and storage structures. The frequency with which the defences of community forts were modified can only have been caused by episodic insecurity. The entrances to at least two of them were soon strengthened by projecting hornworks. Caches of slingstones and fire damage hint at a threatening external landscape.

One of these community forts took shape on a hilltop thirteen miles east of Stonehenge, where the River Test and one of its tributaries partially framed a block of high land overlooked by a prominent, 140-metre hill. The place-label people used for this dramatic location is lost, but on Ordnance Survey maps it's 'Danebury' (a name given later, when *denn* was a woodland pasture and *burh* a fort). It was a site that appeared to have erupted from the earth, pitched impregnably against the sky. You could see it from miles away, smoking on its summit and girdled by fields and trackways. The defences were formidable. Major construction had first occurred on this prominent hilltop several centuries earlier (maybe around 800 or 700 BC), when an area of over 16 hectares had been enclosed within a ditch and a discontinuous bank related to a linear earthwork apparently connecting the hill to the valley of the Test. In common with the other hilltop enclosures of this time, it was probably designed for controlling livestock. But early in the fifth century BC, part of the hill was remodelled by the construction of an earth rampart enclosing 5.3 hectares. External timbers formed a vertical facade to the rampart, and within it, a settlement of fifty or so roundhouses accumulated together with a variety of smaller structures that included storage pits cut into the chalk bedrock. The pits were ideal for keeping grain: once sealed, carbon-dioxide build-up within the pits restricted germination and limited grain-loss to around 2 per cent. By storing one-fifth of each harvest as seed-corn for the following year, with an expectation of a yield of five to one, the inhabitants would have been producing sufficient grain to support around a hundred people. It seems that the pits also had a symbolic presence; once their role as grain silos had run its course, these chalky voids were used for burials and the placement of offerings. Over time, 3,600 pits were excavated within the ramparts at Danebury.

Danebury was one of many similar forts. These ringed communities were physically and socially bounded, with formalised interiors. In its early iteration (c. 600 BC), the gateway of the community fort on Crickley Hill opened to a thoroughfare formed by flanking longhouses. At Danebury, the main, eastern gateway opened onto an internal pattern of lanes that branched out like limbs of a tree to various parts of the settlement. The 'trunk' of this lane system passed through the fort and out of the western gateway, bisecting the settlement. The ordered interiors of community forts like Danebury were indicators of the social use of space. The encircling rampart restricted and intensified that space, demanding access routes, food-storage facilities and dwellings that could function only through the co-operation of the majority. Danebury was among the most developed settlements in Britain. Meanwhile on the continent, Athens (population c. 40,000) was reaching a magnificent zenith; by 432 BC, the great trading *polis* on the Aegean was adding the final, sculptural flourishes to an enormous, new temple of stone – the Parthenon.

The end of the cold wet phase turned out to have been anticipatory rather than real. Although the seasons had become progressively warmer and drier from around 650–550 BC, conditions did not entirely recover. For a century or two, the multitudes of streams and rivers dissecting Britain had been less active, and bogs less sodden. But in around 400 BC, the island was thrown back into colder, wetter seasons, and a return to valley flooding, swollen rivers and diminished food supplies.

This was the seasonal backdrop to a technological development that had begun in around 700 BC, as bronze was gradually overtaken by a new metal from the east. Unlike the ores needed for the manufacture of bronze, iron ore deposits were widely spread through Britain, from the Weald to the hills of the Forth and Clyde. Not only was iron accessible, but it was far more effective than bronze for tools and weapons. In the early centuries, it was a local, roundhouse craft, each family or community smelting enough for its needs. The process required ore, water and wood. Ore would be collected from opencast pits, and then washed to remove soil and then dried by roasting, which also made it more porous and therefore easier to shatter prior to smelting. Mixed with charcoal, it was heated in a bowl furnace to temperatures of around 800°C, when gases from the burning charcoal chemically reacted with the ore to produce a spongy lump of iron as the melted, unwanted minerals – slag – drained to the bed of the furnace. The final stage was the most physical, as the iron 'bloom' was repeatedly heated and hammered to remove the last residues of slag. From its raw state, the lump of iron could be forged and turned into tools or weapons, or into ingots for trade or exchange. A finished bar of iron weighing a kilogram required about 100 kilograms of charcoal and 25 person-days to produce, so a group of three or four people working hard could produce more than enough for a batch of tool-heads in less than a week. With time, communities switched to iron sickles and axes, hammers and gouges, saws, files and adzes, their homestead furnaces marked by scatters of slag, charcoal and bits of ore. Iron was an accelerator of change. Working the land became more efficient for more people, and so did armed conflict. Among the devices that iron unleashed was the war chariot, which was likely to have been manufactured in Britain from the late fifth or early fourth century BC.

The enclosed, fearful Britain of 400 BC was a land far removed from the open farmscapes of 1400 BC. In many parts of the archipelago, 'home' had become a defensive carapace, a place of safety in an otherwise threatening habitat. The perceived risks from a diminishing population of wild animals had been replaced by a real and present threat from an insidious predator capable of extreme violence and total destruction. The reaction was written across landscapes from the Shetlands to the Solent. Britain was learning to live with architectures of defence. Early foragers had survived through hyper-vigilance and an all-round awareness of frequently encountered, low-level risks. Their defence of last resort was always mobility. The siting of shelters – and roundhouses of later millennia – had been dictated by highly developed strategies of locative paranoia: views, clustering, and proximity to food, timber and water sources all reduced the exposure to external threats. But during the first millennium BC, the psychological side-arms that Britons had relied upon for thousands of years translated into massive, defensive structures. Landscapes had joined the arms race.

Two thousand years of elemental battering and it still looms resolutely above the wave-torn shore of a small island in the north Atlantic. From the water, it looks solid, like the stump of a massive column that once stood as tall as a mountain. I'd come north by ship from the estuary of the Thames, coasting past the Farne Islands and Duncansby Head to the Orkneys and then to the Shetland port of Lerwick, where I'd joined a small bus to the stone pier at Sandsayre and boarded an open boat across the sound to Mousa. The tower's topmost level has disappeared, yet its greater part defies the tempests that scour this low, treeless island; winds that boom and whine yet fail to remove this drystone bastion from its remote, oceanic outcrop. It is 13 metres tall. As you approach across the island's grassy plinth, the tower seems smooth-walled, circular and slightly tapered. There are no windows. Inside, a stair curls upward between the inner and outer shell, passing six intramural galleries. The entire structure is assembled without mortar or cement, and is held in shape by placement and gravity. It's a 3D, drystone jigsaw. At the top of the stairs, the view fills with a huge, blue void. The tower on Mousa island is the finest surviving example of a broch, one of the most complex stone structures to have risen from the surface of prehistoric Britain.

Brochs probably appeared on the ragged edge of the far north from early in the fourth century BC. Eventually, at least a hundred of them existed. Whether climate change had a hand in their form is impossible to call with any certainty, but it is the case that Europe was plunged into a deep solar minimum that reached its low point at 325 BC. It wasn't until 200 BC that solar activity restabilised, by which time brochs were well established. So appropriate were brochs to life on Britain's margins, that they were still being used in the second century AD. In form, they grew from the simple, stone roundhouse, circular in plan yet ascending far above existing contemporary buildings, their apparent mass concealing an internal structure that made as much use of thin air as solid stone. Within the curved, drystone exterior was an ingenious system of cavity-walling. The outer shell encircled an inner shell, and between the two there was a void. Rainwater penetrating the outer wall would dribble downward within the void, allowing the inner wall to remain dry. Internal space within the broch was maximised by placing the stair within the void, where it also acted as a structural support between the two walls. Scarcement ledges of stone supported the timbers of intermediate floors and the tower was capped by a roof. In the more architecturally advanced brochs, the principal internal space may not have been on the ground floor, but on a raised first floor of timber, where a hearth could have been placed on an insulated base of clay or stone. This arrangement would have created far warmer quarters. Livestock stabled on the ground floor would have provided underfloor heating, while heat from the hearth would have risen up the tower, drawn in part by the cooler downdraught within the wall void.

Building a broch required communal dedication on a scale comparable to the monuments of earlier millennia. They were massive enterprises. The tallest, broadest brochs contained around 2,000 tons of stone and would have demanded substantial quantities of timber, a rare and valuable resource on the islands. Very roughly, the number of worker-days required to build a tall broch would have been around 10,000. One hundred people working flat-out for three months could complete the task within a single summer season. More probably the number of labourers available was less; a workforce of only ten might take three years. In a more refined scenario, a couple of broch-specialists overseeing 24 unskilled workers might put up a tower in 200 days if all the stone had been sourced and readied in advance. The location of brochs was governed in part by outcrops of suitable building stone; the broch on Mousa was built beside a low cliff of Old Red Sandstone, which is hard, yet splits easily into portable slabs that suit assembly into drystone walls.

So much effort, but to what end? What were brochs for? Wear-patterns on surviving stairs are not sufficient to suggest that tramping up to the wall voids or to the top of the broch was a regular routine. Perhaps the stairs were there to aid construction and repairs. Perhaps they were used for drying blocks of peat. In a cold, wet, windy climate, brochs made sense as dwellings. A first-floor communal living area would have been far drier and warmer than a ground-level round-house. Without any windows, it would also have been draught-free, and dark. Well insulated and efficient to heat, brochs were ideal homes in landscapes where fuel was a limited resource; time saved sourcing firewood and peat contributed to the resilience of the community. Had they been elaborate, low-carbon houses, each broch would have been capable of accommodating an extended family unit. The brooding, eyeless towers of the tallest survivors – Mousa, Dun Carloway and Duns Telve, Troddan and Dornaigil – are suggestive of withdrawal. They're exercises in anxiety, designed to repel the weather and perhaps outsiders, too. That, anyway, is what the Vikings seemed to think when they eventually came across brochs, 'broch' being a likely derivative of the Norse word 'borg', which translates as fort or stronghold. Brochs were virtually impregnable. Assailants would have been unable to get anywhere near the walls without moving within range of defenders, and every tower was built with a single entrance: a low passageway that passed through both walls to the inner chamber.

And in most brochs, visitors had to pass a cell or guardhouse set into one inside of the passageway.

Whatever their role (and it would have evolved over the centuries), size probably mattered. On Orkney, the external diameters of brochs ranged from 12 to 23 metres and the internal diameters from 7 to nearly 14 metres, a fourfold difference in enclosed volume that might have reflected social status or family size. Variations in the height of a broch, the quality of its stonework and detailing might also have been measures of social standing. The brochs on Orkney appear to have been nuclei for families of lesser structures which gathered around their footings, so that the tower became the central feature of a stone hamlet; the adult guardian surrounded by a clutch of diminutive, stone infants. At Gurness on the north-west coast of Orkney's main island, no fewer than fourteen ancillary buildings clung to the skirts of the main broch. Overlooking the sea on the other side of the island, a tower-cluster of six dwellings conformed to the same basic pattern. In both of these Orcadian 'broch-villages', a gated entrance to the village opened to a pathway which headed for the broch's doorway, where it split into two, each gated, narrow, paved passageway curving around the base of the tower to provide access to single-storey dwellings fitted with hearths, ovens, stone-lined tanks, storage cells and stone furniture. Smaller and less regular than conventional roundhouses, these gated complexes were effectively a shared structure, huddled together for warmth and security. At its peak, the total number of inhabitants gathered into the 'village' at Gurness may have been as many as 200 to 400.

Mid-millennium Britain was an anxious island. The brochs were just one example of paranoid construction. Across the far north of Britain, the traditional stone roundhouse had been superseded in places by massively built homesteads. The 'dun' (a Gaelic word for 'fort') had a considerable internal floor space of up to 370 square metres. Like brochs, they were drystone constructions; their walls could rise three metres or so above the ground and sometimes incorporated internal cells or cavities. They were intimidating structures.

The pattern of strongly defended homesteads extended south along the Atlantic facade to the Lizard and Land's End. Except for an interruption between the Solway and Dee, the entire western fringe of Britain was strewn with domestic fortlets. Around the lower slopes of Snowdonia, on the island of Anglesey and on the tapering tail of the Lleyn peninsula, a variety of defended buildings rose from coastal sites. The gradual evolution of a hilltop site down near the tip of the Lleyn peninsula may have been typical. Here, an open settlement of several timber-framed roundhouses developed in the fifth or fourth century BC. Later, a part-finished timber palisade was built around the huts, and still later, the settlement was rebuilt in stone and surrounded by an earthen bank around 75 metres in diameter. In a final development, a second earth bank was raised ten metres or so inside the first one, to create a secure space, possibly for protecting livestock. The long-term result, effected over several centuries, was to create a settlement of considerable resilience. Defensive homesteads became a common feature along the fertile coasts of mid-western Britain. A variant – perhaps reflecting the status of its owners – was the rectangular enclosure, again protected by a high earthen bank. The site at Bryn Eryr began as a solitary clay-walled roundhouse set within a timber palisade. Later, a second roundhouse was added, and the pair became the subject of a planned settlement, surrounded by a rectangular bank-and-ditch earthwork and incorporating storehouses or granaries.

At the southern end of the Atlantic facade, an entirely different but equally definitive form of walled settlement modified the lands of the West Country during the last three or four centuries of the millennium. They were created by digging massive earthen ditches and banks, usually round or sub-rectangular in plan. In general, the internal area was seldom more than 1.5 hectares, so they were far smaller than the old hillforts of the chalklands. Some of these walled hamlets had second or even third sets of surrounding banks, the open spaces between the banks likely to have been used as stock enclosures. Locations tended to be low-lying and fertile, with immediate access to fresh water; in some cases the outer banks of the settlement were angled to protect the local spring or stream. Given that the inhabitants were pastoralists, it's tempting to see these walled enclosures as successors to the moorland settlements that were abandoned when the climate turned. Multi-walled settlements also appeared in the territories along the northern side of the Bristol Channel; these too were at locations chosen for their proximity to springs and rivers. Regional variation in architecture was nothing new, but the scale of expression was ramping up. The stone brochs, duns, forts and enclosures of the north and west had grown so big and structurally solid that they were virtually impervious to weathering.

Architectural anxieties achieved their most spectacular expression in the rich, populous, violent, well-connected lands of the south. During the fourth century, many of the existing hillforts in the central chalklands were abandoned, perhaps as a result of attack, and there was a shift of focus to a lesser number of strategically sited settlements, whose defences were massively embellished. Ramparts around these 'developed hillforts' were radically reworked. The defiant verticality of timber-faced ramparts, and their conspicuous consumption of a valuable resource, had been an unsustainable expression of territoriality. They looked imposing, but didn't last. Timber ramparts were quick to flame when under attack; they were prone to rotting and their footings tended to slump into the adjacent defensive ditch. Constant repair and replenishment was necessary. By the fourth century BC or thereabouts, the authorities who controlled defensive forms in the south and east – where stone was rare and timber getting rarer – had turned back the clock and reverted to soil-shifting. Biodegradable palisades were replaced by a sloping rampart of earth set immediately behind a ditch. The combination produced a formidable, continuous slope that rose from the bottom of the ditch to the top of the rampart, which was crested with breastworks of flint or timber. The passage of time actually strengthened the defence: silt that collected in the ditch was periodically scoured and dumped outward, which created an additional, defensive counterscarp. Among the hillforts to be upgraded like this were South Cadbury, Poundbury and Hod Hill, and further north, Breedon-on-the-Hill and Hunsbury. At Mai Dun (Maiden Castle's Celtic name translates as 'great hill' or 'principal fort'), the tilted earth wall behind the ditch rose for over 25 metres at an angle that would have been all but impossible for an attacker to scale. An elaboration of the form saw sections or circuits of ramparts and ditches being added to improve defence, or to enclose greater numbers of livestock. Some may have been added for reasons of status. The number of defensive circuits could reach six. In the west and north of Britain, several forts were embellished with cunningly placed anti-equine barriers: stones (or timber) erected in patterns to deter sudden assault – or in places simply to create 'no entry' zones. Mai Dun grew to be the most spectacular of its kind, eventually accumulating three banks and two ditches, with an additional bank and ditch on its southern side and two opposing entrances which appeared to be competing for complexity. Some of the sites surrounded by these impenetrable walls were so effective that they remained in use for the best part of 500 years. Permanence had been achieved through mass. Fifteen miles up the Test, Danebury was modified, too: the western entrance was blocked off and, soon after 400 BC, a second circuit of ramparts was added. Visitors now approached the settlement's single eastern entrance through a curved, artificial defile beneath a command post well provided with clay slingshot and pebbles from the Test.

Places like Danebury protruded from the landscape like the hubs of massive wheels. They were a lot more than forts. They contained vast food-storage facilities, where basics like corn and wool could be hoarded and protected. They were production centres, where raw materials from distant sources were collected for conversion into high-value products; bronze and iron, salt and shale funnelled for processing into tools, weapons or cured meat. Some of these strongholds appear to have functioned as foci for religious congregation. They were densely and continuously occupied. The spokes that radiated from these hubs trembled with the exchange of forces as people, goods and services, and information passed to and fro between centre and periphery. Each commanded its own territory. The fortified hubs of the south-central chalklands developed on sites that were roughly equidistant from neighbouring centres; each was within a mile or so of a permanent water supply and each was adjacent to open down-land grazing. Concentrations of population and activity like this were unprecedented, and so was their economic reach. Danebury dominated a territory of some 400 square miles. Developed hillforts were adaptive settlements, their basic form and location sufficiently suited to the evolving culture of the time that their primary structural features – the ramparts and gateways – could be reworked as new re-quirements and architectures provoked upgrades. The constant amid all of this defensive gentrification was that height mattered.

It was during this beleaguered era that Britain began to peek out of prehistory. The archipelago had been a place known to its people for millennia. Rich traditions of oral storytelling sustained and updated a narrative that reached back through time like verses of remembered song. The temporal linearity of the island's past was complemented by personal and social geographies that gave each new generation a pro-found awareness of place; a sense of distance, relation and interaction. Intensely felt and intimately mapped, these were well-known lands. The view of Britain from afar could not have been more different. To most Europeans, Britain was a rumour. But as the first millennium moved beyond its median, impressionistic whispers about new places began to coalesce in the imaginations of Mediterranean thinkers.

Greek geography was fuzzy at the edges. They knew well the shores of their own sea, and had produced countless maps. But knowledge dropped with distance. 'I have never,' wrote the historian Herodotus, 'found anyone who could give me first-hand information of the exist-ence of a sea beyond Europe to the north and west.' There was a very good reason for being interested in these distant, northerly latitudes, for it was the people 'on the circumference of the inhabited world' who produced 'the things which we believe to be most rare and beautiful'. What little Herodotus had learned was a mix of myth and hearsay passed on by transcontinental intermediaries. Gold was known to come from the far north, and was said to be procured by one-eyed Arimaspians, who stole it from griffins. Amber was rumoured to come from the river of Eridanus, which flowed into the 'northern sea'. And Herodotus had heard that the Greeks got their tin from 'the Tin Islands', but was vague about their whereabouts. This precious metal came from 'what one might call the ends of the earth'.

ELEVEN

Oppidum

400 BC–AD 43

The 'ends of the earth' were put on southerners' maps a century after Herodotus died. While Britons lived in clusters of roundhouses, the Mediterranean clamoured with Greek cities. On the island of Sicily, Syracuse had become the first Greek city to top 50,000. Aegean islands were dotted with towns. In around 320 BC, a seafarer from a Greek colony in the western Mediterranean sailed to the circumference of the inhabited world and found many of the answers Herodotus had sought.

The adventurer's name was Pytheas, and he made the first recorded circumnavigation of Britain. Pytheas returned home to write up his findings on a papyrus scroll. *Peri tou Okeanou – On the Ocean –* was an extraordinary feat and a great read. The book has not survived in its entirety, but fragments were passed down the ages, woven into works of authors who looked upon Pytheas as an original source. To a seafarer from the virtually tideless Mediterranean, the exposed, coastal waters of Britain, snagged with tide-races, ferocious capes, sandbanks, fogs and fickle winds must have been challenging at best, terrifying at worst (at one point, he did note Britain's shape-shifting tidal habit of transforming islands to peninsulas with every ebb). Pytheas must have used pilots who knew Britain's waters. Or perhaps he chartered British vessels. The sense of extremity can only have been accentuated by the climatic downturn of the early 300s; Pytheas sailed north just a few years ahead of the deep solar minimum of 325 BC.

One of the first places Pytheas saw was 'a promontory called Belerion', whose inhabitants were 'especially friendly to strangers' and who had 'adopted a civilised way of life because of their interaction with traders and other people'. These were the tin workers of south-west Britain. He learned that the island was known as Albion, and that it was 'thickly populated' by the Pretani – 'the painted ones' – despite a climate that was 'extremely cold'. Readers of Pytheas didn't learn a lot about Pretani architecture. The only descriptions of agricultural structures to survive from Pytheas's account were 'roofed buildings' used to store the harvested heads of grain. Pretani houses were built mostly 'from reeds or timbers'. Compared to the urbanised fleshpots of the Med, Pretani lifestyle was 'modest', a deficiency Pytheas attributed to being 'beyond the reach of luxury which comes from wealth'.

Pytheas was no lone explorer. Long-distance maritime exchange along the 'Atlantic facade' had been occurring for millennia but it wasn't until the third century BC, or thereabouts, that sheltered havens on British coasts began to be significantly adapted for shipping. Midway along the Channel coast, where the muscular bulk of the Purbeck hills wrapped a protective arm around one of Britain's largest natural harbours, a pair of timber-piled moles were built to facilitate the loading and unloading of seagoing ships. Poole Harbour's ancient moles can claim to be the earliest-known cross-Channel port facility in Britain. One was 55 metres long, and the other stretched to 160 metres, with stone-laid decks 8 metres wide. They were enormous construction projects and their form resembled many that were already ancient in the Mediterranean. Although rising sea levels eventually rendered this particular port inoperable, it appears that overseas exchange in this era left distorted echoes in at least one coastal place-name, on the Thames.

At the eastern end of the Channel, climate change had created another superb natural harbour. The rising sea levels that had breached

Doggerland had also severed the tip of south-eastern Britain from the mainland to form a small, chalk-walled island which provided a seamark for mariners approaching from the continent. Between the island and the mainland, a narrow, mile-long passage of water twisted and turned between tidal flats, providing a short cut between the Channel and Thames estuary. A shingle spit formed a natural break-water most of the way across the south-eastern end of the passage, thwarting wind and swell from disrupting one of the most important natural harbours on Britain's south coast. On the other side of the island was the most significant cape in south-eastern Britain; the tran-sition point between the Channel and the Thames and North Sea. It carries a place-name that appears to precede Celtic, the Phoenician 'Y TNT', the 'isle (of) Tanit', derived from their goddess Tanit and given also to the island they colonised near the mouth of the Guadalquivir. The Isle of Thanet was blessed with good soil and the harbour-cum-shortcut between it and the mainland spread its reputation far and wide. Thanet was the saltwater gateway to interior Britain.

From the indomitable brochs of the north to the gigantic ramparts of the south, social stress clamped the archipelago as resolutely as the climate. There may have been many contributory causes, but one in particular had the capacity to inflict a universal ill. After 8,000 years or so of continuous modification, the downward curve of productive land may have met the upward curve of population growth. Increasing competition for good land would help to explain the steady rise since the end of the second millennium of a warrior culture. The sharp solar minima centred on 750 BC and 325 BC cannot have helped. But by 200 BC, solar activity had stabilised. The weather systems of the North Atlantic shifted into a positive mode, bringing dry conditions to the lands of the Mediterranean and Anatolia. Glaciers retreated in the Alps and water levels in lakes receded, releasing new tracts of fertile land. A reassuring rhythm of moderate winter precipitation balanced by dry summers encouraged settlement to multiply in Italy and the central Mediterranean. In 121 BC, expansionist Roman forces crossed the Alps and a slice of Celtic territory became the Roman province of

Gallia Narbonensis. Forty years later, the Celtic parts of northern Italy and Liguria were taken over and became Gaul-this-side-of-the-Alps, *Gallia Cisalpina.*

In Britain, the kinder climate unlocked the landscape. In the two centuries since the solar minimum centred on 325 BC, a less severe pattern of weather systems became established. Sea-ice withdrew from the fringes of Greenland. As the Mediterranean was relieved of its warm, wet climate, Britain became less cold. Stable, balanced rainfall and warm seasons increased the take from fields. Improved varieties of barley and wheat raised cereal yields, while the cultivation of flax, beans and peas became more widespread. More land was put to the plough, helped by the development of iron-tipped plough-shares capable of ripping through the heavy clays that characterised so much of lowland Britain. Spin-offs from the basic wheel encouraged a technological revolution: in roundhouses from southern vales to northern glens, home-workers took to the lathe, the rotary quern and the potter's wheel. On tracks and foot-trails, traffic increased with the interchange of local and imported goods. Britain had never been as busy, or as crowded. The population crept towards – perhaps even passed – the one million mark.

Into this burgeoning land dropped a small golden disc. Slightly larger in diameter than the widest part of a thumb, it was decorated on one side with a disarticulated horse not unlike the leaping beast of the chalk downs. On the other side was the full-lipped profile of a figure trailing a cascade of golden tresses. The first coins to reach Britain had been die-stamped in the lands between the Rhine and the Seine and made their way across the Channel to the estuary of the Thames and then inland to the territories of the south-east. Other coins came ashore on the south coast. These tokens of exchange between the rich and powerful represented the increasing worth of cross-Channel networks. Other coins followed, of other metals, other designs. Coins were a message from another world; a world of markets, mints and, in particular, the settlements Romans knew as *oppida*, a word that seems to have come down from the third century BC, when it was used to describe a circus enclosure: 'a place surrounded with *pedis*' – a wooden palisade. By the first century BC, the Gallic *oppidum* had become a more nebulous label and place: a large defended settlement; an enclosed space of economic significance; a place where raw materials were processed, coins struck and products made, stored and traded; the kind of place where merchants and the military could stock up. Some *oppida* had a political role as a centre of local government. They were settlements with a long reach.

Coinage and the proximity of Gallic *oppida* were just two factors that contributed to seismic adjustments north of the Channel. It was as if the landbridge was being reconstructed, sod-by-sod, across the shallow sea separating the continent from its largest island. And increased connectivity brought profound disruption to ancient, insular ways of life. Products from the Mediterranean were funnelled through multiplying portals. Flare-ups between expansionist Rome and indigenous Gauls unsettled waves of migrants. Boat people came to Britain; people like the Belgae from Gaul, who – according to Caesar – 'came to raid and stayed to sow'. The forces of change had little geographical or temporal continuity. The rate of landscape-change accelerated dramatically in the contact zones of Britain's south and east, where insecurity was recorded at places like Danebury. Renewed interest in military engineering saw the eastern entrance remodelled as an unassailable labyrinth, with gated, outer hornworks built beyond two inner hornworks. Confined to a 46-metre-long flint-walled approach corridor, unwanted visitors would have to face continuous fusillades of missiles from a cunningly sited sling platform (it's possible that the strengthening of Danebury's defences was a response to the reoccupation of a neighbouring hillfort which appears to have been used at about the same time for constructing war-chariots and training horse-teams). Fifty miles to the south-west, the eastern entrance of the great hillfort at Mai Dun was also reconfigured on a massive scale, with ingenious, overlapping earthworks intended to confine attackers within range of batteries of slingstones. An arsenal of 22,000 stones was left by one of the gates.

Densely settled hillforts were not restricted to the continental contact zones. In the west of Britain, in the lands between the Dee and Severn, a number of huge, walled enclosures were constructed in extreme locations. The sheer scale and inaccessibility of these enclosures speaks strenuously of deteriorating security. Visit them today, and it's the wind that brushes the gateless portals of these mournful sanctuaries, but once they must have clattered with activity. Two of these human eyries are on the Lleyn peninsula, where several mountains rise like vertebrae on a reptilian skeleton. On one of these vertebrae, a spiky peak called Tre'r Ceiri, gangs of labourers shifted thousands of tons of rock to form a straggling perimeter wall around the summit. Stone-walled huts clustered within the walls. On a second peak, in sight five miles to the south-west, similar walls were constructed around the summit of Garn Boduan. At a third location, further to the north, overlooking the sands at the mouth of the Afon Conwy, a three-sided defensive wall was raised, the fourth side being protected by a precipice. All of these mountaintop bastions were large settlement sites, enclosing between twenty and eighty stone-walled huts. It's possible that some or all of these densely settled, defended hilltops were not permanently occupied, but functioned as seasonal settlements for people who lived in the valleys during winter and climbed to herd sheep and goats on the uplands in summer.

Distinctive peaks attracted comparable settlements in the north, too. The upland south of the Forth and Clyde included a strange, triple peak which offered enormous views across the upper levels of the Tweed valley. This weathered volcanic stump had attracted builders since around 1000 BC. On the summit of the northern of the three peaks at least three building phases eventually led to an area of some 16 hectares being demarcated in piled stone. Less a rampart than a boundary, the stones were an evolving territorial statement. Over time, 296 hut floors were cut into the interior surface of the enclosure, although the total number of building plots may have been double that figure. Even if only half the plots had been occupied at any one time, it might have been home to over a thousand people, an enormous settlement by the standards of the age. It was the largest hillfort north of the Solway and is likely to have been the chief centre of the people occupying lands north of the Cheviot Hills; effectively a regional capital. Twenty-five miles to the east, and inter-visible on a clear day, a smaller hillfort covering five hectares crowned the summit of Yeavering Bell, a peak on the edge of the Cheviots. Symbolism may also have played a part in the form of the 'wall' surrounding this summit, too, for it was constructed of piled andesite, a rock that glows a striking pink when freshly quarried. Long before this livid circuit marked the hilltop, a large burial cairn had been raised on the highest point. Up here, within their pink wall and surrounding an ancestral cairn, as many as 500 people might have lived at any one time.

Albion's people began to come down from the hills. During a period likely to have spanned the early decades of the first century BC, the enormous, ditch-and-banked hillforts of the south were progressively abandoned as foci shifted to settlement-sites with better communications and services. Typical locations included the sides of valleys and important river crossings. One habit at least was brought from the heights: some of the new centres were surrounded by massive protective earthworks. Within a few decades, parts of the south-east had undergone a remarkable transformation as hillforts were usurped by lowland settlements. From around 70 BC, some of them began minting the first British coins. A suitable noun for these places is hard to find. A visitor from Gaul may have thought of them as primitive *oppida*. They were concentrations of population, some more nucleated than others; they were economic hubs; they were mints; some may have been political and religious centres. There's little doubt that the inhabitants of these well-connected, physically defined settlements saw themselves as socially separate from the planters and tillers living 'out' among the fields and woods. The labels came later, of course, but maybe these closer-packed people saw themselves as vaguely 'urban' and the thoroughfares between their blocks of houses as 'streets': roads or ways whose primary function was to provide access to the buildings attendant on them; communal spaces rather than connecting routes.

Two early examples of these proto-urban agglomerations – marked on modern maps as 'Dorchester' and 'Abingdon' – developed close to each other, right in the heart of southern Britain, on the upper reaches of the Thames. This stretch of navigable river was equidistant from the mouth of the Thames in the east, the mouth of the Severn in the west, and the twin mouths of the Solent in the south. From this same, central stretch of riverbank, the upper Thames and Cherwell opened the way to northern Britain. Both sites were route hubs and controlled crossing points on the Thames; they were crossroads in the demographic heart of Britain. Back in the heyday of hillforts, these two spots – exposed on a flat-as-a-coin floodplain – would have been non-starters, but in the emerging, interconnected hub-world, communication was more important than altitude. Security was provided in part by the river. At the Dorchester site, two sides of the 46-hectare enclosure were defined by an elbow in the River Thames, and a third side by the course of a tributary river, the River Thame. The fourth side of the settlement was protected by a massive earthwork. Moated on three sides and monumentally banked on the fourth, nobody arriving at this settlement could fail to have been impressed. Over half of the enclosure – some 25 hectares – was densely occupied at some time during the settlement's history. (It wasn't the first time people had been drawn to this spot; at the eastern end of the site was a quintet of barrows containing the mortal remains of those who had farmed here centuries earlier.) Eight miles upstream, the neighbouring Abingdon site occupied a semi-moated, 33-hectare enclosure at the confluence of the Rivers Ock and Thames. Where insular centrality was a feature of the pair of Thames settlements, other major lowland hubs took their locative cues from the coast. Far to the east, an important agglomeration coalesced close to the sea on the banks of the Colne (not the Heathrow Colne, but the Essex Colne which flows through modern Colchester). South of the Thames estuary, three more sizeable settlements developed close to the North Downs. By around 60 BC, there were over a dozen enclosed *oppidum*-style settlements in the south-east.

Critical to the functioning of *oppida* were places on the coast where goods could be trans-shipped. By the first century BC, the harbour-moles behind the Purbeck Hills had been overtaken by rising waters and by an even better harbour, ten miles along the coast at a place known now as Hengistbury Head. This too was a large, sheltered inlet protected from prevailing winds by a promontory; a headland lined on its south-western side with cliffs. Geologically, the cliffs were a series of layered sediments ascending about 40 metres above sea level, from hardened sand at the foot through to layers of sandy clay, river gravels and windblown sand on the top; the kind of ingredients that can be demolished with remarkable rapidity by the sea's pounding tides. But buried within a 15-metre band of brown, sandy clay were countless boulders some one to two metres in diameter. As the boulders dropped from the cliff, they collected along the beach and acted as sea-defences. And these were boulders with a difference. Each was a great lump of ironstone, the raw material for metalworking. Hengistbury was a superb entrepôt. On the other side of the peninsula, a wide, sheltered estuary acted as a freshwater portal to Britain's interior by way of the rivers Stour and Avon, and the beach on the southern shore of the estuary provided hardstanding for flat-bottomed vessels. Gravel offered a handy source of ballast. The estuary was well placed for the Atlantic trade routes which linked the Garonne and Loire in western Gaul with the Armorican coast and Britain. Incoming cargoes included dried figs, purple glass and amphorae of north Italian wine. Placed midway along Britain's south coast, the estuary was also perfectly placed as a conduit for exports: tin, lead, silver and copper from the south-west, corn and cattle, salt, iron and shale from the immediate hinterland. Britain's coast was ringed with thousands of natural havens which had been used by seagoing vessels for millennia, but Hengistbury was an entirely new order of development. A massive double dyke was built across the neck of the peninsula to protect the port from landward interference and, at its peak, activity stretched for over half a mile. Gathered on the foreshore were hard-standings for beached vessels, pits excavated for ballast, smelting and metal refining zones and the roundhouses of a substantial settlement. Hengistbury was a port.

Cause-and-effect has been carried away by countless tides, but there may have been a link between this booming port-of-trade and the abandonment of southern hillforts. The demand from an expanding Mediterranean empire for raw materials and slaves, and the flooding of southern Britain with luxury products from faraway lands – economically lubricated by coinage – must have disrupted insular habits. The drift from the hills was one-way. Settlements nucleated in response to economic activity, their myriad functions sensitive to Europe's trading networks and their political roles exposed to expansion from the empire lurking to the south.

It was probably during the first century BC that people in the far north of Britain began to come down from the heights, too. On the islands of the Hebrides, brochs began to be abandoned in favour of low-level structures. The sheer cost in time and resources both to build the elaborate drystone towers and to keep them standing ceased to be justified by the status they conferred and they were either neglected or remodelled into single-storey structures that put domestic utility before defensive posturing. The tradition of a roughly circular ground-plan was continued, but typically, a number of stubby stone piers now protruded a short way inward from the encircling wall. The piers had a dual function, both supporting a roof that often had to span a diameter of ten metres or more, and to subdivide the interior of the house. The 'wheelhouses' of the Hebrides proved to be a remarkably enduring design.

Coin-by-coin, a new layer of human geography of the island began to be recorded in Britain's silts and soils. Patterns on the dies associated coins with the people controlling the metalworking operation. As these tiny pieces of treasure passed from hand to hand along the trading webs centred on each mint, many slipped through fingers and purse-holes. Some were buried as hoards, or tossed into waters as votive offerings. Redundant coin moulds and furnace debris marked the sites of mints, while the reach of each mint was defined by the limits of its currency. The earliest indigenous British coins marked out a territory that stretched across the chalklands of the Kennet, the Avon, Test and Arun, from the Solent north to the Thames – lands associated with the migrant Belgae from Gaul, people who became known as the Atrebates (the name means 'settlers' or 'inhabitants'). Coins scattered further west, between the Atrebates and the wetlands of the West Country, roughly demarcated the lands of the Durotriges. By 60 BC, mints were operating in the east of Britain, too, with three distinct spreads of coins identifying the territories of the Trinovantes and Catuvellauni north of the Thames, the Iceni in the far east, and the Corieltauvi between the Trent and the Fens. With its own currencies and its proto-urban agglomerations, south-eastern Britain was closer to the *oppida* of Gaul than to the brochs and wheelhouses of the far north.

Meanwhile, Rome had moved closer. A series of wars from 58 BC gave proconsul Julius Caesar domination of Gaul, and by 55 BC his legions were in sight of Britain's white cliffs. Underestimating the difficulties inherent in amphibious assault across a body of water coursed by strong tides and fickle winds, Caesar's invasion attempt in the late summer of 55 had to be followed up the next year with a far larger fleet of 800 vessels, among them specially designed landing-craft that could be powered by oar straight onto a beach. After conquering the south-east, Caesar settled for annual tribute and an undertaking from the Catuvellauni that they could not attack their neighbours, the Trinovantes. He then withdrew to the continent with his hostages. For Britain, the writing was on the tablet: the island had proved vulnerable to invasion, and although the legions had left, kingdoms of Britain now paid tribute to allies across the water.

The link between tribute, submission and trade was sketched by the great traveller and writer, Strabo, or 'Squinty' (it was his eyes). Writing several decades after Caesar's adventures in Britain, he described an island that was mostly flat and 'overgrown with forests' (Caesar of course had only seen the south-east). Exports included grain, cattle, gold, silver, iron, hides, slaves and dogs; imports ranged from ivory, chains and necklaces to amber, glass vessels and 'other petty wares of that sort.' Strabo took the view that the British kings had been so thoroughly Romanised that they had 'managed to make the whole of the island virtually Roman property.' It was conqueror's bollocks, but Caesar turned a trickle of trade and tribute into a cross-Channel conduit that had a transformative effect on Britain's swelling urban nodes.

I'm writing this on an iPad at the heart of a ghost town in Hampshire. It's a warm dawn and the air is trembling with the urgency of spring. Around me is an open, grassed enclosure perhaps a kilometre in width. It contains no buildings and is bounded by a distant, uneven embankment topped intermittently by what appears to be stonework. The long, straight farm track I'm beside bisects the site, which occupies a level eminence; not a hill so much as a barely perceptible plateaux or a low spur. It's marked on my Ordnance Survey map as 'Calleva': Celtic for 'the wooded place'. Back when it was named, the woods had yet to be felled. It was a sound site for a settlement, a twenty-minute canter south of the Kennet, with wide views across the woods and fields to the south and east, well-drained gravels for building and a water table close enough to the surface to be reached by shallow wells. Back then, Calleva lay deep in the territory of the Atrebates. In the decades following Caesar's visit, it became the busiest place south of the Thames.

A visitor entering Calleva's gates in 15 BC would have been confronted by a busy hubbub. Streets connecting houses, workshops, stables, granaries, metalworking shops (including a mint) and temples tinkled with Celtic dialects and silver coinage. Wealthier households prepared meals with imported olive oil and fish sauce. They enjoyed wine, too. Calleva had trading links with France and Italy, Spain and the Mediterranean. Several of the timber buildings were rectangular.

Anybody visiting from the outlying districts would have been amazed at the bustle of the place. And its enormous scale. The area within its earthen ramparts amounted to some 32 hectares, ten times greater than the largest of the old hillforts like Danebury. Instead of desire paths branching between wattle and stake, some of the streets ran as long and as straight as spears, parallel with each other, crossing other streets, which were also parallel. You could stand at one end of the main street in Calleva, and see right through to the heart of the settlement.

Over on the east coast, near the head of the Colne estuary, was another thriving *oppidum*-type centre. On a sprawling site defended by woods, valleys and rivers, the Trinovantes had dug an astonishing fifteen miles of dyke to surround their principal power base. Beside the dykes, excavated soil had been used to raise banks up to 25 feet high. Attacking chariots, horsemen and foot-warriors had to cross lines of these obstacles to gain access. The site seems to have changed hands several times between the Trinovantes and their neighbours, the Catuvellauni and the Cantiaci; appropriately, the Trinovantes named the place after the Celtic god of war, Camulos. Camulodunon, the 'fortress of Camulos', was a very successful settlement, minting its own coins and trading through neighbouring territories. The man-power needed to create the settlement's enormous defences suggests that it was densely inhabited and one of the several spectacular burial sites has released from the soil artefacts that may have accompanied a king. It seems probable that Camulodunon was a large royal estate whose incumbents had developed a thriving market and port. Camu-lodunon – now Colchester – had the makings of a capital.

The ripening of Britain's proto-urban hubs was slow, as if the necessary sunlight had faded with distance from its Mediterranean source. In around 10 BC, place-names began to be recorded in the soil. First remembered was a place abbreviated on Trinovantian coins as 'VER', 'VIR' and 'VERLAMIO', a sprawling agglomeration some sixty miles west of Camulodunon. Camulodunum came next, with its CAM-struck coins. Within a decade or so, Calleva's name

– 'CALLEV' and 'CALLE' – began to circulate on coinage from the dies of the Atrebatic ruler Eppillus.

It had been a long road. Britain had been continuously occupied by humans for nearly ten thousand years. Since the felling of the first tree, some 250 generations had inscribed their signatures on the land surface of Europe's largest island. By about AD 40, a dozen or more proto-urban settlements were dotted across south-eastern Britain and the principal components of a modern, anthropic landscape were in place. Fully fledged field systems covered much of the island. Settlements now existed that performed the functions of a central place: manufacturing, distributing; operating as a node for the surrounding district. In that sense, the landscape had acquired a new distinction, between rural and urban; between 'town' and 'countryside'. Four thousand years after the first flint mines had been sunk into the verdant chalk grassland of the South Downs, extraction sites were commonplace. Mines and quarries now provided everything from lead and tin to building stone and gold. The wilderness was still there, in pockets, but it was shrinking fast. And the habit of unsustainable exploitation was well established: the extinction of species and complete removal of resources like woodland had become established as the flipside of development.

Like extreme flood events, invasions exploit weak points. Fingers have been pointed, more often than not towards the Atrebates. In AD 43, another Roman emperor looked again towards Britain with covetous eyes. The Greek historian Dio Cassius identified the Atrebatic ruler Verica as the man who persuaded Emperor Claudius to invade. Claudius needed little in the way of invitation. To keep ahead of the game in Rome, you had to stay alive, control succession, reward your clients and win glory. Claudius was weak, vain, fearful and lacking political nous and military experience. He needed a military feat to consolidate his power back in Rome. Britain, the island in the Ocean, had been a symbolic prize since the heady days of Julius Caesar. Taking Britain would show the world that no land lay beyond the reach of the Roman people. There were additional benefits: by sending an army to

Britain, Claudius could reduce the forces available to any traitorous commanders thinking of striking at Rome. And he had inherited the most formidable fighting force Europe had ever seen: a standing army of some 150,000 soldiers, organised into at least twenty-five legions. With auxiliary troops, the entire Roman army probably numbered close to 300,000.

New territory meant a larger army. Claudius created two fresh legions. He also reformed the invasion fleet, creating the *classis Britannica*, so that he could ship his legions across the Channel. In another reform, citizenship was promised to auxiliaries if they served twenty-five years or more, a benefit that would have implications for the colonisation of Britain. By April, AD 43, the fleet was ready to sail. The initial invasion force numbered some 40,000 soldiers, a huge army by the standards of the day. Commanding the troops on the ground was Aulus Plautius, whose record included governorship of the troublesome Danubian province of Pannonia and a celebrated road-building project across the neck of the Istrian peninsula on the Adriatic coast.

The men who embarked for Britain were a fighting force and a building machine. Among the legionaries were carpenters, masons, engineers, metalworkers, architects and surveyors together with logistics specialists: the clerks and book-keepers. Along with their helmet, armour, shield, javelins and short stabbing swords, each soldier carried a pair of sharpened stakes and a *dolabra*, a long-handled tool with an iron head that functioned as both an axe and an adze. A legion on the move built a marching-camp each night, excavating a rectangular ditch and then lining the berm on the inside of the ditch with their stakes. Drilled work-squads could build at speed an extraordinary range of structures, from forts and barracks to causeways, bridges and roads. Britain was about to be civilised by an army of psychopathic builders.

BOOK TWO

PART FOUR

Ruin & Renewal

TWELVE

Urbs in Rure

AD 43–200

Reconstructing lost landscapes can sometimes demand gymnastic feats of imagination. Old maps help, but there were not any for Rutupiae, circa AD 43. We cycled there on National Cycle Route 1 and as we bumped over the level crossing on the ditched lane from Sandwich, I tried to picture this blurred tarmac backroad as a saltwater channel deep enough to take seagoing ships; deep enough to take the troopships of a Roman army who knew the Wantsum Channel was Britannia's Achilles Heel. After a while, the lane sloped up the old shoreline to an English Heritage sign-board indicating the entrance to the founding monument of Roman Britain.

Beyond the ticket office (tea machine, flapjacks, indispensable EH guide) loomed a cliff of flint and rubble three times taller than a rugby player, a fort from the third century that we'll come to in the next chapter. The landform that makes this site so special is not quite as theatrical and can be found over in the far corner of the fort. Steeply angled and carpeted in turf are the twin ditches excavated by legionaries immediately after they had hit the beach and claimed Britain for Rome that summer's day in AD 43. There are few more thought-provoking spots to chew a flapjack than the edge of a ditch dug two thousand years ago by dolabra-wielding soldiers with salt on their sandals. Whether we like it or not, they gave us a good duffing. The British landscape was never the same again.

The inner ditch was 2 metres deep and 3.5 metres wide. Two metres in front of it, they dug a second, slightly smaller, parallel ditch. Behind the ditches, they raised a rampart topped with a timber parapet. This three-tiered defence ran for well over 600 metres, protecting the beachhead from counter-attack. Partway along the rampart, a timber tower was built above a 3.5-metre-wide gateway that opened onto a causeway across the ditches. Protected by the rampart, shipping could be beached and unloaded, stores sorted, troops marshalled and tents erected. The two short sections of ditch that remain are the earliest-known earthworks in Roman Britain: a pair of ruler-straight incisions, going nowhere. They introduce to the story of Britain's landscape the habit of ascribing landmarks to known individuals; today, these fairly ordinary-looking grassed trenches are known as the 'Claudian Invasion Ditches'.

It's possible that there was more than one invasion beach. The only surviving account of the landings comes from the Roman historian, Cassius Dio, who compiled it from secondary sources 150 years after the event. Dio states that there were three divisions at sea, 'in order that they should not be hindered in landing – as might happen to a single force'. He also says they were sailing west. Perhaps the invasion force split; perhaps one of the three fleets was a feint. The most likely contender for a second, western beachhead is one of the sheltered natural harbours on the Solent, where a rich collection of Roman detritus accumulated in and around a tidal inlet below the South Downs. The name the Romans gave to this place was Noviomagus, the New Port, now known as Chichester. Along with coins and various bits of military equipment were shards of glossy, red pottery unlikely to have been used after AD 45. In the creek lay a bronze legionary's helmet, dislodged perhaps from a soldier who had taken a dunk trying to disembark after an interminable voyage up the Channel. On the waterfront, a grid of piles were driven into the ground to support a building – probably a granary – over 30 metres long and 8 metres wide.

While there are doubts about the beachhead, the goal of the initial campaign was clear: to cross the Thames and take Camulodunon. The Romans believed in their army; the people of Britain believed in their sea, their woods and rivers. In the event, the legions landed without opposition, the defenders misled perhaps by rumours of mutiny in the Roman ranks, or misdirected by uncertainties concerning the true beachhead. Faith in natural hazards was pitted against experience in overcoming them. Dio records that the Britons 'took refuge in the swamps and the forests, hoping to wear out the invaders in fruitless effort'. The long estuaries of the south-east, with their margins of soft mud and strong tidal streams had always been significant barriers. But a detachment of Batavi auxiliaries swam the first river in full armour and routed the Britons, who had 'bivouacked in rather careless fashion' thinking that 'Romans would not be able to cross it without a bridge'. (The river would have been the Medway if the Romans were advancing from the east or the Arun if they'd been coming from the south.)

The Britons fell back on their great river, the Thames, which they forded at a secret crossing, 'near where it empties into the ocean and at flood-times forms a lake'. Plautius split his force. Again, the Batavi swam, while the remainder crossed further upstream by bridge, 'after which they assailed the barbarians from several sides at once'. With Camulodunon in his sights, Plautius sent for his emperor, who was waiting for the call, together with 'extensive equipment, including elephants'. Local farmers who had never heard of any beast bigger than a cow must have freaked to see swaying, grey, tusked monsters crossing the Blackwater. But it seems unlikely to have occurred. Shipping war-elephants across the Channel, then marching them from Rutupiae and through the wetlands of the Thames would have been taken more time and trouble than it was worth. Claudius arrived in Britain in mid August, triumphantly took possession of Camulodunon, stayed for a couple of weeks or so, then left his legions to build a new Roman province.

The grid crashed onto Camulodunon with all the subtlety of a stamping elephant. The Romans had got rectilinear streets from the Greeks. It wasn't a new idea, even then, but the urban visionary who reasoned that social control could be achieved through spatial planning was Hippodamus, who lived in the Greek coastal city of Miletus during the fifth century BC. Grid-planning was especially convenient for the military. Rectilinear plots were the most efficient way to lay out a camp or fort. Repeated plots of equal size were simple to build and repair; they were easy to navigate and number; they were flexible, in that they allowed for larger blocks to be created without disrupting the overall pattern. Perhaps most importantly, given the military need to avoid confusion, a grid of streets allowed for the most efficient route between a random start and finish point. Because grids were easier for attackers to overrun than random accumulations of curving alleys, they required formidable defences. Grid perimeters were designed to eliminate weak points.

So Camulodunon got a fortified grid, right in the heart of its regional *oppidum*. The screening bank at the entrance to the central precinct was butchered and incorporated into the fortress walls, which saved some digging and humiliated the Britons. The long axis of the fortress stretched for nearly 500 metres and its shorter sides for 400 metres. It took twenty minutes to walk once around it. Externally, the fortress was protected by a V-shaped ditch 2.5 metres deep and 5 metres wide. Behind it reared a 4-metre-thick rampart faced with blocks of clay mixed with sand. Midway along the east wall, against the river, was a gate, and another was set midway along the west wall. Aligned on the grid inside the fortress were at least sixty barrack blocks together with stables, storerooms, workshops, a hospital, latrines, a *principia* (headquarters building) and a *praetorium* (legionary commander's house). East of the fortress they built an annexe which may have included a bathhouse. Camulodunon was Romanised into Camulodunum. Sixty miles to the west, the old Catuvellaunian *oppidum* at Verlamion was rebranded – according to Tacitus – Verulamium. Much later, it would be relabelled St Albans.

The breakout from the south-east subjugated remaining centres of opposition and seized vital mineral resources like lead and tin, copper, gold and silver of the west and north. The course of this pitiless conquest was tracked by intermittent die stamps on the landscape, each one an instant gridded camp. The materials came from the local landscape: turf, timber and earth hacked with the versatile, two-headed *dolabrae*. In hostile territory, the military reckoned on marching ten to twenty miles a day and carried with them a mental manual of fortified settlements that ranged in scale and permanence, but all conformed to orthogonal discipline and efficiency of construction. During training exercises, troops rehearsed for the front line by building rounded corners, gates and straight sections of ditch. The purpose of camps was less to repel an attack than to provide a secure staging zone for troops to muster ahead of battle. Sizes varied according to the unit being accommodated and they could be built in a matter of hours. To avoid surprise attacks, sites too close to woodland were avoided, the ideal being an open space rising gently above a plain. When it came to breaking camp, the *fossa* would be backfilled to prevent the camp being occupied by unfriendly forces. No eyewitness accounts have survived of these fortifications being built in Britain, but the historian Josephus – who had the misfortune to fight the Romans in Judaea – was as mesmerised by Imperial speed as he was appalled by their brutality:

> Nor can their enemies easily surprise them with the suddenness of their incursions; for as soon as they have marched into an enemy's land, they do not begin to fight till they have walled their camp about; nor is the fence they raise rashly made, or uneven; nor do they all abide in it, nor do those that are in it take their places at random; but if it happens that the ground is uneven, it is first levelled: their camp is also four-square by measure, and carpenters are ready, in great numbers, with their tools, to erect their buildings for them . . .

The regimented 'four-square' form was alien to British landscapes. Up until the time of Tiberius, the perimeter of military installations had been dictated to large degree by the contours of the land, leading to a variety of polygonal shapes, but under Claudius, plans became a uniform square or rectangle, with a gateway midway along each side. More permanent forts and fortresses obeyed the same basic ground-plan. Auxiliary units were generally quartered in forts of 1 to 2 hectares. An intermediate size of fort – the 'vexillation fortress' – was used during the early stages of the conquest of Britain, and these could occupy 8 to 12 hectares. The largest installations of all – fortresses such as the one built at Camulodunon – housed a legion of 5,000 and occupied as much as 20 hectares. The surrounding ramparts were earth, turf and timber, and the buildings inside were timber. The interior of forts and fortresses had also undergone a Claudian rethink. Where the main street – the *via principalis* – used to pass longitudinally down the main axis of the fort, it now crossed the shorter axis. A second street – the *via praetoria* – ran from one of the fort's short sides to join the *via principalis* in the centre of the fort, which was also the location of the headquarters building. The most important building within each fort was the granary, placed right in the centre beside the commander's house and headquarters building. Many of these forts functioned as compact towns, feeding off the surrounding countryside and long-distance supply lines, sometimes by sea. Unsustainable in the long term, they were impregnable in the short term.

Camps and forts thudded into Britain. The legions moved fast, and none faster, it seems, than Vespasian's *legio II*, the battle-group most likely to have built a fortress nearly one hundred miles west of Camulodunum, less than a year after his troops hit the beaches. It was probably during the autumn of that first season of campaigning that *legio II* selected a garrison-site for their first winter. They had reached a crossroads, the place where the road from the east, from Rutupiae, Verulamium and Camulodunum, crossed the road from the south, from Noviomagus and Calleva. These were old routes, in use long before Vespasian's time. It was a strategic intersection, right in the heart of southern Britain, equidistant from the south, east and west coasts. The crossroads lay on the western edge of territory belonging to the subjugated Catuvellauni. Topographically, it was an unremarkable site, with restricted access to water and without the benefit of height; it was low-lying, on gravel, and the legionaries had to bring water to the fortress by way of an excavated gully – the earliest-known example in Britain of an artificial supply of flowing fresh water. A thousand metres east of the site, an isolated, prominent hill acted both as a signpost to the fortress and as a viewpoint (during a later war, this hill at the heart of southern Britain was adopted by the British Army for their Central Ordnance Depot, Bicester). Vespasian's gridded fortress covered some 10–11 hectares, and its construction – coming at the end of a year that had seen the legion play a pivotal role in the invasion and conquest of the eastern kingdom – must have been tiresome. They were back in the fortress the following winter, when an annexe was added, bringing the total area to 14–15 hectares.

The storm-surge of Roman colonisation was checked by higher ground and dissipating energy. Vespasian's *legio II* did more than any, fighting no fewer than thirty battles, taking more than twenty *oppida* and successfully mounting a seaborne assault on the Isle of Wight. *Legio XIV* took the centre and *legio IX* pushed past the eastern fens to the Trent. *Legio XX* appear to have been based at their new fortress at Camulodunum. Auxiliary units supported a military campaign notable for ruthlessness and displays of exemplary force against Britain's divided peoples. 'Indeed,' recalled Tacitus, 'nothing has helped us more in fighting against their very powerful nations than their inability to co-operate.' British kings either surrendered or were physically defeated. By the time Plautius's term of office came to an end in 47, his forces – and those of his client-kings – controlled a triangle of territory that reached from Rutupiae to a line connecting Lyme Bay on the Channel to the Humber on the North Sea. Two sides of the triangle were delimited by sea, and the third by a military road – the Fosse Way. As a region, it had a neat geographical integrity, corresponding with Britain's lowlands, and with the part of the country that had – in Roman eyes – a less-repellent climate. Acknowledging its geopolitical significance, Tacitus referred to it as *proxima pars Britanniae*, 'the nearest parts of Britain'; that part of Britain closest to the continent and nearest to Roman Gaul. Once again, the south-east was the default development zone.

Britain was backward. The webs of footpaths and tracks serving settlements were an ill-fitting jigsaw of local and regional networks which were difficult for outsiders to navigate and mired by a climate condemned by Tacitus as 'wretched' and beset by 'frequent rains and mists'. The province needed a road system to serve the swift exchange of intelligence, the rapid deployment of military force and the heavy volumes of wheeled transport anticipated for resupply and trade in an expanding territory. The imposition of an integrated road network laid out on simple, direct lines was also of immeasurable utility to military and civil incomers whose grasp of British geography was at best confused. A pristine, rectilinear web reduced the chaos – and fear – of the unknown to a few straight lines. Fitted with roadside milestones, Roman roads functioned like rulers, providing their users with routes that were easy to memorise as mental maps.

Each section of road was surveyed in advance, the route walked, obstacles noted, sightings taken, and the most direct line between places was marked at inter-visible points. Where the bends in Britain's existing roads had evolved over hundreds (or thousands) of years to link settlements, sources of raw materials, strongholds, religious sites and so on, the bends in Roman roads were inscribed by surveyors as they were forced to skirt obstacles or to adjust direction in order to keep the overall alignment as true as possible. Changes of road alignment at viewpoints were not uncommon. Preparatory ground clearance included the felling of trees, the burning of scrub and stripping of turf. Streams and rivers were bridged and ditches were cut to carry water away from the route. There were variations of practice, but typically a ditch would be dug along each side of the proposed route and the material piled between to form a causeway or *agger* which raised the road-bed above the level of groundwater. The *agger* would be laid with a base layer of large stones or pebbles, and then with a top layer of gravel metalling. Repeated use of the road by hooves, feet and wheels compacted and crushed the metalling into the base layer, creating a firm surface which was resistant to rutting and pitting. A camber helped water to drain into the roadside ditches.

Junctions were hotspots. And in the early years of the conquest, one junction rose above all as the key interchange on the entire network. Geographically, it optimised continental and internal connections; a best-fit junction on the Thames that brought to one place the roads from the south coast ports of Rutupiae and Noviomagus and the road from the provincial capital, Camulodunum, at a spot that also offered a superb natural harbour in the heart of the south-east. To minimise the road-distance between Camulodunum and Rutupiae, and the river-distance from open sea, the place chosen was the lowest bridging-point possible, where a couple of islands – or eyots – protruded from the south bank, dividing the river's stream and offering 'stepping stones' across the river. Facing the eyots on the north bank were low hills which provided a vantage point and dry ground for a junction. The river here was wide and tidal; the hillier north bank was thickly wooded and the low-lying south bank was a maze of creeks, eyots and marshland. It wasn't a section of the Thames valley that had attracted substantial settlement, but the damp subsoil was scattered with artefacts lost or placed by people who had found the broad wetland good for foraging and hunting; hidden from sight were mattocks made from antler, flint axe-heads and adzes, arrowheads, bronze weapons and pottery beakers. At the time of the conquest, the nearest civic centres were at least twenty miles away, well back from the river's floodplain. The Romans knew this part of the Thames as the place where, four years earlier, the invasion had momentarily stalled as the legionaries struggled through wood, floodplain and river to outflank, outwit and outfight the retreating Britons.

The initial crossing would have been some kind of ferry linking the eyots with the north bank, but it would have been replaced by a roadway across pontoons, or a fixed, timber bridge on piers. All were well within the capabilities of Roman troops; nearly a century earlier, Caesar's engineers had put a bridge across the Rhine in ten days, countering the strong current by driving piles into the riverbed at an angle, then connecting them with braced beams which supported a roadbed of poles and bundled sticks. The Thames in front of the hill was some 350 metres wide, although the bridging challenge was reduced by the presence of the two eyots. Less satisfactorily, the low-lying nature of the eyots and the saturated marshland of the south bank would mean that deep footings would have to be built for the roads approaching from the Channel. At least there was plenty of construction material available: timber in the surrounding woods, gravel and sand from the river, and brick-earth – a hard, sandy clay – overlaying much of the gravel on the north bank.

Stand on Cornhill today, and you can't see the river for glinting tower-blocks. But in the winter of 47–48, the hill was wooded with oak and hazel, beech and alder descending in leafless stands to a foreshore of sedge. Near the top of the hill, on its forward face above the river, a T-junction was planned. (The actual spot is marked today by the junction of Gracechurch Street, Fenchurch Street and Lombard Street; an intersection whose corner-plots are currently distinguished by New Look, M & S, Boots and T.K.Maxx.) The road from Calleva and Verulamium approached from the west; the road from Camulodunum from the east; the road from Rutupiae and Noviomagus from the south. Surveyors would have marked out the junction with posts. There was an element of monumentality in the orthogonal T, inscribed on the hill above the river-crossing. Perhaps the form of the junction was associated with controlling traffic at the crossing, through tolls, or through a checkpoint.

Trees were felled, turf stripped and streams diverted. Beside the course of the roads, work-gangs extracted gravel, sand and clay. Pits and quarries appeared, and ditches to carry surface water away from the roads to nearby streams. More V-shaped ditches ran parallel with the roads. No trace has been found of a bridge from the north bank to the eyots, but the effort and engineering would have modest in comparison to other projects being undertaken at the time, and the benefits of a permanent crossing at the most strategic hub in Britain would have been immense. In the relatively modest record of British engineering, however, such a bridge would have been the greatest ever built on the island. Smaller bridges were put across the stream to the west of the hill. The approach roads to the T-junction conformed to Roman specs and were built using material to hand. Base layers were clay, silt, sand and brick-earth. A metalled layer of gravel completed the surfacing. The road across the eyots and marshes of the south bank was constructed on a double-layered timber causeway set in clay and pinned with vertical stakes, with sand and gravel spread on top.

The road-builders left a landscape clawed and spewed with pits, quarries, ditches and dumped lumber. They did the job and moved on. Nowhere do they seem to have allowed for future development. No *insulae* were marked out, or building plots, or boundary ditches, or defensive structures. It was a roads project. But somebody at least saw the potential in such a strategic spot. While most of the artificial drainage excavated during the road building took the form of deep, open ditches, an area on the north bank had been provided with underground, boxed, timber drains so that the land above remained available for the construction of buildings. Whoever built those box-drains, founded a city.

It had worked in Gaul, and it would work in Britain. Terror and intimidation would be followed by systematised administration and conversion of landscapes into a model suitable for exploitation. The keystone that would prevent the entire provincial edifice collapsing was the town. Towns would control and inspire the greater mass of Britain's mostly rural population. Conquered territories would be reorganised into administrative districts, *civitates*, with urban centres ranging from small towns, *vici*, to larger, *civitas* capitals. A number of model towns – *coloniae* – would be built and settled by retired soldiers who would be awarded parcels of land to support their families. Constructed on regimented grids and adorned with emblems of authority, they'd be – as Cicero said of the *colonia* at Narbo in Gaul – the 'watchtower and defence of the Roman people'. *Coloniae* were exemplars and enforcers: they'd demonstrate to unconvinced barbarians the benefits of Roman ways but they'd also act as bases for armed reserves and accommodate grizzled veterans who might otherwise mutate into landless outlaws.

Like forts, towns were planned on a grid, the form most suited to efficiencies and expressions of spatial control. Side-streets ran like teeth on a comb from a main north–south street (the *cardo*) and a main east–west street (the *decumanus*), creating rectangular or square blocks, known as *insulae*. (Grids were not a prerequisite for urban success or status; the enormous, and apparently unplanned, town of Lugdunum at the confluence of the Rhône and Saône successfully performed as the capital of the province Gallia Lugdunensis – and birthplace of Claudius.) At the civic centre of Romanised towns was the forum, an enclosed, traffic-free zone where townspeople gathered for meetings, public pronouncements and gatherings spanning a range of administrative, religious, legal and commercial activities. Here, unity, wealth and status were presented for all to see. The key building in a typical Gallo-Roman forum was the basilica, where the council of the *civitas* and justice officials congregated and where civic records and money were kept. Shops, monuments and temples could often be seen in a forum, too. Beyond the centre, development of *insulae* depended largely on private initiatives to add baths, theatres and so on to blocks among zones of ordinary houses. Individual buildings conformed to the principles of civilised architecture encapsulated by one of Caesar's architects, Vitruvius, who described humanity's progression from woods, caves and groves to shelters of green boughs and then mud-walled huts roofed with thatch or oak shingles. In the final, Roman, era, 'they gave up huts and began to build houses with foundations, having brick or stone walls, and roofs of timber and tiles'. Thus they 'passed from a rude and barbarous way of life to civilisation and refinement'. Vitruvian architecture was *firmitas, utilitas, venustas*; solid, useful, beautiful. The overall effect of these concentrations of coded architectures was to enforce new human geographies. From towns, a trickle-out effect would permeate the countryside.

No more than a year or so after a modern communications link was opened between the south coast and the interior, the conquerors established their first *colonia* in Britain. Locals on the eastern Colne must have been bewildered by the pace of change. Only six years or so after the conquerors built their massive fortress within the walls of the old Catuvellaunian stronghold at Camulodunum, they knocked it down. The defences were flattened, ditches filled and ground levelled for a grid of streets. Many of the barracks were demolished, although some survived as homes for war veterans. Colonia Claudia was founded in AD 49 and was more than twice the size of the fortress it replaced. On the largest single plot in the colony, a temple dedicated to Emperor Claudius was constructed, and adjacent to it, a theatre. At the other, western, end of the settlement, a symbolic portal in the form of a monumental double-arched gate rose above the road leading into the provincial capital from the Thames. The invaders had yet to overrun high-quality stone-quarries, so builders had to make do with bricks and tiles fired from clay. Foundations were mashed together from soft, local stone bonded with mortar, and the extravagant decorations on the Temple of Claudius had to be crafted from porphyry and marble shipped from the Mediterranean. When retired *legio XX* centurion Marcus Favonius Facilis died, his two former slaves erected a tombstone which had been quarried far away in Gaul, where high-quality limestone was available near the Moselle. Britain would not be built in a day.

Southern Britain shuddered beneath the pick of armed developers. By 49, lead ingots were being shipped out of the Mendips bearing the inscription of *legio II*. By AD 50, there was a vexillation fort on the floodplain of the Severn and by the mid 50s, a 20-hectare legionary fortress had been built even further west, on the Usk, which effectively marked the cutting-edge of the Roman Empire. Behind the front line, the Atrebatic power-base at Calleva had became an integrated Roman hub, its streets reoriented to new priorities. A capacious timber building set around a courtyard established a new, north–south, east–west alignment for the town, and roofing tiles began to be used. Twenty miles to the north-east, the one-time capital and mint of the Catuvellauni at Verulamium appears to have developed rapidly along hybrid lines as the local elite of this old aristocratic centre hastened to demonstrate their Roman sensibilities by constructing their own interpretation of a 'town'. The precise dates of construction are unclear, but centre roughly on the decade or so following the initial invasion, when the main street, with its side-streets meeting at right angles, became the axial spine of a rectilinear grid. By 60 or so, several of Verulamium's streets had gained imposing timber buildings and although the streets themselves lacked side-ditches and a camber, they'd been roughly metalled with river pebbles. It was probably during this decade that a new road (later known as Watling Street) was built to link the strategic junction on the Thames with Verulamium. It approached the emerging town between two deep ditches set back nine metres from the road.

Speed mattered as much as form. Calleva and Verulamium were propelled into Romanisation by client rulers; Colonia Claudia was planned and inhabited by a ready-made urban population of retired soldiers. At the key road junction on the Thames, a settlement developed which owed more to commercial opportunism than to political expediency. At its core was a small group of settlers whose traces suggest Romano-Gallic origins. Possibly, they'd been sanctioned by the procurator to construct a settlement which could serve the needs of road and river travellers. Possibly their presence was related to the prescient construction of the buried box-drain on the north bank. Whoever they were, they were on the hill by around 51, perhaps earlier, within a year or so of the river crossing being opened. Their first task was to level the mess left by the road builders. The pits, quarries and ditches were filled with rubbish and material excavated from the edges of the site; drains were laid and brick-earth humped and spread on levelled building plots, then compacted to form floor-slabs, wall-bases and hearths. Some of the more basic dwellings were held up by

posts set in trenches packed with gravel or clay, then walled with wattling which was sealed with daub. By Roman standards, the houses
were modest: top-end properties had solid timber frames which had
been mortise-jointed by skilled carpenters. Roofs were thatched or
tiled or laid with thin, wooden shingles. Virtually all the building materials – timber, reeds, mud-bricks, wattles and daub – were sourced
on site.

The people building on the hill may or may not have been under
the control of a central authority, but their desire to emulate Imperial planning was traced on the slopes around the junction. A couple
of side-streets were laid out at right-angles to the main roads, as if
obeying the basics of an orthogonal grid. Immediately north of the
T-junction, turf was methodically removed from an area some 40
metres by 33 metres, which was then levelled and spread with between
132 and 280 cubic metres of gravel carried up from the river. The
gravelling operation alone would have taken 10 labourers between 18
and 38 days. There was a wish-I-was-a-forum feel to this rectangular,
surfaced, open space crowning the T-junction. Its exact functions are
not recorded but it might have served either as a market space or as a
space for mustering military forces and supplies, or both. Its presence
protected a central block of land from further development.

From the start, this was a settlement with more than one focus.
The proto-forum and cluster of rectangular buildings set about the
orthogonal street grid astride the strategic T-junction on the hill
identified this zone as the administrative centre. Lower down the hill,
a riparian zone emerged, settled by people working on the river. Vessels waited out in the Thames estuary for the flood tide to carry them
upriver to the bridge, where a section of the north bank had been
straightened and revetted with posts and planks so that they could
load and unload. Smaller, flat-bottomed boats were able to sit out a
low tide, beached on specially laid hardstandings of chalk and gravel.
Above the tide-line, terraces were levelled to form waterfront work
areas. As imports flooded up the Thames, the waterfront below the
hill became a bustle of dockers, carters, traders and merchants from

the far side of the Channel. Oil, wine and fish-paste from Hispania came up the Thames, and pottery from Gaul, and quern-stones made of lava from far up the Rhine. Feeding off all this activity from their premises close to the river were shopkeepers and innkeepers, craftsmen and fishermen.

There were other outlying zones, too. West of the T-junction, the road towards Calleva and Verulamium dropped to cross a timber bridge over a stream, then climbed again, over a second hill which was used for rubbish dumping and burials. The small roundhouses out here, with their walls of wattle-and-daub, and proximity to refuse and death, were those of the fringe: scavengers and recyclers making what they could of material surplus. Their only consolation was the prevailing westerly wind. Another zone of relative deprivation could be found east of the T-junction, on the road towards Colonia Claudia, where there was a second area set aside for burials. Here, too, there were small, crude roundhouses. Out here, at the east end, the prevailing winds were less kind.

These outlying areas of activity were limbs of the urban body; mutually dependent and connected to the same circulatory systems. In Rome, 750,000 or so people lived in the city itself, but this central urban zone was surrounded by another 350,000 who lived within a thirty-mile radius. Out here were cemeteries and quarries, slaughterhouses and brick-kilns, farms, shanties and a sprinkling of luxurious villas maintained by those in search of space and clear air. This ambiguous girdle was neither city nor country and despite its importance to both, it bore no descriptive term. The progressive shading of its inner and outer limits, and the catch-all nature of its constituent parts, resisted geographical labelling. It was less a place than an attitude; an expression of potential. A city thus surrounded offered an unstructured hinterland of opportunity in which it was possible to explore the best of all worlds. Very occasionally this deep, free-form periphery cropped up in contemporary literature as *suburbium*. It was notable that the road-hub on the Thames had established the first buds of *suburbium* in less than a decade.

The burgeoning hub on the Thames became 'Londinium', a Romanisation perhaps of Plowonidā, a word that may have been used in pre-Celtic times for the lower, wider reaches of the river, where fording was impossible. In this context, Plowonidā may have been referring to a river too wide or deep to ford; a 'Boat River' or an 'Unfordable River', or possibly a river prone to spilling its banks: a 'Flooding River'. Successive linguistic mangling (not least by Celts, who could not pronounce the letter 'p') converted Plowonidā to 'Lōndonjon'. Cognomination aside, this was undoubtedly a place with identity and scale: in the decade since the opening of the Thames-crossing, this wattle-and-daub, roadside-riverside shanty expanded into the most important hub in Britain with a population of perhaps 4,000.

Sixties Londinium boomed. A two-way rumble of hooves, wheels and feet streamed along its roads. Down at the waterfront, flat-bottomed fishing boats mingled with larger seagoing ships in from the estuary. The T-junction jostled with traffic heading to and from Calleva, Verulamium, Colonia Claudia, Rutupiae, Noviomagus and countless hamlets strewn across the countryside of the south-east. Vast quantities of military supplies and spares passed through en route for the front line beyond the Fosse Way. Londinium was like a tidal sandbank, obliged by the passage of powerful currents to exist in a state of continuous modification. Its fluid population was less a community than an agglomeration of competing interests making the most of location. Londinium fed on trade, on import and export, on production. It wasn't a centre of administration, or defence, or religion. It was, as Tacitus put it, a place 'not yet distinguished by the title of *colonia*, but renowned for its very large population, particularly of traders and travellers'. The stuff left in the earth told of multifarious opportunism: a metalworker on the south bank close to the timber causeway; a butcher on the southern eyot and a grain merchant, blacksmith and cattle farmer working on the northern eyot. Overland travellers heading into Londinium from the west passed a tavern or eating-house, a miller and a baker, a carpenter and a merchant selling dry goods like spices and Samian bowls. At night, the glow from lead and copper foundries cast orange puddles on the rutted gravel of the roadway. On wasteland beyond the built-up areas, shadowy figures scrabbled in the refuse, sifting for lost coins and looking for glass to recycle. Colonia Claudia and Verulamium may have been the earliest settlements in Britain to display the planned virtues of a 'town', but the polyfocal, opportunistic, messy accretion on the Thames was a better exemplar of urban energy.

Physically, Londinium was beyond control. Unconstrained by any defensive walls or palisades or ditches, or by an over-arching plan, it sprawled across two hills, the waterfront, two islands and the south bank. A total area of around 40 hectares was occupied by over 700 buildings which formed a street frontage of five miles or so. New houses, warehouses, sheds and shops appeared every year. Civic administration was still concentrated around the T-junction, where an aisled hall – the largest building in Londinium – now stood on one side of the rectangular open space. Close by, there was a grain store. In this core part of the town, infilling and ribbon development reduced the gap between roadside frontages and the road itself to around five metres, and in parts this strip had been laid with boards or gravel; Londinium had pavements. Londinium could not have been more different to neat, planned Colonia Claudia. The settlement on the Thames had no marbles, or bath house, temple or theatre. Its urban form was led by local geography and cultural flux. The notional planning that had directed the original layout of the T-junction and early roads had been overtaken by a runaway population which took its cue from chance. Londinium's urban trajectory was unlike any other in the Empire. It was Britain's first genuine town. But by AD 61, it was a smouldering charnel-house. During the firestorm, temperatures of 1000°C fused pottery and reduced its buildings to a carbonised shroud over half a metre deep.

Led by Boudica, the Iceni, Trinovantes and others 'not yet broken by servitude', killed – according to Tacitus – nearly 70,000 Roman citizens and allies. The historian was exaggerating, but it was probably the single most horrific act of mass-murder ever to have occurred in Britain. Colonia Claudia, Londinium and Verulamium were reduced to silent drifts of ash and bone. Retribution, Roman-style was swift and savage. Boudica and her barbarian tribes were routed in a pitched battle on the edge of a range of hills which has never been conclusively identified. It was, wrote the historian, a 'glory ... equal to that of our older victories'. It was also a massacre beyond the scale of anything the legions had achieved during the opening phases of the 43 invasion. Against Roman losses of 400, the Britons lost 80,000, a figure that may have been chosen by Tacitus to suit an Imperial rather than British narrative.

The Roman machine took the best part of a decade to recover momentum, but it did so with revitalised vigour. Interior decoration took off in the south. Mosaics and *opus sectile* – the decoration of walls and floors with inlaid pieces of material such as glass or stone – were expressions of artistic sensibility, spending-power and Imperial permanence; the crude ground of Britain broken to fragments and relaid in Mediterranean patterns. The main breaking-ground was a short section of the south coast between the Isle of Purbeck and the Isle of Portland where cliffs and outcrops could be relieved of suitable material: red and yellow burnt mudrocks; hard white chalk and 'marble' from Purbeck; cementstones that came in a range of colours from pale yellowish grey to yellowish black. The volumes required were considerable: one floor of 25 square metres would have required around 200,000 tesserae; half a metric tonne of finished stone. The fashion spanned a wide spectrum of construction sites, from legionary fortresses to fancy villas, masonry town houses and more ordinary, timber-framed dwellings. The number of mosaics being laid suggests that owners were being subsidised by the state or had access to favourable loans; enticements perhaps to speed the Romanisation of native aristocracies. Quarrying, processing and shifting so much stone to so many sites can only have been achieved through organisation and the availability of several workshops or teams of travelling craftsmen. Over in the west, the bathhouse at *legio II*'s fortress on the Exe was given a mosaic floor in around 60, and in the far east, a villa on the Medway had its *frigidarium* floored with a variety of tesserae which included grey cementstone, white chalk, ceramics and red mudrock.

Towns destroyed in the Boudican firestorm rose from the ashes with invigorated ambition. Verulamium was rapidly rebuilt, and substantially enlarged. Enclosing more than 40 hectares, a new defensive ditch was excavated, rectilinear in plan, with the fourth side being fronted by the river. Among the new buildings latched onto the grid of streets was the bathhouse and – occupying the central *insula* – a huge forum-basilica complex measuring some 70 by 64 metres, with walls of mortared flint faced with painted plaster, a landmark project and possibly the first major masonry forum to be built in Britain. An inscription bearing the date 79 recorded its completion. Over to the east, on the River Colne, Colonia Claudia was reinvented on a dramatic scale. Debris was cleared, the former street grid was re-established and new buildings were erected on some of the original plots. And in a spectacular expression of intent, defences of the most formidable kind were erected to protect the rebuilt settlement. Almost a perfect rectangle in plan, a freestanding wall 2,800 metres long and 6 metres high was built around the colony. It was 2.4 metres thick and rested on a foundation trench 1.2 metres deep and 3 metres wide. In a part of Britain bereft of good building stone, the engineers appear to have been tipped off about stone nodules which could be extracted from the local beds of clay. The nodules (they became known as *septaria*, from the Latin *septum*, 'partition', due to the cracks within them) were available in vast quantities. They were used with mortar to fill the foundation trench; they were used for the core of the wall and were dressed for the inner and outer faces of the wall, where they were laid between intermediate courses of brick. A total of six gates were built into the wall, two on each of the longer sides, one at the eastern end and the other at the western end, where the Londinium road passed beneath the original double arch, now enlarged into an even more impressive monument by the addition of adjacent pedestrian arches (a surviving one is still being used by Colchester's commuters). The stone circuit was punctuated at intervals by rectangular towers. The effect was to create a territorial monument. Not uncommon in Gaul and Germany, it was the first of its kind in Britain.

Fifty miles down the road, the Boudican catastrophe provided the chance to resurrect the town of Londinium along more formal lines than its earlier, chaotic evolution had allowed. Workmen moved into the ruins, their identity hinted at by bits of armour and leather tent they left, and by a riverside beam bearing a branded mark which may record the presence of Thracian auxiliaries. Their camp, complete with granary, cookhouse and latrines, was protected by massive double ditches and a rampart which obliterated the road east towards the lands of the Trinovantes and Iceni. Charred mud-bricks and scorched timbers were recycled and used for the core and strapping of the rampart. In the upper part of the Walbrook valley, fire debris was levelled, new buildings with mosaic floors were constructed and timber boardwalks were added to the roadside. An ingenious bucket-and-chain machine was installed to provide the reconstructed town with drinking water. By 64, a massive, new wharf was in operation down on the waterfront. It wasn't till the 70s, over a decade after the Boudican firestorm, that the construction of public architecture began to restore a sense of civic presence. On a levelled terrace down on the riverbank, public baths were built, with a hypocaust-heated hot-room and plunge-pool approached by a suite of ante-rooms. Five hundred metres to the north, close to the corner of the fort, a timber amphitheatre with seating for 7,000 was built. In around 75, work began on the construction project intended to place Londinium firmly at the heart of the province. Along the northern side of the forum, the ground-plan of an enormous basilica was marked out. Fifty metres long, it would be by far the largest building in Britain.

The campaigning continued. By the late 70s, the Roman army had pushed far beyond the Fosse Way and a full-size legionary fortress had been built on the upper reaches of the River Severn, on the threshold of western Britain. In the north, legionary fortresses were built each side of the Pennines: Deva (the future city of Chester) in the west stood on the Dee and Eburacum (York) in the east controlled the Ouse. Both rivers provided navigations to the open sea. Eburacum occupied a prime site in the centre of a huge, fertile vale set between the Pennines and the Wolds, and between the lands of the Parisi and Brigantes. Long corridors of conquest extended further northward, marked by forts, like rivets in an iron bar. They'd been built by order of governor Gnaeus Julius Agricola, and were so well placed, designed and supplied that Tacitus could boast that none of Agricola's forts that had been built 'on a site of his choosing was ever taken by storm, ever capitulated, or was ever abandoned'. From the safety of its military bases, the army sallied forth on punitive missions, then withdrew to its walled sanctuaries. Sufficient grain (usually wheat) to feed the garrison for an entire year – sometimes longer – was stored in venti-lated buildings. The granaries not only made it possible to withstand prolonged siege, but allowed the army to maintain a four-season pres-ence in the most inhospitable latitudes. It was the failure of Rome's occupying forces to withdraw south at the onset of winter that left the natives – if Tacitus is to be believed – 'baffled and in despair'.

There were now two borders within Britain: the Fosse Way sep-arated the civilianised south-east from the militarised north, while the frontline of military expansion now ran across the indented neck formed by the Solway and Tyne. Beyond here stretched undefeated 'barbarian' territory. In 84, Agricola administered what many Romans hoped would be the final gladius-thrust into the throat of British re-sistance by breaking a force of Caledonians who had taken a stand on a hill south of the Great Glen. Calgacus appears to have fought far more effectively than Boudica, killing nearly as many Romans (360) at a cost of 10,000 Caledonians, just one-eighth of the losses suffered by the Iceni queen. Tacitus put the Caledonian army at 30,000, im-plying that some 20,000 left the field of battle. The bloodbath at Mons Graupius concluded the forty-year conquest of Britain, a period that saw between 100,000 and 250,000 native Britons killed; or between one-twentieth and one-quarter of the entire population. To celebrate Britain's submission to 'the terror of Rome', Agricola despatched ships of the *classis Britannica* to sail around the north of the island and back down the west coast to the port they'd departed from on the south coast, thus completing the first Roman circumnavigation. The encircling of the whole island of Britain by the Roman fleet was symbolic and illusory. Agricola neither controlled nor knew the entire island.

The landmark that embodied the illusion was built on the beachhead at Rutupiae within a year or so of Agricola's claiming 'command of Britain'. It was a four-square monument so massive that its foundations of layered flints and clay were 10 metres deep and so tall that it could be seen from far out to sea. Beckoning ships towards the spoils of conquest, its panels were embellished with military scenes and inscriptions, and the decorative work included Carrara marble, sculptures, fluted columns and bronze. Its form was *quadrifrons*, with four arches above two axes which intersected at the centre. Above the arches was a solid attic storey, and each of the two axes acted as a passageway, both real and symbolic. One passage led up a flight of steps from the waterfront and continued through the monument to emerge at the beginning of the new road towards the heart of Britannia. The other axis was oriented along the coast. Here, where the armour-clad legions of Claudius had disembarked to take the land beyond the Ocean for the Empire, were the two intersecting, controlling lines which formed the gateway to Britain. It was one of the largest monumental arch-structures ever built in the Empire. Through the arches lay conquered Britannia, a vast, rural island rich in raw materials, manpower and sources of taxation.

More than ever, Britain was a land of extremes. While brochs still spiked the coast of the far north, a spectacular palace was being built on the coast of the far south. At what is now Fishbourne, gardens, courtyards, baths and floors decorated with black-and-white geometric mosaics graced a complex composed of four wings and an entrance worthy of a king: a towering pedimented facade raised upon six columns some eight metres high. Inside, the entrance hall stretched for over 30 metres. At the time, it was one of the most dramatic – and architecturally advanced – structures in Britain. Its evolution over several decades symbolised the gradual transformation of the south coast from a front line into a region of unlimited aspiration. Back in the early days of the conquest, the tidal inlet outside the harbour-town of Noviomagus had been a military supply base, with a granary and various other store buildings. Within a decade or so, a timber-framed villa had been built here, upgraded in the 60s to a stone villa with a veranda, a bathhouse and mosaic flooring. Work on the Fishbourne palace complex began between 75 and 80, but it was so ambitious that its gigantic footprint couldn't be accommodated on the land occupied by the old villa. Around 30,000 cubic metres of clay and gravel had to be shifted in order to create an artificial terrace large enough for the four wings. The fabulous palace upwind of Noviomagus enjoyed one of the most secure locations in the province: within earshot of the lapping waves which linked Roman Britain to Roman Gaul, and well within the pacified heartland bounded by the Fosse Way. While the hacked corpses of Caledonians seeped into the gritty soil of Mount Graupius, the owners of the palace outside Noviomagus were strolling in colonnaded gardens.

The 80s were good if you lived on the safe side of the Trent and Severn. Past masters at conquest follow-through, the Romans were feeding the Brits with – in the words of Tacitus – 'schemes of social betterment'. In Gaul, it had taken just one generation for a Celtic civilisation to become Romano-Gallic. The defeated aristocracy were Romanised by being offered positions in the army which allowed them to retain some level of status, and higher grades of citizenship were used as enticements to those who demonstrated a loyalty to Roman culture, in particular by helping to create towns and raise public buildings. This strategy worked in Britain, too, as Tacitus described:

> Agricola had to deal with people living in isolation and ignorance, and therefore prone to fight; and his object was to accustom them to a life of peace and quiet by the provision of amenities. He therefore gave private encouragement and official assistance to the

building of temples, public squares, and good houses . . . And so the population was gradually led into the demoralising temptations of arcades, baths, and sumptuous banquets. The unsuspecting Britons spoke of such novelties as 'civilisation', when in fact they were only a feature of their enslavement.

The emerging centre of hedonism was a hot pool on the western Avon deep in the territory of the Dobunni. It had been a popular spot long before the legions showed up. Britain was not a land favoured for geothermal springs. Groundwater was normally 10 to 12°C, but there were half a dozen places where water rose to the surface warmer than 17°C. The spring on the Avon produced water at an astonishing 46°C. To the Dobunni, it was a sacred place belonging to their Celtic goddess, Sulis, whose enchanted, curative waters drew pilgrims from high and low. A causeway of boulders and gravel had been built to improve access to the spring and from around 35 BC – when the Dobunni began minting their own coins – the pool accumulated a glittering silt of silver staters bearing the triple-tailed horse. The first impressions of *legio II* are lost, but the presence of these hot springs in a cold province quickly elevated this place to a unique status in war-torn Britannia. It wasn't in the most accessible part of the province, but it lay within the militarised zone defined by the Fosse Way, at a crossing-point on the Avon. It was also within carting distance of some of the best building material in the conquered territories. On the slopes above the spring were outcrops of oolitic limestone so evenly textured that it could be cut in any direction without splitting, a 'freestone' that was ideal for sawing into blocks and for ornamentation. And barely an hour's ride to the west were a range of hills – the Mendips – whose limestones were veined with lead for pipework and ducts. By the 60s or 70s, the spring had been upgraded for a wider, Romanised clientele by the construction of a massive stone reservoir lined with lead, which provided a head of water for an adjacent suite of baths and hypocaust rooms. With time, the curative waters of the Dobunni were rebranded Aquae Calidae – Hot Waters – although they became better known as the Waters of Sulis – Aquae Sulis.

In this variable-speed Britain, urban development raced ahead in the south-east. At Verulamium, a guiding-light among Romano-British towns, the 40-plus hectares enclosed by its defining ditch were occupied by a mix of dwellings aligned on its neat grid and by open space for paddocks, gardens and production or craft zones. Solidly founded rectilinear houses rested on footings of flint and mortar, roofed with tiles. One private house on the eastern edge of town by Watling Street even had its own hypocaust. The population by 120 was at least 4,000 and perhaps double that figure.

Acting as urban models for the *civitas* capitals, another two *coloniae* were designated by the end of the first century. Colonia Domitiana Lindensium (now Lincoln) was founded on the site of the old legionary fortress overlooking the River Witham. Colonia Nervia Glevensium (now Gloucester) was founded at about the same time on – or close to – the legionary fortress beside the River Severn. At Calleva, a rectangular street grid was superimposed on the earlier *oppidum* layout. Timber-framed houses were demolished and re-placed by new buildings of masonry floored with mosaics bordered by tile tesserae. All three of these towns were defended with massive, banked earthworks. Other towns defended by earthworks during the first century included Verulamium, Isca Dumnoniorum (Exeter) and Venta Belgarum (Winchester). The stone wall surrounding Colonia Claudia can be excused in part as a reaction to the Boudican fire-storm. But the enormous encircling earthworks around these other towns spoke of more nuanced motives. There was clearly an anxiety concerning social order. Defensive circuits protected settlements from lawless bands and facilitated the screening of incoming and outgoing travellers. But they were also spectacular demonstrations of economic and social firepower; dramatic symbols of civic authority, pride and status. Defensive urban circuits were not widespread on the continent. The walling of towns seems to have sprung – at least in part – from an insular tradition within Britain, a regenerated urge to seal nucleated settlements from the surrounding countryside. These were not permeable settlements. Walls and banks removed towns from the wider landscape, separating town from country; townspeople from country-people. To pass through a town gate was to transit between one world and another. Walls interfered with a town's ability to function as a central place by imposing a fixed boundary within the urban–suburban continuum.

Londinium was different. Unwalled and open, this key Roman settlement had not only recovered from the Boudican rampage, but its street grid now draped over both hills on the north bank while a town-sized suburb filled the eyot on the south bank. By 120, Londinium's population probably topped 25,000, far in excess of any other settlement in Britain. It was by now demonstrating that peculiarly urban habit of episodic regeneration. Fifty years after they'd been built on the riverbank, the public baths were enlarged with two additional hot-rooms and various other facilities. In around 120, the temporary military base that had been built near the forum was replaced by a square stone-built fort on the north-western edge of the town. Covering some five hectares, it was a formidable structure with gates midway along each side, watchtowers on each corner and walls over a metre thick and 4.5 metres high. Inside, rows of barrack blocks housed up to 1,000 soldiers. Domestic architecture included numerous courtyard houses, many with mosaics. Londinium now sprawled over 120 hectares, along 1.2 miles of riverbank.

By the middle decades of the second century, there were at most twenty-four major towns in Britain and around a hundred lesser urban settlements, a density far below the Roman provinces in Gaul, Spain and Africa. And most of Britain's urban settlements were concentrated south-east of the Fosse Way. Urbanisation in Britain had not been a widespread success, but the couple of dozen large towns of the 150s were a world away from the meagre scattering of pre-Roman *oppida*. Most had a forum and basilica, and public buildings such as temples and baths. Streets were lined with shops. Some towns had amphitheatres and many of their buildings had masonry walls and tiled roofs. Interiors were decorated with polychrome paint and elaborate mosaics. But these urban trappings masked insecurities: all the *coloniae* had been walled before 200, and most of the *civitas* centres, and nearly twenty small towns. In most cases, the circuits were earthworks, or walls set into existing earthen ramparts.

Beyond the towns, the southern 'lowland' zone was dotted with rectilinear buildings and villas. There seems to have been a regionalism of rectilinear style, with a swathe of villas running from the Exe to the Wash. In the parts closest to the continent – the old lands of the Cantiaci, Catuvellauni and Trinovantes – layout tended to follow a model popular in Gaul, in which the villa was constructed as a long strip of rooms, linked perhaps by a corridor. Further west, in the old lands of the Dobunni and Durotriges, the cultured, rural elite developed a penchant for mosaics which showed off their knowledge of classical mythology. In the East Midlands, large, aisled buildings were popular. And in parts of the south, the roundhouse survived as an architectural form.

Aside from military and urban centres, one of the most 'Roman' places in the south developed around the hot, curative waters at Aquae Sulis. Right beside the lead-lined pool, a spectacular temple was built on a raised podium mounted by a flight of steps rising to a porch supported by four gigantic, fluted columns crowned by Corinthian capitals and an enormous triangular pediment. There was nowhere like it in Britain, and the ready availability of superb local building stone meant that it could only get bigger. During the second century, the Sacred Spring was converted into an all-weather facility by the construction of a huge, vaulted chamber which sealed it from the vagaries of wind and temperature. The temple podium was extended to create a circulating ambulatory around the temple. Shrines were built. With its crisp statuary and classical dimensions, Aquae Sulis was an exhibition of the possible, admired each year by the rich and influential who would return to their estates and towns with reinvigorated architectural ambition.

Never had landscapes in Britain been subject to so much change in such a short period. Only four generations separated the beachhead ditches at Rutupiae from a web of towns linked by engineered roads. The cohorts of carpenters and brick-makers, stone-masons and soil-shifters who stormed ashore on the south coast in the year 43 were the vanguard of a force that brought forts, temples, baths, amphitheatres, villas, mosaic flooring, hypocausts and urban drainage. In eighty years, Britain had been force-marched towards a Mediterranean vision of civilisation.

In the militarised zone beyond the Fosse Way, landscapes were on a lesser trajectory. Episodic resistance to Roman rule tied up a disproportionate number of Rome's soldiers; the 40,000 to 50,000 stationed in Britain during the second century represented over one-tenth of the Roman army's entire strength. While some of the legionary fortresses had been reconfigured as *coloniae* in the south, the north was still a war-zone, the ebb and flow of successive campaigns recorded by the rectilinear imprints of yet more marching camps and forts. Many still reek with melancholy. Located for reasons that were lost once the soldiers left, these rectangles in the turf are often preserved in places bereft of modern settlement. Gusting moor and bleak col want for house and garden. On the west side of Hardknott Pass in the Lake District there's a rocky spur that runs down from the mountain known as Border End. Where the spur becomes less steep for a few hundred metres the rectilinear walls of a 1.2-hectare fort stand chest high around the remains of the commander's house, the headquarters buildings and the granaries (you can still see the platform entrance where carts could be unloaded). Some way below the fort, on the side of the spur, are the ruins of the bathhouse: a furnace room for heating water, hot and cold baths, and a circular *laconium* where knackered soldiers could sweat out their aches. With steep drops on three sides and a mountain behind, it's well defended, but it's also exposed to westerly winds and stands at over 200 metres above sea level. Being stationed here in winter must have been a truly horrible experience. Unusually for Roman forts in Britain, the parade ground is exceptionally visible. As big as the fort, it was cut by hand – presumably by the poor bloody soldiers – some 200 metres uphill from the fort. The remains of an inscription from the south gate records the presence of the Fourth Cohort of Dalmatians, men raised close to the Mediterranean coast. There were ten cohorts to a legion of 5,500 men, so 550 soldiers may have been stationed on the pass. You can still stand beside the remains of the tribunal ramp which was used to review the troops as they trained on the parade. They must have suffered in the cold and rain of lakeland. I was there a few months ago, at the end of a balmy September weekend on the fells. It had been a day of Apennine clarity, silent and blue, and as the evening sun sank over Eskdale, falling shadows inched across the interior of the fort, bringing with them the first chill of darkness.

Britain's situation as an Imperial province meant that political events at the epicentre on the Tiber had seismic implications at the periphery. The death in 117 of Emperor Trajan and succession of Hadrian replaced an expansionist warmonger with a consolidator. Hadrian was the kind of emperor who liked to share scran with the squaddies. He'd done time as a tribune in *legio II Adjutrix*, a unit that had been recruited by Vespasian from an auxiliary force of marines. Recorded by his biographer as a cheerful consumer of 'such camp-fare as bacon, cheese and vinegar', Emperor Hadrian would march 'as much as twenty miles fully armed', wear ordinary clothing, carry an unadorned sword, visit sick soldiers in their quarters and appoint to the rank of tribune only those with full beards who were old enough to confer authority. He was an inveterate traveller who personally selected the sites for camps and understood how the military could cooperate with landscapes. Significantly for Britain, Hadrian – along with many of his military commanders and governors – had drenched himself in Greek culture. The 'Greekling' had an acute grasp of geography. There's little doubt that he'd absorbed the stories about the 'long walls' constructed to connect Athens to its port and similar walls that had been built across necks of land at Corinth and Gallipoli. So when Hadrian reached Britain in 122, he arrived with a long-wall solution to the most troublesome frontier in the entire Empire.

He divided the island. The idea, according to Hadrian's biographer, was 'to separate the barbarians from the Romans'. The place he chose for his Greek-revival 'long wall' was the most obvious geographical pinch-point in Britain's mid-latitudes: the sixty-mile constriction between the mouth of the River Tyne in the east and the estuary of the Solway in the west. Ethnically and politically, it was a divisive cut, following neither the edge of Brigantes territory, nor the current limit of the Roman province. Between the two seas, Hadrian's surveyors made use of a succession of rivers and crags to strengthen the purpose of the wall. Their most spectacular natural ally was a series of volcanic intrusions protruding in cliffs, scarps and ribs from the bleak grass-lands along the intended course of the wall. Dolerite belonged to the family of rugged volcanics that had stood firm as resilient headlands and eye-catching crags from the West Country to the Orkneys (and had contributed the bluestones for Stonehenge) and its sombre, in-timidating aspect contributed to the wall's inviolable verticality. Raw materials like timber, turf and stone could be sourced locally; labour was provided by the army. At the time, there were three legions in Britain, and all were involved in construction. Plans changed after work commenced. Initially, the western thirty miles of the wall were to be built from piled turf, and the eastern section from stone. Along its entire seventy-three-mile extent, gates (milecastles) were to be set at one-mile intervals, with two observation towers (turrets) between one gate and the next. In the west, a fifteen-mile chain of forts contin-ued along the southern shore of the Solway Firth to the important port at the mouth of the River Ellen. Long before the job was complete, it was decided that a far larger garrison was needed for the wall than could be accommodated in the milecastles and turrets, and so forts were built astride the wall, with gates to north and south so that units of soldiers had secure bases from which they could mount offensive operations in both directions, the perceived threat being as likely to come from the south as from the north. A second modification was the excavation of a deep ditch flanked by two banks, set back on the south side of the wall. The subsequent addition of the huge *vallum* along the southern side of the wall restricted access to the frontier area and reduced the number of crossing points from eighty to around sixteen. Movement and passage near and through the wall could be tightly controlled. It was a zone of control rather than a defensive structure; a set of sluices rather than a dam. For indigenous Britons, free passage between north and south was but a memory.

Emperor Hadrian's long wall was fiendishly successful. There was an attempt, within months of his death in 138 and the succession of Antoninus Pius, to shift the control-zone a hundred miles further north. Another linear barrier was built, stretching the forty or so miles across the narrow isthmus between the firths of Forth and Clyde. An exercise in military make-do, this one required the digging of a massive ditch fronted by an upcast bank and turf rampart on its northern side. By the time the rampart was complete, nearly 400 hectares of turf had been stripped. Some thirteen forts were built along the barrier, virtually all of them timber. But Antoninus's was a wall too far, and after a couple of decades there was a new emperor in Rome – Marcus Aurelius – and the garrisons were back on Hadrian's Wall, this time for keeps. The original ditches were cleared of twenty years of debris. Walling, milecastles and turrets were repaired, and a new military road was constructed between the vallum and the wall, creating both a coast-to-coast communication artery and a means of linking all the forts.

Landscapes north of the wall were barely marked by Rome. Substantial occupation lasted for no more than forty years or so and had strict geographical limitations. Mountainous terrain neither suited the Roman preference for fighting in open ground nor came close to providing sufficient quantities of food and military supplies to sustain garrisons. During the most extensive episode of occupation – the Flavian years from 69 to 96 – there were perhaps 25,000 troops north of the Solway. The amount of grain they would have needed each year, together with the needs of cavalry horses, was probably between 16,000 and 19,000 tons, some of it undoubtedly imported to the front line. And this army of carnivores would have also slaughtered annually – allowing for their predilection for beef and pork – around 950 cattle, 1,200 pigs and 1,000 sheep. Wear and tear on leather was huge, and this peripatetic army needed constant renewal of footwear, clothing and tents, which might have required as much as 2,000 calf-hides per annum. Sacrifices at festivals took out another 1,500 animals a year. The combined cavalry units needed over 500 new horses each year. Demands on this scale could not be met in the Highlands. The limited scope of military occupation north of the Solway was determined in great part by productive agricultural land.

The walls and forts were a war-zone legacy written for eternity in the landscapes of Britain. And so too were the communications that sustained them. It had taken the best part of 150 years to roll out a comprehensive road system, but by 180 or so the Romans had pushed highways into the parts of the archipelago they most needed to control. From the principal hub, Londinium, roads radiated outward. In the south-west, they petered out beyond Isca, apparently thwarted by Dartmoor's misted bulk. In the west of Britain, a road ran between the Severn estuary and Dee estuary and another parallel one ran further west from Carmarthen Bay to Snowdonia. The Fosse Way still functioned as a diagonal route across southern Britain. In the north of Britain, a ladder-like network ran along each side of the Pennines, connected laterally with 'rungs' across the hills. The most northern section of the network ran along a ridge towards the fearsome waters of the Tay, but this road petered out in the river's flood-prone valley.

Some of the most important traffic was information being carried by the *cursus publicus* – the Imperial post. To boost transmission speeds, stables – *mutationes* – were built every four miles or so along the main highways, where riders could change horses. Information could be pushed along the highways at a rate of about fifty miles a day, although far greater speeds were possible in times of need (on the continent, Tiberius once covered 200 miles in twenty-four hours). The main highways were provided with rest houses – *mansiones* – at intervals of 11–13 miles, depending upon the terrain. This was not the kind of infrastructure system that would have evolved had Britain's indigenous people been allowed to upgrade their own desire paths. The Roman road web was a tool of Imperial control, the means by which people and information could be shifted speedily between centres. Its stakeholders were political and military. This had a profound effect on the geography of the system: this was a network with a capital bias (all roads led to London) and it was a network whose reach was limited by the scope of military conquests. In the south-west, Roman roads petered out at the River Exe, and in the north they ran out of momentum just beyond the Clyde. The roads of old had evolved through the millennia to work with natural features, following the scarp slopes of downlands or wending their lowland way from ford to ford, their primarily local purpose being to connect adjacent communities. Roman roads stretched as taut as twine between termini which were hundreds of miles apart. Maintenance of the highways was regarded as being of the highest importance.

The road system increased the take from land. Mines, quarries and farmland became more accessible, more open to organised production and distribution. Coal from the seams of the Tyne, Wear and Tees was carted to forts and various other military sites in the north. The coalfields flanking the Pennines and others each side of the Severn estuary fuelled central and southern markets. Salt was hauled from the flatlands south of the Mersey. The brine springs on the River Salwarpe – renamed Salinae, 'salt-works' – became a major settlement and the hub of four radiating roads. Stone was in demand for buildings and ornamental work, and could travel astonishing distances; at least fourteen varieties of marble were used for decorative work in London, their sources including Gaul, Italy and islands in the Aegean. Inferior to marble, but workable nonetheless, stone from various quarries in Britain was also sent to London. Along dusty Roman roads, carters hauled white lias from the West Country, chalk from the North Downs and shale from the Weald. From quarries on the Medway, huge consignments of ragstone were shipped downriver and then up the Thames on a voyage of some fifty miles to Londinium, where the off-white rock was used for the city walls. From the

Cretaceous beds on the Isle of Purbeck came a limestone (one piece ended up on the bottom of a London plunge bath) that could take a polish, later known as 'Purbeck Marble'. The oolite quarries that had built Aquae Sulis provided high-quality freestone that was also carted east to Calleva, for use on the baths and forum-basilica, and north-west – possibly via the Avon – to the legionary fortress at Caerleon.

The quest for metals exacted its own costs on the landscape. Tin, copper and lead were all worked, but in terms of volume, the big one was iron. Iron bound the whole Roman enterprise together. It made military equipment and weapons, fittings for buildings, ships, wagons, horses, surgical instruments, locks and keys, tools for farming and mining, metalworking and plastering and processing everything from leather to cloth, for woodland management and carpentry. The list was virtually endless. The first major source of iron ore to be ex-ploited after the invasion was the Weald, but the breakout opened access to fields in the Forest of Dean and others to the west of the Fens. Enormous slag heaps accumulated (just three of the Wealden slag-heaps would half-fill the Albert Hall). With so many sources of ore, the number of small-scale iron-smelting sites was huge, and with time various production centres developed, controlled by Roman authorities or by private contractors to the state. Slag heaps could cover more than a hectare.

Agricultural land also faced unprecedented demands. Food was wrested from field, pasture, wood, wetland, moor and mountain in unprecedented volumes. Some 40,000 hectares of cultivable fields were required to keep the forces of occupation supplied with corn alone. Productivity increased and several new crops were introduced, among them cabbage, asparagus, celery, beet and carrots – the latter long favoured by Romans for healing wounds, treating tumours and dissolving spots before the eyes. Orchard crops appeared too, among them apples, plums, cherries and medlars. Among the new fruits were vines, the 'one asset', according to Pliny, that allowed Italy 'to have surpassed all the blessings of the world'. Fields bore harvests unseen a century earlier. Less dramatic than it might have been had Britain

been invaded in Caesar's time was the systematic partition of agricultural land. Long before the Romans arrived, Britain's own people had
devised field systems suited to the local peculiarities of topography
and climate. Centuriation, the division of conquered territory into
square blocks (equivalent to 706 metres along each side), had been
a widespread practice in the conquered territories of the western
Mediterranean during the latter part of the first century BC, but had
ceased to be a common land-management tool by the time Britain
was invaded.

The Romans had been helped in their exploitation of the Empire's
most northerly province by the climate. During the three centuries
from 100 BC to AD 200, the climate of northern Europe had remained
exceptionally stable and warm. The volcanoes of the northern hemisphere had behaved too: from 40 BC to AD 150, there was an unusually
low level of eruptions, and corresponding absence of 'volcanic winters' – although the years 75 to 93 did see a cooling, perhaps as a result
of the Vesuvian eruption of AD 79. Alpine glaciers had retreated and
Britain's southern neighbour, Gaul, had enjoyed a stable and balanced
precipitation, with regular river flows and reduced flooding. In Britain, mean July temperatures were at least one degree Celsius higher
than they would be in the mid twentieth century; warm enough
for vineyards to become a common sight on the British landscape.
Rome's northerly adventures were buttressed by unusually favourable
conditions beyond the southern fringe of the Empire. Between 30 BC
and AD 155, rainfall in the uplands of Africa fed the Nile with copious
volumes of water, increasing the frequency of annual floods in the
river's massive delta and bringing a long succession of huge, depend-
able grain-yields to the Imperial breadbasket.

At its most effective – say between 150 and 200 – the reach of
Romanisation in Britain was partial and indicative of an Empire at
the limit. Incomplete domination exaggerated Britain's geographical
inequalities. While the elite of Fishbourne padded about on polished
tesserae, people of the edgelands subsisted in settlements that had
barely changed for centuries. In the north and west of Britain, the

roundhouse remained the dominant style of architecture. In the lands of the Cornovii (now Cornwall), there was a cluster of roundhouses whose walls would last until the twenty-first century. Assembled from piled stone, the form of these houses appears to have become more inward-looking and defensive since the Roman occupation. Surrounded by massive, protective walls parted by just a single gateway, they faced away from the prevailing winds. Small living-spaces were built into the inside of the encircling wall, with thatched or turf roofs supported by a central, vertical post. Each enclosure might have two or three of these rooms, which opened onto a central space in the enclosure. Rudimentary drainage systems carried rainwater out of the house, but these were desperately primitive dwellings compared to the Romano-British villas of the east.

In the far north, a vast redoubt of traditional habitation was isolated beyond the war-zone that prevailed between the Forth–Clyde lowlands and Hadrian's Wall. Military conquest followed by civil rule had brought towns and villas to the south; knowing neither, the north had become a living archive of indigenous landscapes. Brochs were still being inhabited and maintained in the Shetlands, on the Orkneys, on the north-western mainland, the Western Isles and to a lesser extent south beyond the Forth and Clyde. Duns still clustered among the Highlands and islands of the west. Wheelhouses still seeped smoke on the Shetlands. Stone-lined, subterranean chambers were still being used as food stores. In the war-zone itself, the reins on development were tugged even tighter by outbreaks of open conflict such as the episode in the early 180s described by Dio, when 'the tribes in the island crossed the wall that separated them from the Roman army and did a great amount of damage, even cutting down a general together with his troops.' A punitive campaign was mounted, but the incursion hinted at longer-term troubles.

THIRTEEN

Scatter of Tesserae

200–550

A long time ago I lived beside a Roman fort. Most evenings I'd go to the river. The route I followed was an inner-urban rat-run, down steps, up steps, along elevated, concrete walkways, past tower-blocks and a medieval bastion to a length of weathered Roman ragstone and then over the swilling traffic of London Wall to the amphitheatre and the place where they'd discovered the water-lifting engine. I'd pass the bathhouse on Cheapside and then Broken Wharf where I'd turn upstream past the temple complex and the tidal grave of the Roman barge. In ten minutes I could run two thousand years.

In around 200 AD, the fort outside my flat was demolished. Or at least, most of it was. They kept two of its four walls, incorporating them into a new, much longer wall which stretched for 3,200 metres around the landward side of the town. With the river on the south side, and gates for access, Londinium was wrapped in rock and water. The surprise is less that the wall was built, but that it had taken so long to get around to it. By 200, the *septaria* and brick circuit around Colonia Claudia had been standing for over a century. The wall around Londinium enclosed an area – 133 hectares – half as big again as Corinium, the next largest town in the province, and three times larger than Colonia Claudia. It was the biggest settlement enclosure ever built in Britain and probably took two or three decades to complete. Around 85,000 tons of ragstone had to be shipped from the Medway in 1,300 barge-loads. To provide the necessary stability, the wall's base was 2.7 metres thick, with two facing courses of squared ragstone rising each side of a core of rubble and mortar. Additional strength was contributed by layers of flat, clay tiles laid at intervals in the ragstone courses. The finished wall was six metres high and turned this notoriously unkempt town into a place that appeared homogenous, huge and well defended. Londinium became an island behind formidable white cliffs.

The wall concealed a shrinking town. Much of the enclosed area was open ground. In part, this was due to inconvenient topography and unruly growth. Londinium straggled over two low hills and several streams, with rising land to the north and an important suburb beyond the walls on the far side of the river. Encasement was necessarily clumsy, but Londinium's population had been declining for several decades. This was a settlement created for – and sustained by – traffic, and for various reasons, flows were easing. By nature, river ports existed in a state of constant degradation. Silting and falling tidal levels required quays to be periodically – and expensively – extended so that ships could continue to berth. Londinium was also bleeding trade. The episodic rumblings in northern Britain kept army units on station at key defences like Hadrian's Wall, diverting military supplies from the Thames to northern ports. One of the larger outbreaks of trouble occurred in the first decade of the 200s when renewed aggravation north of the Wall culminated in the governor of Britain, Alfenus Seneccio, appealing to Rome for reinforcements. Emperor Septimius Severus showed up in person, with two sons and sufficient troops to mount a four-year campaign which led to the eruption of a new string of marching camps – some of them huge – way beyond the Forth–Clyde line. In terms of holding that line, the Tyne was now more crucial than the Thames: it was at this time that the Tyneside fort at the eastern end of Hadrian's Wall was enlarged and converted into a major supply base. Military buildings were demolished and replaced by thirteen granaries, with another six being added later.

The Severan campaigns of 208–211 also boosted Eburacum's role as Britain's northern hub. Well over a century had passed since *legio IX* had built their great fortress on the east bank of the Ouse and its walls were now surrounded by a sizeable civilian settlement, a *canabae*, while a self-contained settlement now occupied the facing, west bank. It was to his *domus palatina* – imperial residence – on the Ouse that Severus returned each winter from his shock-and-awe campaigns further north and it was here that the emperor died in February 211 as he was preparing to ride north yet again. The decision was taken in Rome to restrict the size of military forces available to any one commander, by dividing up provinces. Eburacum became the capital of *Britannia Inferior*, a new province stretching from the Wall south to a line from the Wash to the Dee. Londinium became capital of the remainder of southern Britain: *Britannia Superior*. The Thames capital remained the senior of the two provinces, with a consular governor against Eburacum's praetorian governor, but this politically motivated split reduced Londinium's administrative duties. The civilian town of Eburacum joined the veteran *coloniae* of Camulodunum, Lindum and Glevum by being elevated to the status of honorary *colonia*.

In a land with a rural, civilian settlement pattern characterised by dispersed farmsteads and occasional nucleated clusters of buildings, there was little of permanent record being left on the landscape. The most spectacular building projects were logistical and military. At three points on the southern segment of the east coast, huge, stone-walled trans-shipment points were built. The most southerly was the one built no more than eight miles from Rutupiae, at the northern end of the Wantsum Channel. Another one appeared at the mouth of the great estuary formed by the Bure, Yare and Waveney, while the third occupied the northern shore of the old Iceni lands at the mouth of the Wash. All three were well placed to act as entrepôts for productive hinterlands and located at points convenient for inland vessels to trans-ship materials into larger, seagoing ships. Architecturally, these fortified ports conformed to the usual fort-plan: rectilinear, with rounded corners and a gateway midway along each side. (After your picnic on the grass at Reculver, take the short walk down the path to the lane and you'll see the tarmac follows the third-century wall, whose core of flint and concrete underpins the hedge; the most likely location for the Roman harbour is now a caravan site.)

With increasing logistical demands there was a coastal surge later in the century as many more fortified ports appeared. At least seven more were built at logistical hotspots along the coasts of the south-east from the Solent to the 'Great Estuary'. This second generation of forts was built with projecting towers. There was an ominous circularity to this late-third-century bout of fort-building. One of the twelve or so that can be said to belong to this system of logistical trans-shipment centres was built on the beachhead where it had all begun in AD 43. Rutupiae had risen and fallen. It was still, as Ammianus Marcellinus put it in the fourth century, 'a quiet haven on the opposite coast'; the southern end of the Wantsum Channel was still the closest haven to the continent. The enormous monumental arch that had greeted arriving sailors for two centuries was demolished and its marble cladding smashed and incinerated in kilns then converted into concrete to bind the fort's walls. It was a massive affair, with masonry 3.3 metres thick, and rising to 10 metres and enclosing an area of 2.5 hectares.

Walling with stone had extended by the end of the third century to most towns as they replaced or upgraded earlier circuits of earthworks, programmes that probably had as much to do with distinction as defence. By the fourth century, many of these walls had been modernised further with the addition of projecting towers. Prominent turrets seem to have been related to status rather than function, the outcome perhaps of Imperial concepts of what a modern town should look like, or the vanity project of urban elites. At London and Caerwent, turrets did not accompany the full circuit of walls and it's highly unlikely that the manpower existed to mount a credible defence. It was as if the whole Roman project was calcifying, becoming less flexible. The hundred or so towns that dotted the southern part of

Britain may have been home to around 240,000 people, give or take 50,000. Given that the total population of Roman Britain was probably between three and five million, around 95 per cent of the entire population was 'rural'. Towns may have been the engines of Roman civilisation, but they were enveloped by tracts of field and wood, fen and mountain which must have seemed interminable when viewed from freshly chiselled ramparts.

The encasing of towns with stone was accompanied by an apparent loss of vitality within. Support for public buildings and urban monuments withered as successive generations of wealth-withholders spent their money on capacious stone town-houses and countryside villas. The custom of embellishing towns with look-at-me construction projects, sculptures and inscriptions died out. Some key towns lost buildings which a century earlier would have been regarded as essential to civic existence. In about 300, Londinium's magnificent forum/basilica complex was demolished, and at around the same time, the amphitheatre ceased to function. The London mint closed in 326, but was briefly revived between 383 and 388 under the emperor Magnus Maximus. In some towns, the central forum-basilica space was put to new uses: in Calleva, for example, metalworkers moved into the basilica before the end of the third century. Hearths and furnaces of metalworkers were also set up in the part-demolished basilica at Venta on the Severn. And at Glevum, the forum ranges were demolished and the enlarged open space resurfaced with cobbles. In Viriconium (Wroxeter), the forum and basilica were damaged by fire and not rebuilt. The same happened in Ratae (Leicester). Out in the far west, the inhabitants of the old garrison town of Isca on the Exe undertook restoration work on their forum-basilica around 350, but the site was cleared by the end of the century and taken over by metalworkers and then, fittingly, it became a cemetery. Large public baths fell out of use. The great urban project had effectively failed. Few towns had the productive resilience to endure beyond the departure of Roman administration. Britain was unready for a city state.

The urban death throes of the fourth century were accompanied by a terminal floruit of rural extravagance. While the wider world fell apart, those with money went mad on extensions and interior decoration. In the least insecure districts, country houses sprang sun-trap courtyards and revamped reception rooms. Additional steam-rooms were suddenly essential. Mosaics were laid as fast as tesserae could be supplied. They cared so much for appearance and yet their identities proved no more enduring than the tiled roofs they paid others to lay. Among the country-villa crowd may have been urban exiles escaping the increasingly dreary confines of town walls and the expectations that they should fund urban projects. Some may have been rich immigrants from the continent, shifting their wealth to an offshore haven as fear of eastern barbarians outweighed attachment to homeland. Some may have been families who had made a stack from local production, land or politics.

The anonymous developer who poured a small fortune into a western Cotswold villa during the first half of the fourth century was intent on creating a talking point, the kind of rustic palace visitors couldn't forget. The site was delightful, bedded into a side-valley of the Frome, equidistant from Corinium Dobunnorum and Glevum (Cirencester and Gloucester). Up at the top of the steep scarp above the villa lay excellent sheep country, and down the valley lay the flatlands of the Severn. Within a five-mile radius, there were at least seven more villas. The existing villa was far from poky. It already had two courtyards, the smaller one being 27 metres square, with cloisters on three sides. But a third courtyard was needed, and with it came more cloisters and suites of new rooms, one of which was spectacular. By the time it was finished it can't have been anything other than the talk of the Cotswolds. The room stood on the central axis of the complex, at the northern end. It measured 15 metres along each side, and its soaring roof was supported on four pillars standing within the room. Beneath its floor, enormous hypocausts provided heating. And on that floor was the largest mosaic pavement to be commissioned in Britain. Around 1.5 million tesserae were used to create a geometrically patterned, square frame which held three concentric circles and an octagonal centre-piece themed on the evergreen story of Orpheus, the paradisiac, lyre-strumming pop-star, seen on the villa floor charming processions of beasts and birds around the floor in concentric circles. It was the kind of story that could be read as an allusion to the owner's Mediterranean learning, and to the richly endowed game-park lying beyond the room's doors. It was an image that turned the chaotic living world into a peaceable procession. When it was completed, the villa, with its three courtyards and sixty-plus rooms, covered over two hectares and had nearly 1,000 square metres of mosaic flooring.

The wrecking-crews of the Roman Empire recruited from far and near. Transformative forces included climate refugees thousands of miles away in inner Asia, where vast steppes traditionally supported nomadic pastoralists. In the latter part of the fourth century, these fertile grasslands were seared by one of the worst droughts for two thousand years. It began in 338 and lasted forty years, long enough to drive the nomads – the Huns – westward, over the Don, to collide with the Goths, who fled into a weakened Roman Empire, which ultimately led to one of the worst military defeats the Romans had experienced. For Rome – and for Britain – it was demolition by multiple hammer-blows. Campaigns launched against Roman territories on the continent affected the Empire's ability to defend Britain and disrupted trade routes between British and continental markets.

From the early 360s, Britain's coasts were preyed upon by seaborne raiders: the *Picti* came from the far north; the *Scotti* from Ireland; the *Attacotti* (of unknown provenance, but possibly the Western Isles); the *Saxones* from the far side of the North Sea. In 367, these assorted plunderers coordinated their attacks. Parts of Britain as far south as the Thames were ravaged by the Picts, the Scots and the Attacotti, while Saxons and their neighbours, the Franks, attacked the shores of Roman Gaul. Not settlers but looters, they snatched booty from an unguarded land with, as the contemporary Roman historian Ammianus put it, 'cruel robbery, fire, and the murder of all who were taken prisoner'. Londinium was 'plunged into the greatest difficulties'. Hadrian's Wall was overrun and defences on the North Sea coast, too. Simultaneous strikes on multiple fronts, by land and sea, were unprecedented. The invaders of 367 were eventually repelled and surviving marauders wiped out, but the malaise was irreversible.

In the midst of chaos and uncertainty, a new cult accelerated the abandonment of old ways. The intermittent replacement of paganism by a religion practised by followers of Jesus Christ did little to reinvigorate religious centres, although it may well have delayed their demise. By 320 there were Christian bishoprics at London, York and Lincoln, although evidence of formalised assembly buildings is blurred. By the 360s, Christianity had been adopted as the religion of the Empire and by the 390s, all public displays of paganism had been banned. Tombstones ceased to be commissioned and pagan monuments disappeared, but whatever success Britain's Christians enjoyed was not reflected in the construction of enduring stone monuments or the resurrection of civic presence. Christianity's earliest contribution to townscapes was the *ecclesia*, or church, which first appeared in the public spaces of Romano-British towns during the twilit, wane-of-Empire decades of the late fourth century. In a later age, the remains of what may have been a church and baptismal font emerged from the soil of Rutupiae. In richer villas, the tracing in tesserae and wall-paint of Christian symbols also hinted at spreading beliefs.

The final systematic construction project was a chain of signal stations along the east coast from the Tees to the Humber. Each took the form of a central masonry tower surrounded by a wall or enclosure fitted with projecting towers. Whether their primary purpose was to relay warning of seaborne threats back to Eburacum, or as a screen intended to suggest to potential attackers that the coast was strongly defended, their effect when lit must have been poignant to some: a string of flickering beads along the furthest reach of an Empire breathing its last. On the cliff at Scarborough, you can still see the ground-plan of one of the signal stations, etched with stone into the grass; a rectangle haunted by anxious ghosts. The huge cost of repelling

Britain's raiders and invaders was largely met by ordinary folk whose taxes rose, strangling their ability to buy goods at the marketplace. With demand plummeting, production slowed. Iron furnaces fell silent by the red-bedded streams of Wealden oak-woods. In the suburbs of towns and cities, pottery kilns cooled, and were abandoned. When people needed a pot, they moulded it by hand from local clay and hardened it in a bonfire. With less money circulating, and mints silent, life leeched from Britain's towns. After Emperor Magnus Maximus was killed in 388, Britain's markets had to rely on coins already in circulation, and on imported coins, mainly from mints in Trier, Lyon and Milan. Once the supply of imported coins dried in about 400, the end was imminent. Britain became a land with insufficient coinage, unpaid, rebellious soldiers and neutered governance. By 390, there there no Roman troops in western Britain and by 407 any remaining units in Britain embarked on ships to support yet another usurper in his bid for Imperial power. Without their economic and military scaffolding, the unsupported structures of Roman Britain could only crack.

Scattered across Britain's land surface in various states of disrepair were around 500 military camps and at least 300 forts and fortresses. The island had been cut into three by a long wall and a long bank, and diced into segments by 7,000 miles of Roman roads. Around 2,000 villas had been built, and perhaps as many as 100,000 other farms and rural sites. Maybe twenty or thirty of the villas were extravagant enough to be described as palatial. The largest monuments were the hundred or so towns. The legions of chippies, brickies and plumbers had modified select locations to spectacular effect, but Britain was an island of 23 million hectares and most of the island was unmarked by Mediterranean architectures.

Across much of the north and the west, the collapse of the Roman 'system' left little or no mark on the landscape. The roundhouse still dominated. North of the Clyde and Forth, settlement evolution was removed from the military and cultural inundations of the south. There's little to suggest that the relatively brief Roman tenure had any effect on native patterns of settlement. At some time before the mid-millennium, the broch-settlements ceased to be occupied, and wheelhouses fell out of fashion too – although the remote wheelhouse at Old Scatness on the Shetland Islands continued to be used as a dwelling. Generally, there was a shift from imposing domestic structures to simpler, cellular buildings, sometimes built on the sites of old brochs or wheelhouses. Some dwellings were partially subterranean, to provide shelter from wind and to economise on building materials. Waterlogged ground was an issue, and the only feasible settlement sites were those on well-drained rises. Here and there, sufficient dwellings had accumulated to form small, hamlet-sized settlements. Hillforts and promontory forts continued to be built and occupied, their emphatic presence on the landscape continuing to identify the territories of local aristocracies, an increasingly complex picture with the arrival of the Scotti, people from north-eastern Ireland who established their own centres on the west coast, countering the power-bases of the Picts, who dominated the Highlands and east coast. Most of the forts at the time were small – less than a hectare in area – and protected by timber-laced or drystone walls.

The failure of the Romans to dominate the entire archipelago was not the only reason that the far north appeared little changed by four centuries of the Imperial presence in Britain. Most of these northern landscapes were unfavourable to human settlement and farming. Less than 20 per cent of the land was cultivable and it was in these lower, coastal belts and valley floors that people eked a living: herding small, brown-fleeced sheep, goats and shaggy cattle. Most of the territory north of the Wall was still a wild land of mountain, moor and rough pasture patrolled by wolves, wild boar, wild cats and countless deer. The total population north of Hadrian's Wall was probably less than 200,000. A tantalising vignette of upland Britain is provided by *De Excidio Britanniae*, the longest insular text to have survived from this era. *On the Ruin of Britain* was written in the first half of the sixth century and was a lament in which the British, classically educated author – a cleric called Gildas – bewailed the decadence of Britain's rulers and Christian clergy. At one point in the text, he referred to 'mountains particularly suitable for the alternating pasturage of animals'. It's the earliest known reference in Britain to the age-old practice of transhumance and its significance in the lines of Gildas is that pastoralists were still engaged in an ancient method of migratory subsistence some 4,000 years after the introduction of cultivation to Britain.

The greatest decay occurred in the most Romanised provinces. Writing and literature disappeared along with the coinage and legions. Villas stood roofless to the sky, mosaics fraying, hypocausts silting. Sites that had once fed markets with cartloads of pots and nails, timber and tiles stood silent. Cemeteries were overtaken with wild grasses. Towns withered, their streets holed and houses derelict. Conduits choked. The great road system fell into disrepair; potholes multiplied. Timber bridges went unrepaired; trees felled by gales blocked the highway. The military, political and economic motives for maintaining a long-distance road system no longer existed. In places an old *agger* performed a function as a boundary between farms. Britain's population slumped, and may have dropped to two million or so. Once-productive fields were retaken by weeds, scrub and woodland. Culturally, this had become a no-man's-land: the regimented rows of cultivated crops, a metaphor for Romano-British order, overtaken by an opportunist free-for-all as a variety of invasive species sought out the most fertile pockets of abandoned soil and set down roots.

A significant migration of peoples occurred. Some were already in Britain; some came from southern Scandinavia; some from the lowlands of the Ems, the Weser, the Elbe and the eastern rump of Doggerland. Their ethnic identity was as cloudy as a muddy estuary. Writing a century or so after the first migrant boats lurched into the choppy North Sea, the historian Procopius claimed that 'Brittia' was composed of three peoples: the *Angíloi*, *Frissones* and the *Bríttanes*, each being governed by their own king. A bit more flesh was put on these bones in the early 700s by a Northumbrian monk called Baeda, or Bede, who related that the immigrants had come from 'the three most formidable races of Germany: the Saxons, Angles, and Jutes'. Migrants need a very good reason to leave a homeland and embark upon a long and hazardous sea voyage. Migration required the perceived benefits of an alternative habitat to outweigh the anticipated risks and costs of travel and of leaving the homeland. War, plague and famine were perennial stimuli. Not for the first time, it appears that Britain was the recipient of climate refugees; from around 300, northern Europe's characteristically dry summers had begun to get wetter; by around 350, wet summers had set in, and they persisted until around 450. The effect on the deltas and floodplains of northern Europe would have been dramatic; the timber buildings of at least one settlement on the lower Weser were abandoned in the mid fifth century, most probably due to rising sea-levels. The migrants had much in common with the people of Britain: their homelands shared the shores of the North Sea; they came from beyond the orbit of Roman civilisation and knew little or nothing of towns or coins; they must have been desperate and they were taking a one-way voyage. They were effective colonists because they had to be. Like so many earlier migrations to Britain, this one was spread through time and was less a 'movement' than a succession of sporadic intrusions by small groups taking advantage of resources, weather and tides. They came by boat to the shores closest to the continent and sought land in the south-east. Initially, the immigrants came as family groups and their impact was small, but from around 470 the numbers increased, and then tailed off in the mid 500s. Nobody checked them in at the coast and so the total number of migrants who arrived was uncertain, but probably lay between 10,000 and 20,000, far less than half the capacity of the amphitheatre in Rome.

Their settlements appeared in vales reaching from the Channel to the Thames, Ouse, Nene, Trent and beyond. The draughty, timberless heights of the Cotswolds and the Pennines acted as western deterrents. In the course of the fifth century, around one-third of the area of mainland Britain became dotted with migrant settlements and cemeteries. The migrants didn't build in stone, or lay out streets, or define their places with theatres and statuary. In the early decades, the sites favoured were generally unoccupied, a choice that may have been related to damper conditions which now prevailed in Britain. During the three hundred years of the 'climatic optimum', Romans and then Romano-Britons had grown accustomed to building on low-lying sites, often by rivers. In a wetter Britain, many of these low-lying sites had become untenable. The immigrants favoured landscapes that mimicked their homelands. They liked fertile lowlands and they built in timber on well-drained sites. Compared to the villas of the Romano-British, their rectangular houses were crude yet practical, typically varying from 6 to 11 metres in length, their width being roughly half their length. Earth-fast posts supported roofs which were shingled with wood, or thatched, while their east–west alignment took advantage of the sun's warmth. Interiors were usually a single space, but larger houses might be divided by an internal partition. Suspended, planked floors allowed air to circulate, reducing the creep of rising damp and wood-rot. Midway along each of the longer sides was an entrance. Throwing open both sets of doors allowed through-draughts to dry the interior.

While rectangular, timber-built houses were following a Romano-British tradition, the design of the second type of building favoured by the settlers was brought to Britain from their homeland. Already common on the continent, these ingeniously designed outhouses typically measured some 4 by 3 metres and were erected over an excavated pit which could be up to half-a-metre deep. Conclusive evidence is lacking, but the pit may have been covered by a suspended floor, the void beneath providing air-circulation intended to reduce humidity within the building. Walls and roof were supported on vertical timber posts, set into holes in the ground. Roofs were thatched or shingled. That these sunken-based structures were more than sheds is demonstrated by the effort required to build them, which was equal at least to that required for the adjacent houses. These *Grubenhäuser* (pit houses) were an essential settlers' requirement, suitable for a range of uses from grain storage to craft-working such as weaving.

In places, the loose clustering of houses and outhouses in one lo-cation formed distinctive settlements. They were not 'villages' in the sense that they were usually demarcated by permanent boundaries or aligned on long-term roadways, and they did not sit within a patch-work of communally worked fields. If they had a style at all, it was rural sprawl. Perhaps they're best seen as 'hamlets', although even this suggests that they were more formally arranged than seems to have been the case. One of these settlements evolved on a gravel terrace on the north shore of the Thames estuary. Above the flood level, south-facing and a short walk from the water, the site provided the basics for subsistence agriculture: water, grazing land, fishing, foraging, communication. The presence of pits and graves spoke of settlement here some three or four thousand years earlier. The banks of a fort rose nearby, and there were traces of over one hundred roundhouses which had stood here before the Roman invasion. The terrace had been settled by Romano-British farmers, but they had disappeared before immigrants rediscovered this spot early in the fifth century. Once established, the settlement here consisted of ten houses and fourteen outhouses; enough to support maybe a hundred people. None of the farmsteads was markedly more elaborate than its neigh-bours. The place clung, rather untidily, to its patch of marginal land.

Also dating from the early 400s, but far to the north, were a number of settlements taking root in and around the great vale north of the Wolds – the same vale that had been identified 9,000 years earlier by some of the first people to cross the land-bridge from the continent once the ice had melted. By now, the lake that had been such a draw for fishing, foraging and hunting had shrunk to puddles in the bed of the broad, green vale. One of the settlements grew up by an aban-doned Roman shrine seven miles up the vale from the edge of the old lake. With time, there were about seventy-five people living here, in ten or so extended families. It was an ordinary kind of place, with no structures denoting a conspicuous social hierarchy. Along with the timber houses was the usual scatter of sunken-base outhouses where the community could keep grain and the equipment and tools needed for the multiplicity of tasks from spinning and weaving, to butchery, malting and metalworking. In the absence of a functioning commercial web to convey flows of goods, communities like this had to be self-sufficient.

As usual, attachment to place proved more enduring than the longevity of timber. In Britain's damp climate, wood began to decay the moment it was felled. Immigrant architecture was perishable; timbers raised in the fifth century AD were no more likely to endure than those raised by Britain's first farmers four thousand years earlier. Buildings had to be regularly maintained, or completely rebuilt. Space was recycled: a collapsing outhouse was an opportunity to take advantage of a new pit for depositing rubbish which couldn't be spread on fields: discarded loom weights; broken pottery; snapped tools. At the site on the Thames, where the number of functioning sunken-base outhouses probably didn't often exceed a couple of dozen at any one time, the total number of outhouse pits that had accumulated by the seventh century was 200.

Belonging was expressed through diverse practices. In this fragmented Britain, sprawling, urban cemeteries were replaced by local interments at locations that reinforced claims to land. Very often, people were buried close to ancient henges or to stone circles. Graves were dug into the sides of barrows and beside the tumbled walls of villas and towns. Not for the first time in Britain, one of the adaptive strategies during a period of insecurity and dramatic change was the invention of tradition; the reuse of ancient sites to create continuities with the past. Time may have eroded the original rituals but the act of return – of locative repetition – was one way of elongating dynasties and reinforcing territorial rights.

Colonial misadventures had widened the developmental gulf between Britain and Europe. In the Med, the fulcrum of civilisation had shifted east to the fabulous walled city of Constantinople, a city alive with statuary, baths, arches, colonnades, fountains, squares, bookshops and libraries. The tradesmen's stalls on the main street were faced with marble and a shopper could buy anything from dyed silks to perfumes, spices and slaves. There was a university, and in 537, architects working for Emperor Justinian completed one of the most astounding buildings ever seen in Europe, a giant basilica capped with a dome over 30 metres in diameter, arcing nearly 60 metres above the floor. To Procopius, it seemed 'suspended from Heaven'. Two additional semi-domes lengthened the nave to create an integrated, immense space:

> All these details, fitted together with incredible skill in mid-air and floating off from each other and resting only on the parts next to them, produce a single and most extraordinary harmony to the work, and yet do not permit the spectator to linger much over the study of any one of them, but each detail attracts the eye and draws it on irresistibly to itself. So the vision constantly shifts suddenly, for the beholder is utterly unable to select which particular detail he should admire more than all the others. But even so, though they turn their attention to every side and look with contracted brows upon every detail, observers are still unable to understand the skilful craftsmanship, but they always depart from there overwhelmed by the bewildering sight.

The walled area of Constantinople (there were suburbs, too) was nine times larger than Londinium's and enclosed some 1,200 hectares. A fleet of over 1,000 transport-ships was required to keep the city supplied with Egyptian grain; one of the granaries measured some 87 by 28 metres and was apparently of 'ineffable' height. In Britain, the ventilated, wooden outhouses brought by pioneering immigrants were seldom larger than 4 by 4 metres.

The rebuilding of Britain was not helped by famine and disease. The single worst episode began in 536 with an 'event' that must have struck terror into the minds of people for whom comets and eclipses were deeply troubling. That year the skies over much of Europe turned dark. The cause was either a huge volcanic eruption in central America, or the extraterrestrial impact of a comet or meteorite. So many particles were spewed into the atmosphere that the world became overcast. In Mediterranean lands, the sun scarcely cast a shadow from the beginning of 536 until the end of summer the following year. The dust veiled solar radiation, dragging down surface temperatures. In northern Scandinavia, the average summer temperature fell by as much as 3 or 4°C. In the northern hemisphere, summers remained chilled, on and off, until 545 – and in some locations, until 550. Ever since the days of the forager hunters, cold winters had been far less of a problem than cold summers. For about fifteen years from 536, the summers were so cold that crops failed and trees stopped growing. Famine followed. No records have survived of the havoc wreaked in Britain, but descriptions of the long darkness were left by the Roman official Cassiodorus (who had been writing outside Ravenna, in northern Italy), by Procopius of Caesarea, Zachariah of Mytilene, John of Ephesus and John the Lydian. It may have taken as much as four to seven generations for the landscape to recover. The ten to fifteen years of intermittent, reduced summer temperatures in Scandinavia were accompanied by increased rainfall, which coincided with the abandonment of low-lying settlements for new sites on higher, drier land. The population may have halved, leaving hundreds of abandoned settlements. These long, dark, frigid years of disruption and hunger were probably the source of the word that came to represent the coming apocalypse: *Ragnarök* translates as 'twilight of the gods', or 'darkness of the gods'. In a world dependent upon the gods for good harvests and health, the long eclipse of the sun was betrayal at its most profound.

The dust veils of the 530s and 540s coincided with the first recorded outbreak in Europe of bubonic plague. An already weakened European population was hammered by disease on an unprecedented scale. *Yersinia Pestis* was a creation of globalisation. From its initial reservoir in eastern Asia, the plague was communicated by people moving commodities, rats and fleas to the port city of Pelusium on the Nile delta and then – in 541 or 542 – to Constantinople, where it found the perfect host population of human beings crammed together in unsanitary conditions. Procopius reckoned that it was killing 10,000 a day at its height. From Constantinople, the bacterium spread west – probably by ship – through Italy, Gaul, Carthage and Spain. Then it travelled up the rivers to the interior of Gaul and Germany. In 544 or 545, the plague reached Ireland, and from there, Britain. There are no figures, and no hard evidence that bubonic plague wasn't joined by other, equally devastating diseases. But Britain – ringed by seaports and penetrated by navigable rivers – was highly susceptible to contagion. The plague remained virulent on and off for two centuries, with an estimated mortality of between 15 and 40 per cent. The total number of people killed is unknown, but may have reached between 50 and 100 million, enough to have slowed the development of Europe. Procopius claimed that 'the whole human race came near to being annihilated'.

Through dust veil and plague, Britain – a land that had lost the habit of writing – remains mute. There are suggestions that the dust had such a serious effect in Scandinavia that it was one of the catalysts of the Norse diaspora that eventually shaped so much of the British landscape. It's conceivable, too, that the spread of the plague – which started in the depths of the climatic disaster – was exacerbated by the atmospheric dust. A century and a half after the collapse of the Roman Empire, Britain was a land of scattered tesserae; a townless island dotted with monuments and humble houses of stone or timber. To circles and henges, barrows, banks and hillforts had been added the multiform ruins of a Mediterranean civilisation. On this landscape of fragments and memories it was the dynamic immigrant cultures that were best placed to rekindle development.

FOURTEEN

Nuclear States

550–850

Britain was built with immigrant grit. Within a century or so of Germanic boat-people arriving on the North Sea coast, the denuded human landscapes of the south-east began to be repopulated by kingly halls and palaces, fortresses, ports, churches and monasteries. The 'central place' returned, with resilience. There was no particular birthplace, but let's begin at Basingstoke.

In a year not too distantly removed from 550, carpenters selected and felled some trees close to a stream on the edge of a down six miles south of the ruined town of Calleva. The gently sloping site was well drained, with access to plenty of timber and sweet water. The ancient Ridgeway route across southern Britain lay just to the south and the old Roman road to Noviomagus crossed the valley a few hundred metres downstream. Over a period of maybe a century, a succession of enormous structures rose above the down. One of them was over 22 metres long and 8 metres wide. Like the far smaller, rectangular timber houses put up in the early 400s by the first waves of immigrants, these too had opposed entrances set midway along each of the long walls. One of the buildings on the down had a 5-metre annexe added to its eastern end, bringing its overall length to 21 metres. These were not works of architectural precision, chiselled in limestone and

Purbeck marble, but rough-cut, timber power-houses raised by people with sharp blades and strong arms. They had neither the skill nor the will to engineer the tight joinery perfected by Roman carpenters and their loose-limbed, timber-framed structures were supported by ground-fast timbers set in excavated foundation trenches. Walls were assembled in sections of around four to five metres at a time, each resting on its own wall-plate, a modular method of construction which made it possible to raise huge buildings in manageable sections (although the effect could lead to kinky walls if the sections were slightly misaligned). The gaps between the walls' structural timbers were occupied by panels of wattle and daub. Most of the buildings on the down had roofs so heavy that they required an external row of supporting timbers along each side, angled inward to take some of the load bearing down on the roof-plates. Visually, the buildings appeared to be resting on crutches. By the time the settlement matured, it had accumulated some nine of these super-sized buildings, along with a sunken-floored outbuilding and fenced enclosures. The huge size of the larger buildings, their open interior spaces, and the formalised, symmetrical way they were laid out within the settlement, make it possible to identify them as early *healles*.

At great halls such as those on the down south of Calleva, wielders of power hosted feasts and strengthened social bonds with retainers through gift giving. The largest building would have comfortably accommodated sixty guests. Other buildings on the down are likely to have been used for gathering tribute from local or regional subjects. Similar timber-built 'central places' developed in other parts of lowland Britain. On the Roman road from the Channel port of Portus Lemanis to Durovernum (Lympne to Canterbury), a hall was built with a ground-plan of over 160 square metres. Far to the west, another great hall was built a five-minute walk from the Fosse Way, where the road splashed through the Avon in the lowlands between the chalk downs and Cotswolds. The hall became the nucleus for a settlement which spread along the south bank of the river. These great feasting halls would have required assiduous maintenance in the face of rot and weathering and cannot have lasted more than two or three generations, but they marked the end of a constructional recession which had lasted in eastern Britain for two centuries.

Power builds. By the end of the sixth century, the scattered tesserae of post-Roman Britain were beginning to reassemble into new political likenesses: kingdoms of varying size and permanence whose shifting territories overlapped in places with lapsed Roman provinces or *civitates* and in others represented emergent tribal heartlands. At the smaller end of the scale were groups defined by their landscapes, like the dwellers of the Chiltern Hills, the Cilternsæte, or those of the Peak District, the Pecsæte. And at the opposite end of the political scale were over-kings who exerted power through sub-kings. Cantium (Kent) had survived the great disintegration with its Roman label almost intact and took its cue on kingship from neighbouring Francia, expanding its kingdom west from Thanet to take the Weald, North Downs and crumbled Londinium. Where the reach of Rome had been weakest, old continuities resurfaced: in the far south-west, the lands of the Cornovii assumed an identity that would become Cornwall. However ill-defined the peripheries of these coalescing kingdoms, their organisational centres left emphatic marks on the landscape.

One of the earliest 'palaces' took shape in the upland redoubt of the northern Cheviots at a site rich in ancestral relics. The complex was probably the work of Æthelfrith, the pagan, warlord-king of the Bernicians whose grandfather, Ida, had – according to Bede – founded the dynasty with a reign that began in 547. On a shelf projecting into a valley there was already a henge, a stone circle, a cemetery and a ring barrow. Beneath the turf lay human burials, flints and pottery going back four millennia or more. And behind this compound, ritual landscape, on the crest of a dark wall tilting giddily upward for nearly 300 metres, were the tumbled stone ramparts of the pink andesite hillfort we visited in Chapter 11, now a broken, grey shell containing the ruins of 125 roundhouses (on an Ordnance Survey map of the nineteenth century, the hillfort appeared as 'Druidical Remains').

What more could a king want? The site was sandy and free-draining. Close by ran the clear waters of the river. The shelf caught both the rising and the setting sun and commanded a fine prospect up and down the valley. And located at the fold of valley floor and upland, it offered the full spectrum of liminal benefits: hunting with foraging; herding with cultivation; settlement with connectivity. With its topographic utility, landscape and narratives, it was just the kind of place where associations between place and power could be rekindled. An extraordinary structure was built. Not unlike a miniaturised Roman theatre in timber, it was perhaps a venue for addressing assemblies and it shared a loose alignment with a cemetery, a temple, a kitchen and an enclosure which might have been used either for assembling cattle or ritual feasting at certain times of the year, or both. Later, the enclosure seems to have been a corral for horses (when the edge of the site was excavated during the construction of a railway line in the 1880s, huge quantities of horse bones – including complete skeletons – emerged from the earth). The enclosure may have been used for collecting livestock to be rendered to the king as food rent, or *feorm*. Whatever its exact role, the size and form of the structures at the site, and its presence a century later in Bede's *Ecclesiastical History*, identify it as a notable place known as Ad Gefrin, 'the hill of the goats' (Gefrin eventually became 'Yeavering', although Ordnance Survey maps still mark the knoll as 'Gefrin').

If Ad Gefrin had a role as a livestock centre, much of that four-legged tax revenue might have been herded to related kingly strongholds, and there were two candidates nearby. Less than a day's ride east, a massive outcrop of rock reared upward from the coastal plain. It's now known as Bamburgh. Back in geological time, the original basalt sill had spread above level beds of limestone but marine erosion had subsequently torn at the sill's edges leaving a single raised crag; a spectacular vantage point with immediate access to the shore. The ovate top of the stack was some 100 metres wide and 400 metres long, rising nearly 50 metres above sea-level. Its planed summit was a perfect platform for building. A hillfort had once crowned this crag, which had become known as Dun Duarioi, *dun* being a reference to its being fortified (the *Anglo-Saxon Chronicle* recorded a palisade enclosing the summit, later replaced by a wall or rampart). By the age of Æthelfrith, the stack had become a settlement of such prestige that the king – according to Bede – gave it to his queen, Bebba. A hint of the scale of the settlement here was conveyed by Bede, who described a siege during which attacking Mercians pulled down the neighbouring settlements and carried to the defended crag 'a vast quantity of beams, rafters, wattled walls, and thatched roofs', which they piled on the landward side of the fortress and fired. Although this event happened some fifty years before it was committed to vellum in Latin, the impression given is that Dun Duarioi was a significant 'central place' on the Bernician coast.

Kingship was all about stature. The sites chosen for these power bases were not unlike geomorphological thrones. The palace of Ad Gefrin was built on a sunlit dais buttressed by an ancestral peak. Dun Duarioi occupied a plinth of basalt commanding Bernicia's productive coastal plain (dark and rugged, basalt's presence in the kingdom was also asserted as the sea-girt islands in sight of Dun Duarioi and as the foundation for the most imposing sections of Hadrian's Wall). The coast itself was a source of authority, a space to be commanded by sight and a medium across which to move objects of tribute. Forty miles up the coast from Dun Duarioi, at what is now Dunbar, another fortress occupied a promontory with views across the mouth of the Firth of Forth. This, too, had been occupied long before the Romans arrived, but was then recolonised by incoming immigrants in the sixth century, who constructed rectangular timber buildings here, together with sunken-floored outhouses, the whole settlement being enclosed by a palisade or fence. In the latter part of the seventh century the promontory was redeveloped with new rectangular buildings of timber within a stone rampart, a reaction perhaps to the Mercian raids around 651 or to conflicts with Pictish forces in 685. It's possible that both this royal centre and Dun Duarioi were the ultimate destination of livestock that had been collected at Ad Gefrin.

On an island re-emerging from isolation, coastal havens and navigable rivers offered power-building possibilities beyond the reach of
landlocked centres. Not for the first time, the corner of Britain closest
to the continent enjoyed an accelerated trajectory. By the end of the
sixth century, a vast swathe of territory from the Isle of Thanet to the
Humber was under the control of Æthelberht, a ruler descended – if
Bede is to be believed – directly from the first generation of Germanic
immigrants to arrive on the beaches of Kent. The increased security
that royal oversight brought to the region made it safer for travellers
and markets to operate at a time when the Frankish seaboard facing
Britain had come under the unified authority of Merovingian kings.
Driven by Frankish entrepreneurs, cross-Channel trade had been
increasing since the middle of the sixth century, and by the early
600s it had achieved a regularity that turned several coastal trading
posts into more permanent proto-ports. Æthelberht ruled his domain
from the old Roman *civitas* capital of Durovernum Cantiacorum,
rebranded at around this time as Cantwaraburg, 'the stronghold of
the Kentish people'. The king had built a bridge across the Channel
by marrying the Christian daughter of a Frankish king, a prestigious
union which opened the door to ideas and exchange with Francia.
It was probably Æthelberht who redeveloped Cantwaraburg's Roman
theatre, a structure well suited to assemblies and one that would have
impressed Frankish visitors. The closest tidal havens to Cantwaraburg
lay less than an hour's ride to the east, on the channel separating the
Isle of Thanet from the mainland. Imperial Rutupiae had gone, but
the walls of the fort still stood so tall that they could be seen from
far offshore. Over the intervening centuries, silting had reduced the
width of the Wantsum Channel and it's possible that by the end of
the sixth century the main disembarkation point had shifted from
Rutupiae to a point closer to the open sea. Æthelberht also controlled
the old Roman port of Dubris (Dover), which was a little further from
Cantwaraburg, but lay closer to the continent.

Now known as Ipswich, another trading post was operating north
of the Thames, fifteen miles up the coast from ruined Camulodunum,

where the deep estuary of the Orwell squirmed inland to meet the clear water of a river draining the borderlands of the East Angles and East Saxons. It was a heathland site with good beaching and it covered no more than six hectares or so but through the 600s it gathered several peripheral cemeteries whose higher-ranking graves were accompanied by products that reflected the reach and success of the port: pottery from Frisia, northern Francia and the Rhineland, and imported glass palm cups, along with fancy brooches, from Kent. Outbound cargoes are likely to have included timber, which was in short supply on the Frisian coast, agricultural products and metals such as tin and lead, woollen cloth and slaves. The settlement's several foci suggest that it developed as a collection of market 'patches' exploited by different households or consortiums, while the presence of field boundaries a few minutes' walk from the settlement is indicative of a population that depended upon land as well as water for sustenance.

Change came from the coast. The re-emergence of trading settlements was the response to thriving marine exchange. Led by mints in Kent, coins began to be produced again, and with them, a rekindled monetary economy. Besides the reawakened ports on the shores of Kent and the East Angle/Saxon borderland, another trading station was developing far to the west, close to the shores of the Solent. One bend down the River Itchen from the ruined walls of a small, once-fortified Roman town, trading vessels were being beached at a place that would grow into Southampton. But the most significant coastal portal to return to life was the old Roman crossing-point on the Thames. After a hiatus of more than two centuries, the abandoned Imperial city had spawned a busy little appendage. On the bend of the Thames just upstream of the dilapidated Roman walls, ships were unloading continental wares and a settlement was beginning to coalesce on the slope that would eventually become Covent Garden.

There was another transformational locus. By 600, paganism was enjoying a monumental floruit at centres like Ad Gefrin. In eastern Britain, whatever support churches had received in the fading decades of the Christianised Roman Empire had long ceased to register a presence on landscapes. But in the far west and north, Christianity had been a resistant cult, encouraged no doubt by its promise of redemption in an age of discontinuities. Beyond the reach of pagan immigrants, the tombs and shrines of saints continued to be venerated and churches continued to be built. (Occurrence of the place-name 'Eccles' may record the presence of indigenous British Christian centres, the word having come down from the Latin *ecclesia* [church] through Celtic *eglēs*.) Christianity was a religion waiting to be revitalised. Where paganism took its cues from local spirits and deities, Christianity was universal and missionary; a belief founded on superhuman moral authority and expansion. In an age of jostling kingships, Christianity had political utility. From the mid 500s, missions from Ireland and Francia began to revitalise dispersed pockets of Christianity. Populated by celibates living under communal rules, a new form of expansionist settlement – the *monasterium* – took root on the landscape.

Monasteries became a potent form of 'central place': the hubs for networks of churches and the bases from which missionaries departed to convert communities far and wide to Christianity. The single most active Christian region in Britain was the western redoubt that was beginning by the sixth century to be known as *Cymru*, the 8,000-square-mile tract of valley and upland bounded by three coasts that projected westward between the Dee and the Severn. By the middle of the sixth century, these parts were adopting their own language, Brittonic giving way to Welsh, a word most likely to have derived from an English term referring to people who had been Romanised. The territory itself hardened its identity, taking the name *Cymru* (and variants) from the Brittonic *Combrogi*, 'fellow-countrymen'. *Cymru* had been a receptive haven for Celtic missionaries arriving by sea from Brittany and Ireland, searching for places suitable for their cells – often places that were already holy and therefore venerated by local communities that might be won from their pagan ways. Physically these sacred settlements were often marked by a *llan* or enclosure, and some evolved into *clasau*, monasteries. One of the most successful monasteries in southern *Cymru* developed on the bank of the River Taff close to its outfall into the Severn Estuary. Llan-Taff, or Llandaff, became the hub for sixty or so churches. In the north of *Cymru* on the River Dee, a monastery that became known as Bangor (the word means 'wattle fence', a reference to the enclosure that surrounded the site) was reported by Bede to have had more than 2,000 monks 'all of whom supported themselves by manual work'. The enormous number may have been a composite total that included outlying communities, but it was an indication that some monasteries at least were sizeable farming settlements.

The monastic enclosure picked up on one of the oldest human landforms in Britain: the ditch-and-bank circuit that dated back to the fourth millennium BC. In its latest, monastic iteration, the enclosure demarcated Christian space; the consecrated ground for burial, for building plots and for the church. The monastery founded by St Columba on the Hebridean island of Iona in 563 was delineated by a bank-and-ditch vallum. With its communal church and circular huts, it was a model imported with Columba from Ireland, and then copied across to the Pictish mainland, where an important monastic settlement sprang up on the east coast, at a place now called Portmahomack. The physical relationship between the island of Iona and peninsula of Tarbat is intriguing, for they lie at each end of the Great Glen, girt by ocean, one of them a seamark for voyagers up and down the west coast, while the other performed the same role in the east. It's conceivable that the settlement at Tarbat was founded by Columba himself, who is reported by his biographer, Adomnán, to have travelled in 565 up the Great Glen and past Loch Ness to meet the leader of the northern Picts – who Bede claimed were converted by Columba's 'preaching and example'.

The Tarbat site appears to have suited both the missionaries and the Pictish king who gave it to them. A ridge of uneven ground set between sand and a marsh, it offered a wide view across the Dornoch

Firth and had been a burial ground for a thousand years. It was an old place. Constructing the settlement was a backbreaking task (skeletons in the graves record compression fractures, hernias and miscellaneous injuries of the kind suffered by those devoted to heavy lifting and digging) as the monk-pioneers shifted rocks, levelled the site, built a massive D-shaped enclosure, raised a church and organised their water supply. Rainwater was directed down the hill into a stone-lined cistern, a wicker-lined well was constructed and they excavated a pit and filled it with pounded charcoal to purify rising groundwater. On the slope leading down to the marsh, cereal was cultivated, cattle tended and metal worked. The number of monks living at the monastery varied but may have centred on around thirty or so. Some forty miles north of Iona, another monastery on the island of Eigg was large enough to have accommodated the fifty-two men who were reported in Irish chronicles to have been murdered in around 615.

In the pagan heartlands of the south and east, the establishment of Christian foundations was delayed by uninterest in an imported theology expounded through unintelligible texts that were at odds with oral folk traditions built through the ages around inherited mythology and local landmarks. Not for the first time, transformative pressures were deployed through cross-Channel traffic. Backed by Rome and the Franks, a papal mission landed in 597 in the Wantsum Channel near the ruins of Roman Rutupiae. Tasked by Pope Gregory to convert the wayward inhabitants of the former Roman province, Augustine and his forty or so missionaries were granted an audience with King Æthelberht, who permitted the foreigners to settle at his royal seat within the dilapidated Roman walls of Cantwaraburg. The arrival of Augustine brought the south-eastern fringe of Britain back into the orbit of Rome and opened the way for Romanising centres. After an absence of two centuries, the city on the Tiber had re-extended its reach to Britain.

The missionaries from Rome came with a building programme that had already been rolled out on the continent: a tiered system in which strategically placed episcopal churches dominated a network of baptismal churches and private oratories. Thus would the mission bring this 'barbarous, fierce, and pagan nation' into line with Christianity – and with Rome. Writing over a hundred years after the missionaries landed, Bede described how they established their bridgehead: 'Augustine proceeded with the king's help to repair a church which he was informed had been built long ago by Roman Christians ... and established there a dwelling for himself and his successors. He also built a *monasterium* a short distance to the east ... where at his suggestion King Æthelberht erected from the foundations a church.'

Four years into the mission, Augustine had made very little progress beyond the Cantii heartland. Resistance to Christian proselytising was strong among people loyal to native traditions and Augustine did not have sufficient missionaries. Neither was there wholehearted political support from Æthelberht, whose power risked being undermined if his people were subjected to coercive conversion by foreign zealots. A demand by Pope Gregory that Æthelberht should destroy all pagan shrines was countermanded in a subsequent letter by instructions that they be converted into Christian churches, a process that required little more than the removal of idols and interior decoration. Gregory now took the view that coercion was counter-productive and that the way forward could be found through the creation of social and financial inducements – an approach that had worked well on the continent. The role of the old pagan shrines as social hubs was to be encouraged by allowing people to ritually sacrifice cattle and pigs in the grounds of the new churches. Temporary timber structures – *tabernacula* – could be erected where people could continue to sacrifice animals (the only caveat was that the animals be slaughtered 'to the praise of God' rather than 'to the devil'). Christian celebrations like saints' days were to be morphed into the ancient ritual calendar. Pagan hearts and minds were being massaged rather than bludgeoned. Little by little, they'd get the word of God. Gregory's instruction that the *tabernacula* should be constructed 'from the branches of trees around the churches which have been converted out of shrines' may well have been a devious ploy to rid Britain of its ancient, pagan groves. It wasn't until 604 that a church was established on the Medway at Hrofescae-stir – Rochester – and another on the Thames at Londinium, the old capital of Roman Britain. Dedicated to St Paul the Apostle, the church was built just inside the western gate of the old Roman walls and it gathered around it a clutch of ecclesiastical buildings.

Lubricated by kingship, Roman Christianity took a generation to reach the lands beyond the Humber. Æthelberht's Christian daughter was handed in marriage to the pagan king of Northumbria, Edwin, who had succeeded Æthelfrith in around 616. Bede, a man whose quill quickened whenever there was a Romanising breakthrough to write about, had King Edwin converting to Christianity 'with all the nobility of his kingdom and a large number of humbler folk.' Edwin relocated the ritual centre of his kingdom to the heart of the old Roman city of Eburacum, and in doing so, resurrected northern Britain's urban hub, which he Christianised by putting up a small oratory, 'hastily built of timber', a structure that he soon enclosed within a much larger, 'more noble', square church built from stone. It may have been these events – dated by Bede to 627 – that led to the rebuilding of the palace complex out on the edge of the Cheviots at Ad Gefrin, where the great enclosure was embellished with a timber palisade. Between the en-closure and the assembly structure, two enormous halls were raised, with a connecting courtyard. Beside the cemetery, a timber-framed church was built. According to Bede, one of Pope Gregory's mission-aries, Paulinus, spent thirty-six days here conducting mass baptisms.

And so it spread. The year after Edwin had instructed that a stone church be built at Eburacum, the once-pagan reeve of the old Roman *colonia* at Lindum also built a stone church, 'of fine workmanship'. Many of the new Christian centres were sited in the ruined towns and forts of the Romans. Some were already hubs of population, with suc-cessful churches. In the chalklands of southern Britain, the balance of power between the old deities and monotheism tipped after 634, when the Pope sent Bishop Birinus to work on the Saxons living in the 'completely heathen' Thames valley. Cynegils, the King of Wessex, gave Birinus the old Roman town of Dorcic – Dorchester – to found his see. With time, the long-gone town on the Thames was recolonised and became one of the most important centres in Wessex. In around 634, too, the little archipelago off the coast of Northumbria was transformed by the arrival of an Irish bishop, Aidan, who was given the island of Lindisfarne as his episcopal see. The 400-hectare island was distinguished by a spiked intrusion of dolerite, part of the same basalt system that had guided the course of Hadrian's Wall and attracted early Bernician kings to the coastal crag below the Cheviots. Lindisfarne had been a coastal seamark for millennia.

There was a pattern of changing fortune: the initial conversion programme would win support from kings and courts, and to some extent from their subjects, but it was often followed by an apostolic reversion to the old ways – encouraged by the observation that God could not stop war, bad harvests and plague. Bede wrote of an occasion when an outbreak of plague exacted such a heavy death toll that the East Saxons ruled by Sighere 'abandoned the mysteries of the Christian faith and ... began to rebuild the ruined temples and restore the worship of idols'. Such reversions didn't last; Sighere's people received a visit from King Wulfhere's Mercian bishop, who 'proceeded with great energy' to 'correct their error'. Sighere's churches were briskly reopened and the rebuilt pagan temples and altars abandoned or destroyed. Apostasy was usually followed by a committed return to Christianity; in a world where plague knocked on the door as a matter of course, the promise of resurrection had a persuasive appeal. It was this second coming to the Faith that strengthened Christianity's grip on Britain, frequently commemorated by a transition from timber to stone.

Within three or four generations of Augustine's arrival, all the southern and eastern kings – and their courts, too – had turned from pagan idols to Christianity. Along the rough road to conversion of Britain's eastern pagans, the original ecclesiastical masterplan of Augustine and his continental missionaries was modified to suit the prevailing circumstances. Unlike Francia, Britain did not have a surviving *civitates* which could be recycled into ecclesiastical units. Instead of the episcopal church and subsidiary baptismal churches and oratories, a more *ad hoc* system of *mynsters* – from the Latin *monasteria* – developed in which a church was served by a Christian community rather than by a single priest. Royal or aristocratic patronage provided minsters with security and recruits. By the 660s the way was clear for spectacular, minster-led flowering. The arrival of Archbishop Theodore and Abbot Hadrian was followed by the establishment of a school in Canterbury admired by Bede because it 'attracted a large number of students, into whose minds they poured the waters of wholesome knowledge day by day'. The teaching of Christian music spread from Kent to the furthest kingdoms. New Christian houses were established and old houses expanded. In 656, a minster was founded at Whitby and it was here that Cædmon composed the first Christian poetry in English. One of the Canterbury students was Aldhelm, who went on to be abbot of Malmesbury, where he built churches and wrote pioneering works of prose and poetry. In 664, a church was added to the religious settlement on Lindisfarne. Framed with oak and thatched with reeds, it was upgraded by a subsequent bishop, who stripped the thatch and clad the entire building in sheets of lead. The armouring of Lindisfarne against the inclemencies of Northumbrian gales was a measure of wealth accruing to Christian churches and a signal of their permanence on the landscape.

Romanising centres re-established a physical connection with the city on the Tiber. Among the religious and cultural ways of Rome that Britain's leading Christians sought to import was architectural style. One of the most influential proponents was a Northumbrian nobleman, Benedict Biscop, who in around 674 founded a minster where the River Wear flowed into the sea. Biscop's repeated pilgrimages to Rome and a period spent on an island *monasterium* in the Mediterranean qualified him as one of the best-connected Christians in Britain. Bringing stonemasons and glaziers from the continent, he built a stone church on the Wear. This was the centre of learning – still run by Abbot Benedict – that embraced young Bede, born on the monastery's lands and sent by his family to Wearmouth at the age of seven in around 680. In 682, a companion minster was founded at Jarrow, six miles to the north, on the banks of the Tyne. Jarrow's axial layout, the stained-glass windows, turned pillars and redbrick floors sought the form of Rome, as did the stonework, much of it pilfered from the walls of a Roman fort. With time, the twin-minsters filled with imported paintings, relics and books; so many books (maybe as many as two hundred volumes) that Wearmouth-Jarrow grew to own the most extensive library in Britain. This was the centre of learning that sustained Bede through a lifetime of writing which included his epic description of the Christian mission in Britain. Written in 731, *The Ecclesiastical History* 'of our island and its people' effectively book-ended the conversion of Britain from paganism. The last of the provinces of Britain to accept the Faith of Christ had been the Isle of Wight, and this outpost of paganism had not relinquished idolatry until the year 686. With its excellent communications by land and sea, Wearmouth-Jarrow was well placed for spreading the word. As early as 710, King Nechtan of the Picts requested that 'architects be sent him in order to build a stone church for his people in the Roman style'.

Eternal stone, bearer of authority, bastion against idolatry, rose again, block by block, above Britain's solidifying central places. It was the physical resurrection of a landscape well known to poets of the age:

> Wondrous is this stone wall, smashed by fate.
> The buildings have crumbled, the work of giants decays.
> Roofs have collapsed, the towers in ruin . . .

In Old English, this meditation on the impermanence of worldly glory summoned the imagery of a once-great city, of 'ingenious ancient work' now 'ringed with crusted mud'. As the poem unfolded, the city rose from the lines as a place that may have been more than allegorical. There were identifying features: it had 'many bathhouses'; it had 'stone buildings' and 'flowing water' that 'threw out heat' and a wall that 'entirely encompassed it / within its bright breast, where the baths were, / hot to the core.'

The crumbling city of hot springs and baths would seem to have been Aquae Sulis, whose fabulous stone porticos and steam rooms had once drawn thousands to the valley of the Avon. But it was more than Bath. The poet's city was a civilisation falling. In this post-colonial land, strewn with the ruins of Rome, new stone spoke for the future.

So by the early 700s, Britain was dotted with 'central places'. To the great halls, palaces, fortresses and trading posts could be added churches, minsters, monasteries and Romanising centres from Canterbury to Malmesbury and Wearmouth-Jarrow.

Nucleation was a process: the more settlements were fertilised by population growth, religion, trade and innovation, the greater their human congregation. By the early 700s, some of the south-eastern trading posts had been transformed by the general acceleration of economic activity. The surge occurred in the years centred on around 720, and it was perhaps during the following decades that the six-hectare heathland trading post on the Orwell estuary north of Camulodunum came under the control of a family head called Gip, and became known as Gipeswic, the *wic* suffix borrowed from the Latin *vicus*, meaning a trading emporium. Later, Gipeswic became Ipswich. One of the settlement's cemeteries was overrun by a market area, its trampled earth accumulating scatters of dropped silver. A loose grid of metalled streets evolved, lined with workshops and houses, and a small suburb developed on the south side of the river. On the Solent site that would eventually become Southampton, a settlement of comparable size to Gipeswic had grown up on the tidal River Itchen. This too gained a *wic* suffix at some point after the boom years of the mid 700s. Hamwic was a busy, commercial entrepôt with a planned system of gravelled streets lined by properties set on physically defined plots of land. Hamwic's streets were metalled too, and – as in Gipeswic – expansion was so rapid that streets eventually overran cemeteries. Packed together in their wattle-and-daub

houses, these were communities that put living-space above kin identity. During the course of the eighth century Gipeswic and Hamwic
swelled to cover as much as 40 to 60 hectares and reach population
sizes of between 2,000 and 3,000. Gipeswic and Hamwic were towns.
But Lundenwic was even bigger.

Bede, writing beside the Tyne in around 730, described distant
Lundenwic as the capital of the East Saxons and as 'a trading centre
for many nations who visit it by land and sea'. No other settlement
in his account achieved quite this prominence. It was big, it was well
connected and, in Bede's eyes, it was still bathed in the aura of Augustine's mission: the location first chosen by Pope Gregory to be the
primatial see. The original Saxon settlement on the bend of the River
Thames east of the Roman walls swelled during the eighth century to
a town of perhaps 5,000 to 10,000 by 800, sprawling over 60 hectares
or so – the area covered these days by Covent Garden west to Trafalgar Square. The town dipped down to the beach on the bend in the
river, where a section of the bank had been stabilised with timber to
facilitate the loading and unloading of vessels. If you stand on Waterloo Bridge, you'll be right above the East Saxon waterfront. Up the
slope from the beach (and across the Strand, a street name derived
from *strond*, the Old English for the edge of a river), a dense network
of streets and alleys served houses and workshops specialising in the
trades of the time such as metal- and bone-working, and weaving.
Flat-bottomed ships arrived from the continent with wine, pottery
and glass, millstones from the Rhineland, amber from the Baltic.
Leaving Lundenwic's waterfront would have been cloth, agricultural
produce and slaves. The 'Frisian' in Bede who sold a hapless, fettered
Northumbrian *thegn* called Imma would have been one of the many
foreign merchants operating on the Thames by this time.

So towns returned. After a three-century absence, the urban settlement had finally made its comeback. Driving these markets and
entrepôts were local administrations able to respond to economic
opportunity; a new class of people whose vocation was business:
people who could keep tallies, change money, exchange goods, build

storehouses, victual ships and run wagon teams. Undefended, the *wics* were not bounded by walls, but open to the wider world. While *wics* were hardly glittering cities, nothing like them had enlivened British shores since Roman days.

There was another model of emergent town. Bede's island was centred on a single, dominant kingdom which reached from the Wash to the mountains of Cymru and from the Thames to the Humber. In this Mercian heartland, a succession of rulers lifted the idea of the central place to a new level. Traces of their methods were incorporated in charters dating from the 740s describing how specified engineering works were legally enforceable. The earliest known reference to such obligations is contained in a Mercian charter of 749 which refers to the duty of landowners to contribute to the building and maintenance of bridges and fortifications. The linking of bridge and fortress was a strategic measure, for the two combined could block the movement of enemy forces by water and land. In an era of small-scale endemic warfare, the 'defended centre' supported by landholders was an extremely successful model. With their strategic, riverside locations, these strongholds were regional central places, located where they could be supported by dependent populations. They were, it seems, part of a Mercian 'system'; a centrally orchestrated network of defences that drew on stringently enforced public works and military obligations. The enforcer was King Offa, who ruled Mercia from 757 until 796. Many of these Mercian strongholds developed into thriving communities, with regional markets being held outside their walls. During the first half of the ninth century, banks and ditches were constructed around Mercian centres on the rivers Wye and Tame and the upper reaches of the River Isbourne, an old place whose heights were graced by a gigantic long barrow and the lands of a long-gone Roman villa. Eventually, these three Mercian strongholds became Hereford, Tamworth and Winchcombe.

Offa's vision for Mercia included the physical definition of its western frontier. In an act of earth-moving so monumental that it can only have been a conscious emulation of Emperor Hadrian (and perhaps, too, Charlemagne), Offa's work-teams built a dyke over 80 miles long and around 20 metres broad. In places, it stood 8 metres high, with a gulf-like ditch on its western side. The claim made in 893 by Bishop Asser that the dyke ran 'from sea to sea' may well have been a mild case of Saxon hyperbole, but there's no doubt that the dyke covered enough of the distance between the estuaries of the Dee and the Severn to have had a dramatic effect on the districts through which it passed. The very act of construction was itself an imposition. The dyke slashed through valuable woodland, which was felled and burned, taking with it critical local resources like pannage, charcoal-burning and game. Drainage in pastures and fields was affected, as the bank and ditch diverted streams and run-off. On the larger scale, the dyke drew a clumsy physical contour that roughly followed the boundary between the lowlands and uplands, interfering with – and perhaps severing – interaction between the two. This was a border-land that depended upon both topographies through ancient habits of transhumance, herding and exchange. East–west ridgeways that had carried traffic for centuries were blocked by a north–south bank and ditch. Offa's Dyke enforced a cultural reorientation, from borderland to either upland or lowland. The earthwork hardened an emerging in-ternal border within Britain, helping to crystallise the self-awareness of the Welsh people and leading to a linking of kingdoms that would place the inhabitants of Britain's mid-west under the rule of a single – Welsh – king.

By 800, better-connected parts of the British landscape were be-ginning to display the economic potential that had flowered in the heyday of *oppida*. Economic activity was increasing and so – for the first time since the Roman occupation – was the population.

Much of Britain had become a land illuminated with little centres: a church here, a great hall there; strongholds, monasteries, coastal trading posts, *wics*, defended settlements and the occasional palace. Between these pinpricks of light lay a rural, green backcloth where farmsteads were scattered like handfuls of flung seed.

There was, however, a level of order in this rustic patchwork. In the 'Germanic' parts of Britain, land was allocated in a unit known as the 'hide', which corresponded roughly to the amount of land a peasant needed to support his family. And from at least the middle of the seventh century, several kingdoms were parcelled into administrative units of 50 to 100 square miles. Each block had its own central settlement where dues were paid and justice dispensed. In some locations, there were physical demarcations, too. On this island of accelerating hubs, Britain's oldest human landform had made a comeback. The ditch-and-bank earthwork first made an appearance way back in the fourth millennium BC, when immigrant farmers began building massive social enclosures. This most basic of earthworks had enjoyed many iterations, from linear-bank and hillfort to vallum. It reappeared from around AD 400.

Typically, the latest generation of linear earthworks or dykes consisted of a deep ditch and a bank, topped perhaps by a palisade. At up to two metres high, they could run for as little as a few hundred metres or many miles. On the southern side of the Fens, a number of parallel dykes crossed a band of farmland which had developed between the woodland growing on heavy clays and the waterlogged fenland. Since this band of drier land was the route used by the pre-Roman Icknield Way, the dykes would have been a means of restricting passage along the road, while also demarcating discrete plots of land, each with its own access to woodland, pasture and water. Some or all of these dykes may already have been in place since pre-Roman times, and were simply refurbished. To the east of the Fens were more dykes. Others could be found at various points in the Pennines, in the hills north of the Tweed and in the West Country. The linear earthworks were built over several hundred years. They probably served various purposes at different times, ranging from symbolic walls to the uniting of otherwise heterogenous peoples. They were all building works requiring substantial commitment from the local population, perhaps in lieu of taxation. All, however, were responses to territorial imperatives and hardened existing cultural divides.

An oblique glimpse of rural life in Wessex appeared in the law codes of King Ine, ruler of the kingdom between 688 and 726. It was the responsibility of every commoner to fence, hedge or wall their own land and if they failed to do so, no claims could be made against the owners of any animals that might stray onto the plot and damage the crop. Plots on common meadows had to be bounded, too, and compensation had to be paid to anyone whose crops or grass were eaten by stray beasts. Hedges were encouraged to thicken with tangles of brush such as hawthorn. Fines were to be paid for pigs that trespassed onto another's 'mast pasture'. The penalties for allowing a beast to stray onto corn land were especially harsh; the animal could be slaughtered by the plaintiff, who would be allowed to keep the best flesh before returning the hide and any leftovers to the owner. Encoded in these laws was a tacit grip on environmental protection; as land-use became more particular, the maintenance of dedicated zones of exploitation – whether arable, pasture or woodland – became increasingly important to the resilience of communities. Biodiversity was a security issue. Fines, for example, had to be paid for destroying a woodland tree by fire, or for cutting down a large tree which could shelter thirty swine (although the penalties were less for timber-theft because 'the axe is an informer and not a thief').

The species posing the greatest threat to security was the least controllable. Ine hints at its potential for raising mayhem in clauses dealing with fighting in *mynsters*, stealing, robbery with violence, murder, breaking and entering, cattle rustling and 'belonging to a band of marauders'. It wasn't always wise to walk alone in Wessex. Fundamental to security was a common understanding of territory and ownership exerted through notions of trespass. One of Ine's laws required people from outside the neighbourhood ('from afar,' or a stranger') to shout or to blow a horn if they should travel through woodland off the highway. Any stranger not making a racket in someone else's trees would be assumed to be a thief, and – according to the law – could be killed or put to ransom. The effect was to restrict people to the straight-and-narrow: the highway. The right to roam was a distant memory. In Wessex at least, the landholder was both king and executioner. It was an idea that had particular appeal to West Saxons.

The laws of Ine also contained the earliest reliable documentary reference to a new agricultural landform. Mention of the 'common' field, and the damage that could be done to the crops of a number of people should a cow stray, introduced to our landscape story the large, comunally managed tract of arable land, ploughed into long stips. There were a number of reasons why the 'open field' was spreading through Britain. A farmer might choose to spread his risk by cultivating arable land in a variety of locations; the open fields could have been a method of sharing land among members of a community; they could have arisen through the subdivision of land due to partible inheritance, or they could have been shares in land that had been collectively cleared. And the introduction to Britain of the heavy, mould-board plough may also have been a factor.

The heavy plough had come to the south-east from the continent by the seventh century, but it was a device of the developed world that demanded a substantial investment and a community committed to sharing the burden – and benefits – of its operation. Construction of the skeleton and wheels required large baulks of timber, skilled carpentry and ironwork, and it had to be pulled by a team of oxen. Where paired oxen could haul a scratch plough, or ard, a team of maybe eight oxen was needed to drag and turn a heavy plough. The sight of a heavy plough in action would have been one of the most arresting spectacles people had ever seen on farmland. Where the home-made ard simply scored a line of broken earth, which then had to be scored a second time, at right-angles, in order to produce a reasonable tilth, the heavy plough made a single pass, simultaneously cutting the soil and turning it over. Traditional fields were unfeasibly small for such a cumbersome device, and neither was there a need any more for them to be square. The new ploughs also made it possible to cultivate extensive areas of the heavy clay-lands. Over the period of a season, far more land could be cultivated with a heavy plough than with an ard.

Open fields brought with them a subtle shift in the shape of rural settlements. The operation of heavy ploughs and management of communal fields was more efficiently conducted from settlements that were nucleated rather than dispersed. The actual process of nucleation is unclear, but one model would have a group of farms that were separate and yet not too distant from each other growing into a single, or polyfocal, settlement through an extended period of infilling. Hamlets of perhaps six to eight farmsteads, they could be defined by the elements they did not have; hamlets had no church of their own (although most would have had access to one), or extended central-place function, or predominant trading or industrial activity that would make them more dominant than other clusters of farm-steads. They were small, informal congregations of people whose food security could be increased by sharing an acknowledged neighbour-hood. Glimpses of these farmstead clusters can be read between the lines of Ine's laws, where references to 'common meadow', to 'partible land', 'common crops', the management of woodland, the hiring of oxen, responsibilities to maintain field boundaries, the payment of church dues and so on, all allude to a system devised to avoid conflict in the sharing of common resources.

In the north of Britain, novel structures appeared in the lowlands around the Forth from the mid 600s as the Angles pushed north-ward, introducing to the region their great timber halls and ingenious stores-cum-workshops with ventilated sub-floor structures. Some of the halls may have been built for the same kinds of elites that had set-tled at Ad Gefrin. At about the same time – and seemingly in use from about 600 to 1000 – another new type of building began to appear on the Pictish landscape. Often associated with field systems, they were roughly rectilinear and ranged in length from 10 metres to as much as 30 metres. They were sturdily built with stone footings and some had sunken internal areas that might have been animal byres. These homesteads – or farmsteads – could occur singly or in clusters. In areas of Britain suitable for more intensive farming, such as Wessex, communities were working the land together.

*

The culminating force for nucleation showed up in longships. With so many monasteries and religious houses standing like beacons around the coast of Britain, their presence was far from discreet to seafarers, and many had become rich storehouses of Christian metalwork and related treasures. The earliest recorded raid crops up in the *Anglo-Saxon Chronicle* under the year 789, when 'three ships of Norwegians from Hörthaland' arrived off the south coast. In 793 exceptional flashes of lightning and fiery dragons were seen over Northumbria, portents that were followed by a great famine and the return of the 'heathen' to plunder Lindisfarne, Jarrow and Iona. In these early years of raiding the Norwegians and Danes from the far side of the North Sea were after loot, but by 800, some of the raiders had become settlers. Just as Kent and the Thames had been the open door for Angles and Saxons sailing from the lowlands north-east of the Rhine, the islands of northern Britain were convenient stepping-stones for Vikings departing the fjords of Scandinavia. The sailing distance from the islands at the mouth of Hardangerfjord, to the islands of Shetland, was no more than two days in a wind backing easterly – about the same distance as that from the Rhine to the Thames. The Pictish islands of the far north were fertile and not unlike those that dotted the coast of the fjordland. Northern and western Britain was subjected to a wave of settlement which mirrored the settlement of Angles and Saxons in the south and east. It was a fateful symmetry. By 800, a settlement transformation was well underway, as the cellular stone houses of the Picts were overtaken by far larger, Norse-style longhouses and Norse cemeteries. Orkney, with its low-lying islands, fertile soil and mellow climate became a centre of Norse occupation. Through the early decades of the ninth century, the Vikings spread down the western coast of northern Britain, settling on islands and at sheltered havens on the mainland. Soon they were settling the shores of the Lake District and the Isle of Man.

The countless seaborne strikes recorded in the *Anglo-Saxon Chronicle* revealed a natural vulnerability to hit-and-row raiders. With its long coast, numerous navigable rivers and surviving military roads,

Britain's geography could not have been better suited to surprise attacks by large forces at home on land and water. It took fifty or so years for the Vikings to extend the scope of their raiding to the Thames estuary and to strike at Britain's heartland. In 835 they raided Kent, and for three decades, they returned almost every year, desecrating site after site. Lundenwic was attacked 'with great slaughter' in 842, and again in 851, when the plunderers were back with 350 ships, storming Canterbury and returning up the Thames. After two hundred years of continuous occupation, the largest settlement in Britain was abandoned. What structures remained slowly collapsed and most of Lundenwic reverted to waste or farming plots. At other coastal locations, *wics* that had survived economic downturns were snuffed out by the middle of the century.

Viking attacks revived the utility of protective walls, however decrepit. Old Roman Londinium found a new role as a pre-fortified stronghold; a public settlement surrounded by defensive walls. Within the walled ghost town, the most significant functioning buildings were clustered on the western hill where Bishop Mellitus had founded his cathedral church back in 604. By the mid 800s, Londinium was back in action as a trading centre, albeit restricted: in 857, a Mercian royal charter referred to 'a profitable little estate' and its liberty 'to use freely the scale and weights and measures as is customary in the port'. While unprotected Lundenwic was being overtaken by weeds and Vikings, trading activity had shifted a few hundred metres eastward, within the walls. Much of the old Roman waterfront now lay beneath a metre or more of silt and gravel, and in the absence of quays the most suitable site for beaching ships was a bend in the riverbank where a stretch of gravel-bedded foreshore each side of the mouth of the Walbrook provided a natural hardstanding.

Winchester was stormed by Vikings in 860 and a few years later, Vikings operating out of Thanet 'devastated all the eastern part of Kent'. The numbers of Viking ships engaged in raids had been steadily growing, and so had the number of footholds they had on the archipelagos of Ireland, Francia and Britain. In 865, a force unlikely to have exceeded 1,000 or 2,000 fighting men – the *Anglo-Saxon Chronicle*'s 'great heathen army' – settled into winter quarters on the sandy heaths of East Anglia's brecklands. With horses provided by the pacified East Anglians, the Vikings gained new mobility on land, assisted by the still-functioning network of Roman roads. The campaigns that followed, aided by 'a great summer army' who had arrived by ship, saw the disunited kingdoms of Britain taken to the edge of extinction. The once-powerful kingdoms of Northumbria and East Anglia had fallen; the Mercians and West Saxons were reeling.

Full-scale, mobile warfare reaffirmed the value of strongholds. To prosecute a major war over several years in Britain, the Viking armies needed winter quarters: *wintersetl*. The isles of Sheppey and Thanet were natural fortresses, moated by seawater, but elsewhere, the raiders had to create their own artificial defences. Generally the sites of these camps were selected to suit the needs of a military force on the move. Natural features like rivers and fens were often used to form one side of the camp, creating D-shaped ground-plans. Most of these earthworks were little more than crude banks thrown up to keep irate locals at bay. When the Danish 'Great Army' needed a site for the winter of 873, they took the monastery at Repton on the Trent and built a D-shaped earthwork, with its long side running along the edge of the six-metre-high river cliff. Constructing the fort probably took 200 men five weeks.

In 871, amid the bloody ebb-and-flow of running conflict, Alfred succeeded to the throne of Wessex. He was twenty-three. By 877, he'd lost control of the lower Thames and much of Mercia. The defeat of a Danish army on the northern edge of the Plain early in 878 provided a respite and Wessex found itself pincered between a Viking army based in Cirencester, and a new Viking army that had arrived in the summer of that year from the continent and established a base several miles upstream of Londinium, at a place called Fulanham – Fulham – where the Thames could be forded on exposed banks of sand and gravel. It was at this military nadir that Alfred initiated a strategy of extraordinary consequence.

PART FIVE

Town v. Country

FIFTEEN

Street Plan

850–1150

Viewed by satellite, the rectilinear grid and axial primary streets might suggest a super-sized legionary fortress, and when you alight from the London train and walk across North Bridge the impression of military planning is reinforced by the prospect from the stone cutwaters over the limpid Piddle. The north side of town is obscured by an artificial bank so high that it hides the houses behind. Beyond the bank, 'North Walls' tracks the Piddle for a couple of hundred metres before turning a sharp right-angle and becoming 'West Walls'. On the opposite side of town, there's a street called 'East Walls' and a road called 'Bestwall'. The oldest church in town is known as St Martin's-on-the-Walls. On three sides, Wareham is rimmed by a looming turf barricade and on the fourth, it's protected by the tidal currents of the Frome. The world outside must have been a very threatening place indeed.

The Vikings were good for Britain. True, they carved a wake of cleft skulls and ripped guts, and their fire and plunder relieved a culture of unknowable manuscripts and ecclesiastical treasures, but the sea-pagans also prodded a dispersed, insecure population back down the road towards urbanism. Shepherding his subjects to safety was Alfred, a young ruler who knew that the only course of action for a people threatened by annihilation was withdrawal to the sanctuary of walls.

It's not often in the ups-and-downs of this geographical narrative that a major transformational change can be attributed to an individual, but in this case it's reasonable to identify the son of Æthelwulf, king of the West Saxons, as the principal architect of a shape-shifting urban revival: a second-coming of towns.

Alfred had grown up with walled settlements. The landscapes of his youth were pimpled with hillforts, Roman forts and Viking barricades. Under Offa, the defended centre of authority constructed by enforceable common effort had become a familiar concept in this age of developing kingships; Offa's walled centres of the eighth century had proven their utility in a world beset by small-scale, endemic warfare and had been instruments of Mercia's pre-eminence. And Alfred had also seen the most famous walled settlement in Europe. As a boy, he'd been escorted across the continent to the apostolic see on the Tiber, where he'd been confirmed by Pope Leo IV and no doubt guided through the succession of catacombs, churches and martyrdom sites that had attracted Christian *Romipetae* – Rome-seekers – for centuries. Four hundred years of looting, plundering, earthquakes and floods had pecked and silted this city of shining stone. The suburbs had gone and the patched Aurelian walls now surrounded a shrunken kernel of perhaps 30,000 people. Time was written in every vista. Temples, theatres, forums and bath complexes still reared in eerie, disused splendour. Buildings had been stripped of lead piping, bronze tiles, iron clamps and marble fascias. Lime kilns had been fed with smashed marble monuments. Cattle grazed where senators strolled. Dressed stone had been recycled for churches and palaces. Erosion on the city's hill-slopes and a rampant Tiber had laid carpets of alluvium and earth across pavements and streets. But this gaunt, magnificent city still outshone any place north of the Alps. Alfred would have stayed at the recently rebuilt *Schola Anglorum* beside St Peter's. At over 100 metres long, the basilica's central nave and attendant aisles could accommodate more than 3,000 believers, among them first-time pilgrims gawping at the glittering mosaics, jewelled reliquaries and exquisitely crafted liturgical vessels. Across the river in the old city, there were wonders on every street. In the Forum, Alfred would have seen the gigantic arch of Septimius Severus and the enormous Temple of Castor and Pollux. The entablature of the Temple of Vespasian still dwarfed passers-by and the fluted Column of Phocas – a seventh-century addition to the Forum assembled from recycled stone – tottered taller than any monolith in Britain. And what would a West Saxon boy make of the Colossus? Thirty metres high in gilded bronze, this radiant giant had somehow survived centuries of asset stripping and still stood beside the great Flavian amphitheatre, a structure of outlandish scale, despite its role as a stone quarry. Alfred had seen how this city protected itself from the outside world. Even if he could not comprehend the architecture of the Aurelian Walls, his West Saxon escorts surely did; twelve miles of them, standing taller than six men, studded with 381 towers and rimmed with a moat. Seven years earlier, the Walls had been the deterrent that saved Rome. In 846, the city had been subjected to a *ghazw*, an Arab raiding expedition. The seaborne raiders had sacked the city's port-towns of Ostia and Portus, then looted the churches of St Peter's and St Paul's, both of which stood exposed outside the Aurelian Walls. Rome's response would have been evident during Alfred's visit. Construction teams rebuilt fifteen of the towers and raised new gates. The mouth of the Tiber was protected with new towers and a massive chain now blocked river traffic. And as a precaution against future Arab raids, Leo IV had built a wall around St Peter's and the *Schola Anglorum*, turning this part of the west bank into a self-contained, fortified quarter that became known as Leoniana, the city of Leo. To its visitors from Britain, this walled sanctuary was the *burgus Saxonum*.

Impressions gained during his trip to Rome back in the 850s had been reinforced immeasurably by the subsequent pilgrimages of his elder brother and his father, undertaken by mature adults fully capable of grasping the significance of Pope Leo's response to recent Arab raids. The refurbishment of Rome's Aurelian Walls was a vital precaution against renewed attack and an emboldening of the Christian mission. Of even greater resonance to West Saxon pilgrims would have been the walling of St Peter's and the *Schola Anglorum*. Leoniana was the Christian exemplar, a protective centre of devotion and learning.

So West Saxons were wise to long walls. Faced with Danish armies who seeped across Britain like smoke through thatch, Alfred initiated the construction of a system of garrisoned fortresses in Wessex and central Mercia. *Burhs* were an evolution of Offa's defended centres, strategically sited strongholds that could also function as offensive bases. Not since the Roman invasion over 800 years earlier had military conflict provoked systematic settlement on such a scale.

Construction of the *burhs* probably began in the immediate aftermath of the battle on the Plain, early in 878. Some were built on the sites of pre-Roman forts and others in Roman towns. Some were new foundations. All were spaced in such a way that no part of his kingdom would be more than a single day's ride from sanctuary. In many cases, account was taken of navigable rivers, existing hubs and Roman roads. It was a monumental undertaking. The earthen banks around *burhs* tended to be about 8 metres wide and up to 2.5 metres high. A typical example was hurriedly thrown up where Ermine Street crossed the upper Thames, a strategic blocking-point on the main Viking route from their Cirencester base into northern Wessex. The *burh* built at Cracgelade (the latter part of this place-name, *gelād*, was a reference to a 'difficult river-crossing', an impediment unlikely to be suffered these days by motorists hurtling on a four-lane bypass past Cricklade) had a defensive circuit of just over 2,000 metres, raised to a height of 2.5 metres with an average width of 6 metres. A traditional ditch-and-bank construction, the linear excavations around the perimeter provided some 35,000 cubic metres of spoil for the banks, which were revetted with turf and possibly topped by a timber palisade. If Cracgelade's 1,400 hides had supplied, say, one man per hide, the workforce of a thousand-plus would have been able to complete the entire task in around eight months.

Over on the eastern side of the defensive system, a key *burh* was established on the south bank of the Thames, where the old Roman approach-road from Kent and the Channel coast ran across the low-lying marshes and creeks to the bridging point. *Suthringa geweorche*, the 'fortified work of the men of Surrey', was even closer to the front-line than Cracgelade; in 878, a few hundred metres of tidal Thames separated Surrey's men from Vikings occupying ancient Londinium. The perimeter of the *burh* has been lost beneath modern development, but may have followed the boundary of the Roman settlement on the slightly raised eyot of land at the southern bridgehead. The 1,800 hides devoted to the *Suthringa geweorche* garrison were an indication of the strategic value of this bridgehead on the Thames; Londinium was in a class of its own. All three of the forces that coveted the site did so for the same reasons. Whether Viking, Mercian or West Saxon, each understood that possession of this enormous walled fortress was key to controlling river traffic on the greatest waterway in southern Britain, and road traffic on the vital arterial land-routes that congregated at the lowest crossing point. Old Londinium was a super-hub, bigger than the armies that had brought it so much havoc. Attacked by Vikings in 842 and 851, and occupied by them from 871–2, the city on the Thames had been Britain's single biggest prize. Command of Londinium brought income and goods, power and prestige. Its reach extended up the Thames and deep into what had once been Mercian heartland; it extended downstream along both shores of its gaping estuary; it extended to the prized commercial trading entrepôts of Kent, and it extended across the North Sea and Channel to continental trading stations and ports stretching from Scandinavia to the Rhine and Francia. If Alfred was to relieve Wessex of its Scandinavian visitors and to embrace Mercia within a singe polity, he had to control the lowest crossing on the Thames. *Suthringa geweorche* held the key to the future (surprising perhaps, if you happen to be reading this on a bus in Southwark Street).

Alfred built himself out of trouble. The sheer scale of the building programme and its speed of execution were testament to his ability to mobilise an army of construction workers. It's possible that thirty or so *burhs* could have been fortified by the summer of 879, not much more than a year after blood had dried on the Plain. In a document called the *Burghal Hidage*, thirty-three locations were listed, together with the number of hides related to each. With a garrison for each *burh*, around 27,000 men would have been required to maintain and man the defences.

So impregnable did Wessex appear by late 879 that Alfred was able to dictate terms to the lingering Viking armies. Guthrum's contingent in Cirencester trooped off to East Anglia, and the Vikings who had occupied the fording place at Fulanham boarded their ships and left for easier pickings in Francia. Those based in Londinium withdrew to the new East Anglian lines east of the River Lea. With their departure, Alfred was able to control the whole of Mercia and to reclaim the capital of Roman Britain.

Writing some fifteen years after the *burh*-system was initiated, Alfred's contemporary and biographer, Bishop Asser, mused upon the challenges of reconstruction that had faced his king:

> And what of the cities and towns to be rebuilt and of others to be constructed where previously there were none? And what of the *aedificia* incomparably fashioned in gold and silver at his instigation? And what of the royal halls and chambers marvellously constructed of stone and wood at his command? And what of the royal residences of masonry, moved from their old position and splendidly reconstructed at more appropriate places by his royal command?

Much progress had been made, but the scale of the effort required, had been enormous. Acknowledging the 'mighty disorder and confusion of his own people' during a time of war and hardship, and Alfred's personal difficulties with ill health, Asser mentioned 'fortifications commanded by the king which have not yet been begun, or else, having been begun late in the day, have not been brought to completion'. Those who had failed the king's command had been 'reduced to virtual extinction'. With Vikings loose on land and sea, you had to build or die.

Alfred, the 'king of the Anglo-Saxons', built a new homeland for his *Angelcynn* – his 'English folk'. Constant war had created a wasteland. Alfred's reflection of a time 'before everything was ransacked and burned', when churches 'stood filled with treasures and books', motivated the reconstruction. The land of the Anglo-Saxons would be rebuilt. While defended settlements were key to holding the heathen at bay, the cultural future of the English lay in Christianity and education. Monasteries and churches were restored and schools were founded. Alfred sought immigrants who would populate the monasteries. Priests, deacons, monks, and heathen children who could be groomed for Christianity, were brought 'from across the sea'. Bishop Asser recalled how the two of them would sit reading in the royal chamber, while Alfred collected favourite passages in his 'little book'. Sadly, this priceless commonplace book was lost, but one of the sayings survived in Asser's biography: 'The just man builds on a modest foundation and gradually proceeds to greater things.'

Burhs were a very big idea. When Alfred seized London late in 879, a town was reborn. Within the old walls, a gridded network of streets was laid out for 500 metres or so each side of the Walbrook. It seems likely that this momentous reoccupation was accompanied by the rebuilding or repair of the bridge over the Thames. Linking the old walled city with *Suthringa geweorche*, the new Surreymen's fortification on the south bank, the bridge would have been a critical deterrent to further Viking attacks by river and a vital link for land-based forces moving to protect the kingdom from the expansionist Vikings now rooted in East Anglia. The Thames returned to its role as an arterial conduit serving an enormous, politically contiguous territory. Construction commenced of a port waterfront, sharpened tenon-headed piles being driven into the tidal foreshore then connected with socketed cross-bearers and laid with planking. The first waterfront 'hithe' was christened Æthelred's hithe, named after the Mercian ealdorman (and son-in-law of Alfred) whom the King appointed to oversee the new *burh*. Tidal mud began to accumulate the spillage of trade: *stycas* minted in Northumbria and Alfredian halfpennies stamped with a monogram of the letters: LVNDONIA. By the tenth century, London's mint and river-port were the busiest in Anglo-Saxon lands and the town could be said to have returned to the status of a capital. It was a remarkable recovery.

The urban grid made its biggest comeback since the Romans. Many of the larger *burhs* were conceived as formal settlements and within their walls, streets were laid out on a regular pattern, often in sizeable blocks, or *hagae*, which were usually occupied by the region's high-profile landowners. Their presence in the *burh* added weight, and their stake in these urban manors encouraged them to contribute to the maintenance and defence of the settlement. As centres grew in size and economic activity intensified, *hagae* were progressively subdivided into long, narrow tenements, or burgage plots. Besides providing space for a dwelling, the strip behind the narrow street frontage would be used for various purposes ranging from kitchen gardens and grazing livestock, to store-houses and workshops. A back lane provided access.

Over several generations, the more successful *burhs* evolved into fully fledged towns. Deep in the heart of Wessex, the old royal centre of Winchester had been relaid on a new, symmetrical grid of streets within the original Roman walls and gateways. Assessed at 2,400 hides, this was a major *burh*, and in terms of manpower and wall-length the greatest in Wessex. Plots were developed for homes and churches, nearly six miles of streets were surfaced with around 8,000 tonnes of cobbles and channels were laid to carry drinking water. By the 870s the new town had its own mint and a minster was founded in 901. Alfred's burial there helped to perpetuate Winchester's role as the heart of the Old English kingdom. Without a sizeable river, Winchester's commercial prospects were always going to be limited, but by the early years of the tenth century, the main thoroughfare was known as *ceap stræt*, Barter or Market Street. And by the middle of the eleventh century there were over a thousand tenements in the town and at least four guilds.

Despite their systematic inauguration and semi-formalised street-plans, *burhs* could be idiosyncratic settlements. At a crossing-point of the Severn in the heart of Mercia, Worcester had been a defended centre since at least the Roman occupation, when a gravel terrace on the east bank acquired an oval earthwork enclosing just over ten hectares. It had been a humdrum, small industrial town, practising iron working and cattle-rearing. By the 600s, the town had crumbled back into the soil but the earthworks still stood tall and were chosen as a suitable location for the church of St Helen's, which became the focus for a small religious community. It was a secluded spot, with the clear-watered Severn gliding past the shingle beach and the outside world excluded by ancient banks of grass. It was, in effect, a ready-made location for the new see of the Hwicce kingdom, whose rulers built the cathedral church of St Peter's within the defensive circuit in the late 600s. Worcester remained an ecclesiastical hub until this defended site on the Severn crossing was chosen as a potential *burh*. Deep in English Mercia, it was an ideally placed outlier of Alfred's West Saxon burghal network.

The snag with Worcester was that the interior of the defensive circuit was now a cathedral precinct. This seems to have been the reason that the developer – Alfred's son-in-law, the Mercian ealdorman Æthelred – chose to extend the defensive earthwork upstream to create a new, seven-hectare enclosure. Within this extension, a basic grid of streets was laid out, but this seems to have been insufficient for the *burh* and about fifteen years later, a deal was struck with Worcester's bishop, who relinquished around a third of the cathedral precinct in exchange for 'half of all the rights . . . whether in the market or in the street, both within the fortification and outside . . . land-rent, the fine for fighting, or theft, or dishonest trading, and contributions to the *burh*-wall and all the [fines for] offences which admit of compensation'. The charter probably suited the ealdorman more than the bishop, but losing half of the profits from market tolls and fines was clearly worth paying for 'the protection of all the people'. Exempt from the deal were tolls on cart-loads and pack-loads of salt – 'the wagon-shilling' and the 'load-penny' – passing through Worcester from Droitwich, payments which were already going to the king. These kinds of trade-off were probably occurring at other *burhs*; a state emergency providing leverage for the reallocation of land-ownership. *Burhs* were much more flexible urban spaces than the holdings they replaced.

By 900, Worcester's street grid had doubled its original size. It was not one of the larger *burhs* (not least because the cathedral precinct still occupied nearly half of the site) but it developed into a successful town. By the mid 900s, a community of craft-workers were occupying a suburb outside the walls beside the main road heading off to the south-east. Within the walls, *hagae* were being subdivided as owners broke up their urban manors and rented parcels to craftsmen and traders. Several churches were built. By the time the Normans showed up, Worcester probably had a couple of thousand residents.

But the brilliance behind the *burhs* was less the neat grids than their spacing across the landscapes. Conceived primarily as an integrated defensive measure, they also had the potential to be a network of market towns. Within the walls, the population could trade in safety and many *burhs* were little more than one day's ride from the next *burh*, so merchants could travel from market to market with reduced risk of robbery or violence. The spacing of *burhs* also meant that merchants could schedule their travel so that they appeared on 'market days'. Regularity increased trade and made it easier for the latest prices to flicker through communities, which reduced the amount of time people had to spend acquiring information and increased the likelihood of transactions. There was another reason to trade in towns: a vendor had to prove ownership of goods before selling, and this was done through 'vouching to warranty', a procedure that required trustworthy witnesses. By the time Alfred's son, Edward the Elder was on the throne, it had become compulsory to undertake transactions in towns. Not all *burhs* survived as towns, and some were never meant to be more than emergency sanctuaries, but by 920 or so there were perhaps fifty of them. An urban economy had been resuscitated.

Beyond the *burhs* of Wessex and Mercia, towns were taking shape in Danelaw, the part of Britain in which Viking rule held sway following the treaty between Alfred and Guthrum in 886. Derby, Leicester, Lincoln, Nottingham and Stamford all grew into trading centres, but the largest by far was the old Roman provincial capital of Eburacum, subsequently the river-port of Eoforwic and by the late 800s, the Viking town of Jorvik. The place had suffered multiple ravages since the legions had marched away over five hundred years earlier. Most of the Roman buildings, including the huge fortress basilica, had been demolished and recycled for their stone, much of it ending up in churches. After capturing the town in 866, the Vikings built a bridge over the Ouse and laid out a new approach to the crossing, *mikill gata*, the Great Street, now Micklegate. By the early 900s, one of the streets on the other side of the river – now Coppergate – was lined with tight-packed tenements. Each building in these rows was about 4.4 metres wide onto the street, and 8.2 metres deep, with a central hearth of clay which would have had to provide light as well as heat. They probably functioned both as workshops and homes. Several decades later, they were demolished and replaced by superior houses with semi-basements, planked walls and quite possibly a second storey for living quarters. The excavation of cellars for townhouses may have been an adaptation to demand for storage space at a time of increasing trade.

Burhs were a durable urban template. In many cases, their street plans survived through until our present century. In Wareham, you can walk streets trodden by Alfred's *thegns*. York's Micklegate and Coppergate are Viking shopping streets. In Worcester, the boundary of the riverside *haga* once leased to Ealdorman Æthelred is traced by Copenhagen Street, Birdport and Grope Lane. Back before the Normans, they were Huxstere's Strete, Houndes Lone and Wodestathe Strete. The names changed, but the streets stayed where they were.

*

There is a bike ride I like in the Cotswolds. I put it together by linking a sixty-mile sequence of the smallest roads I could find on Ordnance Survey 1:50,000 maps. On a summer's day, it is close to paradise: a byroad meanders through the borderlands of Wessex and Mercia, making height until the gaping blue vales of the Avon and Severn suddenly appear, hazily rimmed with Brittonic hills. One of the things I like about the ride is the necklace of villages, these congenial clusters of honey-stoned cottages, with their twinkling 4x4s and tumbling petals. Most of them have lost their pubs and their post-offices. And their smithies. But they are still villages, with their churches and parish noticeboards. And they are a world away from the city.

Villages appeared during the age of *burhs*. Unlike towns, however, villages did not spring from a short surge of state-planned settlement, but emerged gradually over three hundred years, beginning perhaps in the middle of the ninth century. In their formative stages, they were, in effect, large hamlets: nucleations of more than six to eight farmsteads. Their breeding ground hints at their economic roots. They were settlements of the arable belt: the strip of lowland running from Wessex, north through the east midlands and up the east coast of Britain, past the Forth all the way to the Moray Firth. This belt of cultivation and nucleated settlement shows a remarkable overlap with the zone within Britain of land receiving less than 600 mm of rainfall per year. In effect, villages developed on that south-central to north-eastern strip that had always enjoyed the best soils and climate. Like the *burhs*, villages were modelled for resilience. The lack of nucleated villages in the south-east may have been related to the continued existence of woodland. They were an economic response to food insecurity; a settlement template that made it possible for more people to live off less land.

Villages were complex systems that could take generations to evolve. And within the 'village belt', timings varied considerably. In the central part of the belt, some villages began to take shape soon after 850, but in the north-east, the process was still underway in the twelfth century. Integrated into the village model was the open field and heavy plough, the plot and the practice that made farming more efficient. Very often, the process of village formation began with the abandonment of dispersed farmsteads and hamlets. Instead of the patchwork of old, each village had two, three or four enormous fields, shared by the community. Each field was divided into 'furlongs', roughly rectangular blocks which were subdivided into 'strips', 'lands' or 'selions': parallel segments. To promote soil fertility, a system of rotation was practised: typically, one field would be put down to wheat, a second to barley, and a third left fallow. The following year, the fallow field would be ploughed and planted with wheat, the field that had been harvested for wheat would be ploughed and planted with barley, and the field that had been used for barley would be left fallow. The entire community had common grazing rights on ploughland that was fallow, and on other land that was deemed 'waste', like the common. Such a system required coordination, organisation and the cooperation of every farmer in the community, and to facilitate this the village would have an assembly or manor court where the overall administration could be deliberated upon.

In the lands of the old Brittonic kingdoms, where extensive districts still lay beyond the reach of regular trade or the Christian building programme, landscapes looked much as they had for millennia. In upland areas, many local economies still revolved around semi-nomadic traditions of transhumance, farmers moving with their flocks and herds to high summer pastures and returning to their lowland settlements for the winter. In the country of Cymru, the landscapes between the Severn and the Dee had changed little since the *civitates* had crumbled. In scattered pockets, cultivation was practised, but the main source of food came from the flocks and herds being driven each spring from the *hendref* (the winter home) up the slopes to the *hafod* (the summer home) and then back again in the autumn. These seasonal migrations, repeated through the decades and centuries, had in some places come to define the boundaries of the *cantrefi*, or hundreds, each *cantref* represented by upland and lowland grazing. The transhumance being practised in western Britain was by now a truly ancient economy, its migration routes engraved deeply into the landscape, its waymarks handed down from generation to generation. The attachment of farmers and herders to their land gave the outlying precincts of Britain a sense of deep continuity. While elites raged and ravaged; while dynasties toppled and political borders shifted, the people who lived out in the clearings continued to tramp the same paths their ancestors had used. In Cymru, even the unfree villeins – the *taeogion* – had rights, and in the more fertile areas, their compact villages were the largest settlements.

In the far north of Britain, shifting forces were amending patterns of settlement. Anglo-Saxon culture had pushed north to the Forth; Britons still occupied the great peninsula between the Solway and Clyde; Scots clung to the peninsulas and islands north of the Clyde; Picts were still settled in the glens of the highlands and along the east coast from the Tay to the Pentland Firth. Vikings had claimed the islands of the north and north-west. The changing dynamics of these groupings could be traced in structures and place-names. The Orkney Islands, already strewn with megalithic monuments dating back over four thousand years, were recolonised and rebranded. By attaching the Norse suffix 'ey' – meaning 'island' – to simple geographical labels, the high island became Háey, or Hoy; the island with sandy beaches became Sanday; the island to the west became Westray, and so on. The Viking word for homestead, *bólstathr*, became the root for countless places dotted across the northern isles. Meanwhile, the eastward expansion of Gaelic-speaking Scottic elites spread the use of *Pit* – piece, or part, or share of land – as the first element of many place-names between the Firths of Forth and Moray. Although the prefix was Pictish, it was invariably followed by a Gaelic element. Eventually there would be some 300 'pit' places, covering a wide range of place types from *pit na h-uamha* (now Pittenweem), the cave place, to *pit-cloichreach* (Pitlochry), the stony place. Some 'pit' places were named after individuals: Pitcarmick was Cormac's share; Pitkennedy, Cennetigh's share. These pieces of land were generally south-facing, elevated above flood-zones and some distance from the coast. They were, it seems, the best sites incomers could find. The shadowy presence of these hybrid sites signalled the gradual merging of northern kingdoms into a Scotto-Pictish territory.

By the eleventh century, the Anglo-Saxon economy was one of the richest in Europe, and also the most advanced. The size of Britain, its location, its climate, its underlying geology, its fishing grounds and farmland had made it one of the most valuable land-banks in the western world. And three hundred years of developments had optimised the methods of exploiting those resources. Britain had discovered market economics: a system in which commodities were exchanged for coins and markets were able to fix their own prices. The physical embodiment of this very big idea was the market town, and by the time the Normans showed up, England had nearly a hundred of them. In the shape-shifting lands of Scotland and Wales, there were no towns at all, and in Ireland, only Viking Dublin could claim to exhibit traits of urbanisation.

Totnes's *motte* protrudes so far above the town's solar-panelled slates that it still provides a clear view to the quays on the Dart. The *motte* was built by a young Norman baron, Judhael, whose reward for a good Conquest had included over one hundred manors in Devon, some of them granted by the king, many of them plundered piecemeal from the shire's *thegns*. In an army of thugs, Judhael was exceptional. To control his assets, he built himself a gigantic fortress on part of the *burh* at Totnes. Walled *burhs* were already over a century old and to conquering Norman lords they were the relics of a failed state. Spilling across the town's existing street-plan, a huge mound of rock and earth was piled on a rocky knoll at the top of the *burh*, then sealed with a layer of puddled clay. Some 20 metres high and 60 metres in diameter, the finished mound was surrounded by the circular ditch that had provided the building material. Within the mound a stone core provided the foundation for a timber tower. Projecting from the mound, an oval-shaped inner bailey over 60 metres long was defended by an earth bank topped by a timber palisade. In the space of a few days, ancient, low-rise Anglo-Saxon Totnes was subdued by an impregnable tower which could observe every moving cart and vessel.

Not since the arrival of the Roman legions had there been such a far-reaching makeover. The first inkling of architectural change had been detected a couple of decades before the Conquest with the accession to the English throne of Edward, who brought with him friends and allies from Normandy and France. With them came a novel structure intended to secure territory. At least three were built in the troublesome border zone between England and Wales. The earliest recorded usage of the word *castel* crops up in the *Anglo-Saxon Chronicle* during an account of military stand-offs in Herefordshire in about 1050. According to the *Chronicle* the Herefordshire castle was so effective that the 'foreigners' (French) who built it 'inflicted all the injuries and insults they possibly could upon the king's men in that region.' Castles worked. Since at least 800 BC, free-standing fortifications had been symbols of power and defiance, but early Norman castles had more in common with Roman marching forts than with tribal hillforts. They could be built with astonishing speed, they were virtually impregnable and their construction had more to do with military expediency than with cultural bonding.

Hastings was Britain's costliest military defeat; a bloodbath then regime change, prosecuted with extreme violence. Slaughter on Senlac Hill was followed by the replacement of Anglo-Saxon elites by Norman lords and barons. Godwins and Ethelreds were elbowed out for Williams and Roberts and the language of the *Anglo-Saxon Chronicle* was eclipsed by French. The Anglo-Saxon mead-hall disappeared; Normans drank wine. Castles were built and cathedrals demolished.

The initial shockwave was followed by architectures of suppression based on the motte-and-bailey model: a massive mound topped by a tower of timber, surrounded by a fortified enclosure to protect the castle's ancillary buildings. No care was taken to consider existing local attachments to a place. Buildings were demolished, streets smothered.

In the early weeks of 1067, an entire pattern of streets within Winchester was obliterated to make way for the castle that was strong enough to dominate the kingdom's main seat of government. In Worcester, Sheriff Urse d'Abetôt allowed the castle to encroach on the cathedral cemetery. In York, the Norman garrison fired the houses adjacent to the castle to prevent them being used as raw material for filling in the defensive ditches. Unfortunately the flames spread, consuming the entire town and monastery of St Peter. Norman mottes could be built with ruthless application: responding to a crisis at York, William raised a new castle in only eight days.

Rebellions in 1068 and 1069 unleashed further onslaughts of castle-building. A clever defensive strategist, William understood the need to hold routes of communication and major towns. But he also knew that the most important sanctuaries for the English and Welsh were the surviving tracts of woodland, fen and upland. A window was opened on the helplessness of the English by a Mercian-born monk, Orderic Vitalis:

> In consequence of these commotions, the king carefully surveyed the most inaccessible points in the country, and, selecting suitable spots, fortified them against the enemy's excursions. In the English districts there were very few fortresses, which the Normans call castles; so that, though the English were warlike and brave, they were little able to make a determined resistance.

By 1100, the Norman invaders had built at least 500 castles.

Architectural cleansing swept away Anglo-Saxon cathedrals and churches. Nine of the fifteen ancient cathedrals in England had been demolished or torched by 1087 and over the next couple of decades, the other six would be reconstructed on continental, Romanesque lines, along with all the major abbeys. They were replaced with buildings that reflected the majesty and taste of Norman imperial status. The cathedral in Winchester, the royal seat of the West Saxons and the place where Edward the Confessor had been crowned in 1043, was pulled down. Masterminding the rebuilding project was Paris-educated Bishop Walkelin, a relative of King William. Walkelin was granted half a hide of land on the Isle of Wight to locate and quarry the creamy limestone he sought. He was also granted four days and four nights to remove as much timber as he needed from the forest beside the road between Winchester and Alresford. In common with many of the barons, the bishop went far beyond the king's grant, and felled the entire wood. When the king rode past the stumps he was enraged to see the 'delectable wood' had ceased to exist. During a confrontation with the repentant bishop, the king was reported to have observed: 'I was much too liberal in my grant, as you were too greedy in availing yourself of it.' This was, perhaps, an apocryphal tale, but it served to illustrate the arrogance of many new Norman landholders, whose general principle was to accept one furlong and steal one hundred. Walkelin's cathedral was the longest in Europe. By the time the Normans had completed their architectural cleansing, no masonry from any pre-Conquest English cathedral remained standing above ground.

Saxon architecture was deemed provincial and contemptible and its traditions base. The relics of local saints were chucked out and the memories of Saxon clergy reviled as 'yokels and idiots'. The latter insult was offered by the new abbot of St Albans, Paul of Caen, who destroyed the tombs of earlier abbots and swept away the Saxon church, building a Norman one using stones and tiles from the old Roman city. But the cultural desecration was selective. Although Bishop Walkelin had insulted his Saxon predecessors by failing to respect the axis of the Old Minster, he transferred some of the Old English saints – and the legitimacy they conferred – to the New Minster. The only major church to avoid demolition was Westminster Abbey, spared because it was the burial place of Edward the Confessor and because William the Conqueror had been anointed there. The church was being rebuilt in the Romanesque style as the invasion took place.

Out in the fields and woods, little changed on the surface. Generally, the manors of the Anglo-Saxons were left intact and the rural economy continued to function through farmstead, village and open field. But the invasion had unleashed an army of land-grabbers who supplemented their king-given fees by dispossessing English owners on a freelance basis. The result was a broken jigsaw of land-ownership. In the aftermath of the Conquest, William commissioned an unprecedented audit of England's resources and their allocation between the king and the various lords to whom lands had been granted. 'The mighty king,' wrote a suitably awed Anglo-Norman archdeacon in Huntingdon, 'sent his justices through every shire, that is, province, of England, and made them enquire on oath how many hides (that is yokes sufficient for one plough per annum) there were in every village and how many animals. He also had them enquire what each city, castle, village, farm, river, marsh and wood rendered per annum. All these things were written in charters and brought to the king.' The audit was entirely dependent on local testimony, albeit under oath, and it was of course out of date as soon as it had been completed. But 'The King's Book' reduced the infinite complexities of the English landscape into a few controllable categories. Later, it became known as 'Domesday', the book of the day of judgement.

Beyond the raising of castles and churches, the Norman invaders had relatively little effect on the built landscape. The underlying forces generating the nucleation of villages and growth of towns were already in place. The surge in urban growth was more closely related to population growth and European economics than to French kleptomaniacs.

While England and Wales were being converted to a Norman model of civilisation, towns began to take shape in Scotland. Over a century after Alfred planted his *burhs* across Wessex, David I included the foundation of *burghs* in a flurry of reforms that included regional markets, the introduction of feudalism and new monasteries. Among the *burghs* to appear at this time were Dunfermline and Aberdeen and St Andrews, which was founded in around 1150 on a fairly ambitious scale, with two wide streets angling back from the site where – a decade or so later – Bishop Arnold would begin building his new cathedral. It was a neat, uniform street plan and with time the spaces each side of the streets filled with *rigs* (burgage plots). The two main streets acquired Anglo-Scandinavian 'gait' names – Northgait and Southgait – and the narrow lanes between them were *wynds*. The marketplace was probably located in the triangle of space where the two streets met, but twenty or so years after the *burgh*'s foundation, a third main street – Mercatgaite – was developing between Northgait and Southgait, with a new marketplace set centrally within the growing town. By 1153, a total of seventeen *burghs* had been established in Scotland.

The presence of town streets from Totnes to Tain marked Britain's urban graduation. This was a level of development far beyond anything that had existed under Roman rule and it connected Scotland to the economic and political webs of northern Europe. But urbanisation was accompanied by a taming of the rest. By 1150, a huge swathe of Britain had been completely transformed by the spread of open fields and villages. The ancient patchwork of wood, copse and tiny field had been overrun. An intricate pattern of hedgerow and ditch, field wall and fence, had been smoothed and rationalised. Vast tracts of native pine still whispered in the glens and straths of Scotland but it was the end of the story for England's wildwood.

Ever since the broadleaves had begun their long march northward nine thousand years earlier, there had been unmanaged woodland in England; places where oak and ash and elm had grown and regrown; where saplings had space. In the thousand years up until the Domesday survey of 1086, England had changed from a half-wooded land into one that was occasionally wooded. By Domesday, maybe as little as 15 per cent of the 27 million acres of land covered in the 1086 returns were wooded, rather less than the proportion of woodland seen today in France. When and where the last English wildwood was felled or burned, nobody knows. The Forest of Dean, perhaps, at some point in the twelfth century. The sound of that last wild oak crashing to the woodland floor was no different to the millions that had preceded it, but its solitary impact was one of monumental finality.

Field systems defined by banks – or reaves – on Mountsland Common, Dartmoor. They appeared from around 1500 BC.

The White Horse of Uffington, on the crest of Lambourn Downs, West Berkshire.

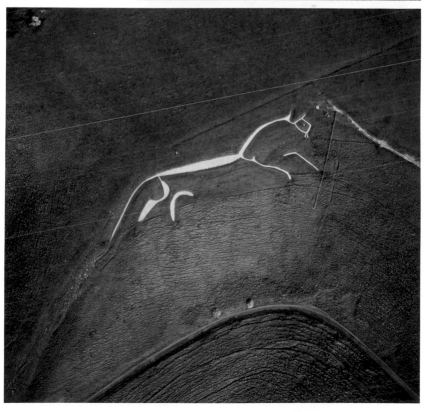

The broch on Mousa island, off the east coast of Mainland, Shetland. Brochs were built from around 200 BC.

Interior of Mousa broch, showing the windowless cavity wall and single, low entrance.

The hillfort known as Maiden Castle, with its maze-like entrance, remodelled around 100 BC.

The Roman fort on Hardknott Pass in the Lake District, early second century AD.

The Roman baths (with subsequent additions) in Bath.

A recreated Anglo-Saxon village on the bank of the River Lark, West Stow, Suffolk.

Offa's Dyke crossing Hawthorn Hill, south of Knighton, close to what is now the Welsh–English border. Offa ruled Mercia from AD 757 to 796.

King's College Chapel, Cambridge. Work started in 1446 and took a century to complete.

Top left The best-preserved *burh* walls in Britain: Wareham, Dorset. Construction of *burhs* began soon after AD 878.

Above left Ludlow, with its Norman castle, market street and grid-like 'new town'.

Left The deserted village of Wharram Percy, North Yorkshire.

Little Moreton Hall,
near Congleton, Cheshire.
Construction began between
1504 and 1508, with later
works completed in 1610.

Britain, then, is the most blessed of islands, rich in crops and
trees, with plentiful streams and woodlands, delightful for its
hunting-grounds of wildfowl and game, and teeming with many
different kinds of land, sea, and river birds. . . .

These lines were written in the mid-twelfth century by an archdeacon
in Huntingdon. The tradition of portraying Britain as an insular Eden
was one that went back to Bede and Gildas. Henry of Huntingdon's
vision was in part an inherited gloss, passed through generations by
word of mouth and vellum, but like his predecessors he contributed
embellishments drawn from his own travels, his reading and his
imagination. Henry's *Historia Anglorum* – *The History of the English
People* – was written, he said, for 'the many – I mean the less educated'.
His Anglo-Norman potboiler was a cracking narrative, packed with
battles, stirring orations and strong views: 'you dregs of the greedy
religious life', he wrote of monks who embezzled tithes and sold
churches.

Historia Anglorum opened with a geographical overview of a
teeming island paradise, with plentiful game, rich grazings, valuable
metal ores and a 'wonderful abundance of fish and meat, of costly
wool and milk, and of cattle without number, so that the wealth of
silver in Britain seems greater than Germany'. And then Henry embarked upon a long passage punctuated with enumeration: Britain
had 28 cities, 5 languages, 2 great rivers, 5 invasions, 7 kingdoms and
35 counties, 4 roads and 4 wonders. Henry's paired quartets were significant. Four was a very old number, analogous to the four seasons,
the four elements, the four winds, the four rivers of Eden, the four
Gospels, the four parts of the cross, the four virtues. Four encompassed the whole. The four highways he described were Icknield Way,
Ermine Street, Watling Street and the Fosse Way. With a little bit of
creative realignment, they reached the extremities of Britain. Henry's
highways represented universal access. The four wonders were there
as a counterpoint: the roads were made by man and explicable; the
wonders came from a realm of Creation beyond comprehension.

The first of the wonders was a wind that exhaled from the caves in a mountain called 'The Peak', a wind so forceful that it 'drives back any pieces of clothing thrown in and tosses them up to a great height'. (Henry's windy cave was the Peak District cavern known today as 'The Devil's Arse'; admission, should you be wondering, is currently £9.25 for an adult and £7.25 for a child.)

Henry's second wonder was a place called 'Stanenges', or Stone-henge (£13.90/£8.30), 'where stones of a remarkable size are raised up like gates, in such a way that gates seem to be placed on top of gates. And no one can work out how the stones were so skilfully lifted up to such a height or why they were erected there.' Henry – who wrote as if he knew the site first-hand – seemed to be under the impression that the fallen stones were part of a second-storey circle which had been balanced on the first. Henry's is the earliest-known record of the name Stonehenge, and the first description, too.

The third wonder was a place called 'Chederhole, where there is an underground cavern which many people have often entered, but although they have travelled a long way over dry land and over rivers, they have never been able to come to the other end.' (Entrance to Wookey Hole, the largest cavern in Cheddar Gorge, £18.00 for adults and £12.00 for children.)

Henry's fourth wonder was rain (no charge). He related that 'in certain places the rain seems to rise up from the mountains and immediately fall on the plains.' Viewed by an archdeacon from the climatically privileged neighbourhood of Huntingdon, the spectacle of rain dumping for hours on end was clearly awesome, but to the inhabitants of northern and western Britain it was less likely to have been regarded as a wonder.

Henry's mother was English and his father Norman, and he'd grown up on the western water-margin of the Fens, in Little Stukeley, three miles up Ermine Street from Huntingdon. He'd been educated at Lincoln by Master Albinus of Angers and by the time he came to work on his thousand-year history, he was reading in Old English, Latin and French. His research covered sources ranging from Bede to the four *Books of Kings*, to Virgil, Martial, Lucan, Statius, Ovid and the *Anglo-Saxon Chronicle*. *Historia Anglorum* eventually ran to ten volumes. As a writer, he was servant to his subjects, but very occasionally he allowed a slender, fleeting ray of sunlight to fall upon his own landscapes. In the poem 'Against himself', readers caught a glimpse of his elegant estate at Stukeley, with its orchards and garden. And in the fifth chapter of *Historia Anglorum* a description of King Edgar's abbey foundations near Stamford and at Thorney and at Ramsey, meandered for a moment into a wetland the author clearly loved:

> These Fens are of wide extent, and the prospect is beautiful; for they are watered by numerous flowing streams varied by many lakes, both great and small, and are verdant with woods and islands.

And in the following chapter, he took his readers through the streets of his own town:

> Huntingdon, that is 'the hill of the hunters', stands on the site of Godmanchester, once a famous city, but now only a pleasant village on both sides of the river. It is remarkable for the two castles . . . and for its sunny exposure, as well as for its beauty, besides its contiguity to the Fens, and the abundance of wild fowl and animals of chase.

Stonehenge, the Fens and sunny Huntingdon were moments of enchantment; fleeting, authorial asides; whispers in ink: 'I was there.' British landscapes were beginning to recruit commentators.

SIXTEEN

Centres of Attraction

1150–1300

One day last summer, we cycled across Wales from the west coast to the border with England. The route cut a transect through a sequence of landscapes from the submerged forests of Cantre'r Gwaelod, up deciduous ravines to breezy, sheep-grazed ranges criss-crossed with ancient ridgeways, then down, down, down, through warm vales printed with hamlets and fields to the castle town of Ludlow. Sitting on a bench in the marketplace, it felt as if we'd pedalled through time, from wildwood to urban hubbub.

Town markets were brought to Ludlow by descendants of Walter de Lacy, a Norman baron from Lassy, in Calvados, who had crossed the Channel in 1066, embedded within the military household of one of the Conqueror's key advisers, William fitz Osbern. After the Conquest, Walter's awards had included 163 manors in seven different counties. One of those manors occupied a strategic confluence of river and road on the Welsh borders in terrain not so different to the *bocage* back home: rolling and wooded with quick streams and good hunting. With it came land for fifty ploughs, twenty hides paying geld, a couple of mills and a church. The de Lacys were energetic colonists and builders (Walter eventually fell to his death from scaffolding on the great spire in Hereford).

Walter built a castle on a rocky spur overlooking the River Teme. It was an impregnable edifice. Four towers and a curtain wall were protected by natural precipices and by a vertical-sided ditch 25 metres wide. Likely to have been conceived as a link in a line of border castles implemented by fitz Osbern, the castle on the Teme was also an out-standing development opportunity. The spur was already crossed by an important road and the manor was well endowed with stone, timber, water, and space for building. A settlement grew below the castle walls, while another expanded along the road, 300 metres east along the spur, the two eventually becoming linked by a market zone and a church. In the late 1100s, the castle gained an outer bailey, and descendants of the de Lacys undertook a planned development on the south side of the spur, laying out two wide, parallel streets. Midway between the new streets, they also laid out an access lane, which gave each property-holder two frontages and therefore the opportun-ity to subdivide – and sublet – their plots. Narrow access lanes like these lubricated urban expansion: property-holders could get higher returns from the subdivided plots, incomers had access to a tier of lower-cost properties, and traffic congestion on the main streets was reduced. The area covered by the grid of new streets probably doubled the size of the settlement. By the end of the twelfth century, 'Ludelaue' could regard itself as a town and one of the key stop-overs for travellers making their way along the borders between Chester and Hereford.

One of many passers-by who saw Ludlow's castle at this time was Giraldus Cambrensis, Gerald of Wales, the son of William de Barri, a Norman knight, and Angharad, a granddaughter of Rhys ap Tewdwr, Prince of South Wales. It was later in life, while Archdeacon of Brecon and occupying himself – as he put it – 'in a sort of happy-go-lucky mediocrity', that Gerald accompanied the Archbishop of Canterbury on a horseback journey around the perimeter of Wales recruiting archers for a crusade to the Holy Land. The journey provided Gerald with material for Britain's earliest-known travel book, *The Journey through Wales*, and for a separate work, *The Description of Wales*, a geographical portrait of Gerald's homeland. Gerald's are the earliest detailed impressions we have of British landscapes.

Long-haul journeys of a month or two were not uncommon in an age of transcontinental pilgrimages and crusades, but many readers east of the Teme would have been surprised by the topographical wonders Gerald revealed. Wales was a land of geographical extremes irrigated by high rainfall. Travel took time. Although Gerald reckoned that Wales could be measured north to south – from the coast of Anglesey to the coast of Gwent – as 'some eight days', the archbishop's party took fifty-one days to complete their circumnavigation of around a thousand miles. As the hours, days and weeks bumped by, an archive of comparable landscapes accumulated, none humdrum, some fantastical. Early in the journey, the easygoing cleric described a lake whose waters periodically turned bright green or blood red; the lake was Llangorse, its alternating colours caused by algae and storm run-off. Then there was a stretch of the west coast where a storm had stripped away the sand to expose tree trunks 'standing in the sea, with their tops lopped off, and with the cuts made by axes as clear as if they had been felled only yesterday'. To this day, storms still strip the sand at Newgale to reveal 'pitch black' beds of peat and tree trunks which shine 'like ebony'. The floating island in Snowdonia is harder to explain. It was so big that flocks of sheep could be carried away by the winds, much to the amazement of their shepherds. Beyond the day-to-day highlights of life on the road, a more generalised picture of Wales emerged. The places Gerald encountered on his circumnavigation came across as isolated strongholds on the Anglo-Norman periphery of a country that appeared to be suffering from over-stretched resources and competition for space:

Those mountain-heights abound in horses and wild game, those woods are richly stocked with pigs, the shady groves with goats, the pasture-lands with sheep, the meadows with cattle, the farms with ploughs. All the things and creatures which I have mentioned are there in great abundance, and yet we are so insatiable in our wicked desires that each in its turn seems insufficient for our needs. We occupy each other's territory, we move boundary-fences, we invade each other's plots of land. Our market-places are piled high with goods for sale, and yet our courts are kept busy with legal cases, the palaces of our kings re-echo with complaints.

Environmentalist, geographer, travel-writer, Gerald seems to have little doubt that his homeland was facing unprecedented pressure:

> This is what we read in Isaiah: 'Woe unto them that join house to house, that lay field to field, till there be no place, that they may be placed alone in the midst of the earth.'

Despite the lack of living space, Wales still had one foot in the wilderness. Along parts of the south coast there was no established road and the archbishop's party had to take to the beach and risk the 'twin hazards of a sandy shore and an incoming tide'. Riders and heavily laden packhorses ran into all sorts of difficulties, including quicksand that took Gerald's mount on the beach near Neath (you can see the spot from the modern M4). Few of the rivers they crossed had bridges, while estuaries like the Dyfi required the services of boatmen. Countering the catalogue of near-death moments and human fallibilities, Gerald had an eye for the beauty of his homeland. While he portrayed the Anglo-Norman rim as a circuit of castles, monasteries, troublesome gulfs and shifting sands (an allegory perhaps of twelfth-century Christendom), the interior came across as a Welsh Eden. Gerald's own county, Brecknock, produced 'a great amount of corn' and was blessed with 'ample pasture and plenty of woodland, the first full of cattle, the second teeming with wild animals'. There was salmon and grayling in the Wye and trout in the Usk. Brecknock Mere (Lake Llangorse) had 'plenty of pike, perch, excellent trout, tench and mud-loving eels for the local inhabitants'.

If a reader had to pick one place that seemed to encapsulate everything Gerald admired about the Welsh landscape, it would be a spot on the 'noble' River Teifi near Gilbert Strongbow's Norman castle of Cilgerran. At a place called Cenarth Mawr, Gerald came across a 'flourishing fishing station' beside a waterfall where salmon leapt 'as far as the height of the tallest spear'. Beside the fishing station and falls was a church dedicated to St Llawddog, and a mill and a bridge over the river, and 'a most attractive garden', all standing 'together on a small plot of land'. This harmonic vignette concluded with a long eulogy to the beaver. According to Gerald, the Teifi was the last river in Wales where beavers could be found. In England, there were no beavers south of the Humber, and he'd been told that in Scotland they existed on just one river, and even there, they were 'rare'. The beaver was teetering on extinction, but in Wales it was building its own monuments: great dams and multi-storey, storm-proof, 'castle-like lodges' which were intricately assembled from logs and willow wands and provided with doorways, lookouts and underground hiding places in the riverbank. All to no avail, for their fur and their scent glands made them a rich prize for hunters. Gerald's was the first – and last – detailed description of living beavers in Britain. He was probably wrong about them surviving north of the Humber, although they were there a few hundred years earlier, when the woods and streams hid sufficient numbers for them to have given their name to settlements such as Bewerley ('woodland clearing frequented by beavers') and Beverley ('beaver lodge'). Thirty miles south of the Humber, Bevercotes ('place where beavers have built their nests') also alluded to their lost habitat.

In Gerald's world, beavers were better builders than men. At no point in his journey did he describe the settlements – towns or otherwise – that had attached themselves to Anglo-Norman strongpoints. All he said of Ludlow was that it had an 'imposing castle'. That day, the last of the ride, the archbishop's convoy had ridden over Wenlock Edge, to Bromfield Priory, then pressed on through Ludlow and Leominster to Hereford before bed; long days like this were not uncommon but the real reason that Welsh towns failed to make the Archdeacon's record was that they were virtually invisible. In the eyes of a cleric who had been educated in Gloucester, studied in Paris and spent time in London, Lincoln and Canterbury, and travelled four times to Rome, a settlement with no more than a handful of streets was hardly a civic spectacle.

The only specifically urban reference the peripatetic cleric made in his record of the journey was to the 'metropolitan city' of Caerleon. Over a thousand years after it had risen from the banks of the Usk as the legionary fortress of *legio II Augusta*, this place was still known, wrote Gerald, as 'the City of the Legions'. No more than a long day's ride down the Usk from his own home at Brecon, Gerald was familiar with the ruins. Caerleon's 'immense palaces', 'gilded gables', 'remarkable hot baths, the remains of temples and an amphitheatre' evoked a lost magnificence:

> Wherever you look, both within and without the circuit of these walls, you can see constructions dug deep into the earth, conduits for water, underground passages and air-vents. Most remarkable of all to my mind are the stoves, which once transmitted heat through narrow pipes inserted in the side-walls and which are built with extraordinary skill.

The derelict Roman 'city' exerted a powerful, physical presence. A 'lofty tower' still stood and the ruins were 'enclosed within impressive walls, parts of which still remain standing'. Denuded by a millennium of plundering and Welsh weather, stripped of its statuary, monuments, roofs and most of its walls, this was still the brightest star in the civic firmament of Wales. Gerald praised the brick, the humble, cuboid, clay brick, that stood for the care and 'one-time splendour' with which this city had been built. Twelfth-century Wales was still in the timber-and-stone age. Drawn by Gerald's devotional quill, the City of the Legions assumed Biblical proportions, a monument to the crisis of its destruction.

Gerald's Wales was a contested land; a land beset by war; a land in which conspicuous construction was expressed in keeps and baileys.

Welsh 'towns' were untidy urchins clutching the skirts of castles. By contrast, urbanisation in Scotland was relatively mature. The seventeen *burghs* that had come into existence by the 1150s had expanded by 1214 to thirty-nine. But the extreme geographies of the north led to an uneven seeding. North of the Forth and Clyde, the highlands acted as a barrier, and to the south, so did the Southern Uplands. Town location was guided by sea lanes and internal connectivity. On the west and east coasts, the seaways that had been so convenient for hunters and foragers became the conduit for urbanisation. Towns that developed along the coast between the Solway and Clyde thrived on their connections with Ireland, while those on the west coast, from Berwick up to Aberdeen, Inverness and Tain, were built on their connections to the Baltic and Scandinavia, the Low Countries, France and England. In this developing landscape, large rivers wielded influence, their bodies as navigable trade-routes and their mouths as coastal harbours. Aberdeen grew at the mouth of the Dee and Perth at the head of navigation on the Tay. Only ten miles from the open sea, Dundee developed around a natural harbour on the north shore of the Firth of Tay. Stirling occupied the head of navigation on the Forth. Edinburgh's lofty vantage point stood back from the coast but the fast little river that darted past the settlement emptied into the sea a couple of miles away at a superb anchorage: Inverlethe, the mouth of the river Lethe. Sixty miles along the coast, Berwick offered a shorter sailing passage to many of the vessels plying the North Sea, but its harbour at the mouth of the Tweed, though sheltered, was far more difficult to approach in bad weather than the generous gulf of the Forth.

In England, markets were breeding like coneys. The right to hold a market was – from the twelfth century – a privilege usually conferred by the Crown. Royal market grants unleashed a virile market economy which fed hundreds of young nuclei in rural parts of the country. The effect on form and function of settlements was profound. Markets were controlled operations, dependent on the collection of tolls and rents for stalls and so the space allocated for them had to be susceptible to control; entry and exit points were commercial portals. Many markets were born in churchyards. Others took over road intersections or part of a settlement's main street. The two consistent elements were location and periodicity: markets were held at the same place and time; fixed on the landscape, fixed on the calendar. At one end of the timescale were weekly markets capturing most of their business from a catchment area reaching to the limits of the neighbouring markets. At the other end of the scale were annual fairs which lasted for several days and attracted buyers and sellers from far afield, often overseas. There was no uniform template. But in this emerging market economy, a fixed location at the centre of an active web of sellers and buyers could only become more established. The regular periodicity of markets hastened their evolution from novelty to habit. They were economically inclusive, working both for budding local entrepreneurs and for itinerant traders who plied their wares from place to place. Competition from neighbouring markets kept prices in a state of continuous flux and increased the motive for regular attendance. Markets were convenient, predictable and commercially addictive. They were also an event, a regular assembly of movers and makers who conveyed and exchanged information. Markets were hotspots on a worldwide web of innovations and ideas.

Physically, markets were messy, idiosyncratic and a far cry from the formal architectures brought to Britain by Rome over a thousand years earlier. And yet their effect on urban topography was infinitely more durable; markets were an idea, a belief system, and their places of congregation assumed a social and economic value far beyond their physical presence. A new market was a non-place, a space. It had no dedicated buildings or permanent structures. It was a topographic void occupied by a transient assembly. Remarkably rapidly, these voids became the community core, the designated place for regular interchange, whether it was commercial, political, cultural or social. Stalls and booths were the gathering ground for all kinds: full-time traders, freelance hucksters, peasant producers, retailers, ale-wives and the swarms of consumers who contributed to the noisiest, most dynamic location in town. It was this absence of physical presence that lay behind the importance of market crosses. These could vary from simple timber crosses to spacious, elegantly carved stone shelters. The market cross was a multiform symbol, a physical manifestation of the town's legal right to hold a market and of the duties required of local officials to regulate fair trading and to maintain order. The market cross was a meeting point and a monument. In Bishop's Lynn, 'Crosmarket near the Church of St Margaret' was a place-name known far beyond the town fosse. The marketplace was the most significant addition to settlement topography since the days of the early *burhs*.

The marketplace came to direct the structure of settlements. The form of these emerging market hubs varied widely, and as ever, each was a composite. There was, however, a common type in which two ribbons of tenement plots developed each side of a single street, which widened midway to accommodate a marketplace. Typically, the narrow-frontage plots were three times longer than they were wide and they were often bounded at the rear by a narrow access lane or footway which also formed the perimeter of the borough. Variations included sites on crossroads or river-crossings, where a town might develop on two axes, and splayed patterns where triangular-shaped marketplaces were laid out with the walls of a castle or monastery along the 'base' of the triangle and burgage plots stretching back from the other two sides. At New Woodstock, the borough grew around a triangular marketplace with one apex at the gate of the royal park. Unfortunately for Woodstock's modern residents, the base of the triangle proved such a popular through-route between London and the Midlands that the A44 now carries 15,000 motor vehicles a day through the twelfth-century town (parish population 3,100), over 500 of them HGVs. In Sussex, the town of Battle developed around a triangular marketplace which took its cue from the abbey. Unusual, fan-shaped patterns of burgage-plots could develop where market-places were semi-circular in shape, a characteristic of trading sites that took their boundaries from the wall of a castle bailey. Richmond is a good example.

The settlements most resistant to being arrayed around a marketplace were new towns developed by the laying out of a planned grid, a model that went back to Alfred's *burhs*, and one that was dependent upon royal will or powerful lords. The street grids at Bury St Edmunds and Ludlow would not have been conceived without the authority and wealth of Abbot Baldwin and the conquering de Lacys. Very few of the early town grids were genuinely orthogonal, the principal axes usually being distorted to fit with features like rivers and existing roads, properties and field boundaries. When a new borough was laid out in 1196 on the west bank of the Avon ten miles from Warwick, the town planner (the bishop of Worcester) was obliged to follow the river's terraces. To this day, the street grid in the central part of Stratford-upon-Avon is a distorted parallelogram. Stratford's plots also had to account for the existing curves of the ridge-and-furrow open fields on which the town was founded. So rapidly were towns growing during this period, that neatly planned kernels quickly became swamped by later asymmetries. Approach roads to many of the larger towns accumulated ribbon suburbs and significant ports such as Newcastle saw a surge of investment in waterfront facilities like quays.

In the corner of the Wash, the port of Lynn was running out of space. Clinging to the rim of Britain's most extensive flatlands, Lynn had taken its name from the old Celtic word for pool, *linn*, a reference to the estuarine lake that had formed where the Ouse flowed into the south-eastern corner of the Wash. Long before Normans crossed the Fens, salt-making had underpinned a string of Ouse-side settlements, among them the North and South *Lena*, and on the west side of the Ouse, West *Lena*. To make the salt, sand from the shores of the *linn* was washed through with seawater and the resultant brine then boiled over peat fires until the salt crystallised. So much waste sand accumulated on the shoreline that land could be reclaimed for building. The Domesday survey recorded 180 or so salt pans close to the *linn*.

Standing back from *Lena* North and South was the estate of *Gaiwde*, now Gaywood, a nice little earner listed by Domesday as having woodland for 160 pigs, 40 acres of meadow, 32 acres of other land, a mill and 21 salt pans. By 1100, the Lynns and Gaywood belonged to Herbert de Losinga, who was responsible for laying out a new town – 'Bishop's Lynn' – on the Ouse waterfront. By the time the third bishop of Norwich was ordained in 1146, Lynn's original planned grid, with its church, Saturday market and waterfront, was too constrained by its peninsular location, so William de Turbe granted a tract of newly re-claimed land for development immediately to the north of the original borough. The bishop's 'new land' was laid out with a grid of streets and provided with a new church and a massive Tuesday market. Even so, Lynn's economy ran ahead of the town: by the 1280s, traders were operating from the Cross-market by St Margaret's on every day of the week except Tuesdays. Both marketplaces were running at full stretch. At around this time, Bishop's Lynn was extended southward, onto more reclaimed land between the dyke known as Millfleet and the River Nar. These days, King's Lynn is under siege from the sea, its quaysides lined with flood defences. It was never an ideal spot to develop a town, but its commercial success created one of England's greatest collections of early urban architecture.

The urban boom of the late twelfth century was accompanied by a revolution in building technology. There were several contributory factors. New tools such as the saw and the chisel became available, making it possible for carpenters to work to closer tolerances than were possible with the adze. At about the same time, the development of the pegged mortice joint made it possible to construct multiple-storeyed, timber-framed buildings. The mortice joint had been used in buildings for centuries, but the pegged mortice locked the two timbers together, preventing the tenon part of the joint from pull-ing out of its socket. If the peg was the correct diameter relative to its length, these joints were formidably strong. Entire frames could be jointed and pegged lying flat on the ground, and then raised into position. Alongside the development of full timber framing came the rapid adoption of solid foundations. Instead of earth-fast timbers set directly against – or into – the ground, walls were rested on a ground-plan of dwarf stone walls or padstones. Solid wall-bases increased the structural integrity of the building and stemmed the perennial problem of rising damp. Wood rot was reduced and ground floors became drier. Properly founded, a timber-framed building would last for centuries. Fire risk in these new high-rises was reduced by roofing them with ceramic tiles. In a remarkably short period of time, a small number of technological advances made a big difference to urban topographies; multi-storey buildings climbed upward to unclaimed light.

At the cutting edge of the building revolution was the largest, wealthiest city in Britain; a city far more than twice the size of the next-largest town; a city whose population may already have reached 50,000. Britain had never known a city of this size or appetite:

> Among the noble cities of the world that are celebrated by Fame, the City of London, seat of the Monarchy of England, is one that spreads its fame wider, sends its wealth and wares further, and lifts its head higher than all others. It is blest in the wholesomeness of its air, in its reverence for the Christian faith, in the strength of its bulwarks, the nature of its situation, the honour of its citizens, and the chastity of its matrons. It is likewise most merry in its sports and fruitful of noble men.

That was William Fitz Stephen, Londoner, clerk and biographer of Thomas Becket, writing in the late twelfth century. The view from the provinces was rather different. He's hardly representative, but here is a monk in Winchester exercising his imagination in a passage offering advice to young Christians planning to cross the Channel from France:

> When you reach England, if you come to London, pass through it quickly, for I do not at all like that city. All sorts of men crowd

together there from every country under the heavens. Each race brings its own vices and its own customs to the city. No one lives in it without falling into some sort of crime. Every quarter of it abounds in grave obscenities . . . Behold, I prophesy to you: whatever evil or malicious thing that can be found in any part of the world, you will find in that one city. Do not associate with the crowds of pimps; do not mingle with the throngs in eating-houses; avoid dice and gambling, the theatre and the tavern. You will meet with more braggarts there than in all France; the number of parasites is infinite. Actors, jesters, smooth-skinned lads, Moors, flatterers, pretty boys, effeminates, pederasts, singing and dancing girls, quacks, belly-dancers, sorceresses, extortioners, night-wanderers, magicians, mimes, beggars, buffoons: all this tribe fill all the houses. Therefore, if you do not want to dwell with evildoers, do not live in London.

This exotic urban tribe occupied a city that appeared to know no bounds. London had been expanding rapidly for over two centuries. Streets were extending, new houses were being built and riverside quays restored. Communities had multiplied so prolifically that Fitz Stephen was able to claim that the city had thirteen Conventual churches and 126 lesser parish churches. The main part of the city was still described by the old Roman walls and the river, although the Thames wall was long gone, undermined by tides and spates. Within this increasingly crowded zone, side-streets ran to north and south from the axial road that entered from the east through Aldgate and ran along Corn Hill to West Cheap and the focal point of the city: the formal gathering-place of the general assembly, Folkmoot, and right beside it, St Paul's Cathedral. West of the cathedral, Fleet Street left the walls through Ludgate, then dropped down the hill to a bridge over the River Fleet. Fitz Stephen did concede that London suffered from a couple of 'plagues': one was the frequency of fires, the other – a nod here to his contemporary, the Winchester monk – was 'the immoderate drinking of fools'. The boozing and burning may well have been connected.

Beyond the Roman walls stretched a dynamic hinterland of wood, pasture and field encroached by advancing development. 'On all sides, beyond the houses,' wrote Fitz Stephen in a line suggestive of ancient Rome, 'lie the gardens of the citizens that dwell in the suburbs, planted with trees, spacious and fair, adjoining one another'. In the west, housing and side-streets now stretched beyond the River Fleet all the way to the Palace of Westminster. On the far side of the river, Southwark was developing into a semi-detached trading station, linked to the city by a long, timber bridge across the Thames. To the north of the Roman walls, another suburb crept along the east bank of the Fleet to Faggeswell Brook and beyond. The north offered the greatest scope for expansion. Unconstrained by the Thames and enjoying a gentle elevation above the floodplain, this well-drained district was a picture of 'pasture lands and a pleasant space of flat meadows, intersected by running waters, which turn revolving mill-wheels with merry din'. Out here too was the Priory of St Bartholomew and a tract of level grazing that once a week (feast days excepting) became a public arena where every class of citizen could mix on the pretext of trade and amusement. The 'smooth field, both in fact and in name' (we know it as Smithfield, today) was the site of the weekly horse fair and country market, a colourful, noisy jamboree where, according to Fitz Stephen, 'Earls, Barons and Knights' mingled with other Londoners as horses were scrutinised, haggled over and raced by boy-jockeys across the hard turf. In a separate area, people from the countryside set out their wares, selling anything from farming tools to pigs, cattle, sheep and mares 'meet for ploughs, sledges and two-horsed carts'.

This belt of mixed land-use girdling the city was also the place used by Londoners for recreation. Fitz Stephen wrote of the 'excellent wells' out here, 'whose waters are sweet, wholesome and clear'. The most popular were Holywell, Clerkenwell and St Clement's Well, visited on summer evenings 'by thicker throngs and greater multitudes of students from the schools and of the young men of the City'. Fields outside the city were used on Carnival day for cock-fighting and ball-games, and during Sundays in Lent, horse-back tournaments were held using lances with their steel points removed. It was on these green spaces too that summer sports were staged, where 'youths exercised themselves in leaping, archery and wrestling, putting the stone, and throwing the thonged javelin beyond the mark, and fighting with sword and buckler'. Spectator sports on winter feast days included gladiatorial contests between bears and packs of hounds, and tusked combat between wild boars. One of the geographical legacies of the Roman occupation had been the formation of a waterlogged area abutting part of the city's northern wall. The cause of the bad drainage here dated back to the building of the Roman wall, which had blocked the natural slope draining towards the Thames. Although culverts had been built beneath the walls, these had eventually become blocked and water had backed up. Known as the 'great marsh' or the 'moor', it stretched from the Aldersgate road east to the Walbrook. When it flooded then froze, this too became an urban playground as crowds of young men came out to slide on the ice or drag their friends on 'seats of ice like millstones'. Those more adept at winter sports would lash animal shin-bones to their ankles and propel themselves across the ice at great speed, using iron-shod poles 'borne along swift as a bird in flight or a bolt shot from a mangonel'.

The suburban belt merged into 'a great forest with wooded glades and lairs of wild beasts, deer both red and fallow, wild boars and bulls'. Woodland like this was critical to the city's economic well-being and to its fire-hearths. Special privileges allowed London's citizens to hunt in the counties of Middlesex and Hertfordshire, in all of the Chiltern Hills, and in Kent as far as the River Cray. It was out here that Londoners used their merlins, falcons and dogs to 'wage warfare in the woods'. London's green belt was far bigger than the city itself.

London was an aspiration. The geography of Heaven could be found on earth, and to men like Fitz Stephen, London was the simulacrum of the City of Jerusalem, walled four-square in the Book of Revelation and transferred to the north bank of the Thames. His elevated descriptions raised the earthly city beyond the comprehension of uptight provincial monks. *Descriptio nobilissimae civitatis Londoniae* was a pitch for recognition: after establishing London as the noblest city in the kingdom, Fitz Stephen constructed its holy credentials with a tour of churches, and then buttressed its aura of security with a paean for its walls, fortifications, seven double gates and – getting just a little carried away – London's 20,000 armed horsemen and 60,000 foot-soldiers. Subsequent sections described the city's superb educational opportunities, international trading connections (all eastward, of course, and reaching as far as Arabia, Babylon, the Nile and China), sporting prowess and its lawfulness. Topping up the Christian measure, he reminded readers that London had been founded before Rome (by the Trojan, Brutus): 'I do not think there is any city deserving greater approval . . .'

A footpath seen from a railway train leads to lost geographies. It's a tangible, surviving link with the 'slow web', that network of paths, roads and waterways, bridges and causeways which used to connect Britain's multifarious settlements. From tiny capillary to primary artery, this was the system that oxygenated town and country. Parts of the system were very ancient indeed. It's likely that some of the ridgeways and valley routes of 1200 had been seasonal through-routes for Britain's pioneering foragers and hunters; many of the fording-places and connecting watersheds that had attracted traffic in 9000 BC were still doing so in the thirteenth century AD. Much of the road network surveyed by Roman engineers was still in service; some sections had been abandoned and others had been grafted with local webs which served the new towns of the urban boom. Developing at locations removed from the Roman system, the Domesday towns of *Couentreu* and *Oxeneford* – Coventry and Oxford – had both acquired the necessary connecting links. In their post-Roman, market-based reincarnation, roads ran from town to town; long-distance, rectilinear, military through-routes were of little use to trade and travel if they bypassed important markets. The routes taken by Roman roads radiating from successful towns like Leicester, Canterbury or Winchester continued to be used but many other sections of Imperial roadway – like those linked to Calleva – had little relevance to inter-town transport.

Measured against the speed of urban and economic growth, the road system functioned well. Britain had no Alpine passes exposed to months of snow-block, nor desolate tracts where water and shelter were scarce. The most numerous obstacles were Britain's abundant rivers and floodplains, which had given rise to an extraordinary variety of bridges and causeways, the design and condition of each being a response to local engineering challenges, and to the source – or not – of funding. Because of its relatively low cost, timber was still the favoured construction material for bridges, but balanced against the savings in construction were high costs of maintenance. The bridge over the Exe carried the main route through the West Country and yet by the twelfth century, it had been reduced to rickety 'clappers of tymbre' suitable only for pedestrians. Horses and wheeled vehicles used the adjacent ford. It took an Exeter worthy, Nicholas Gervase, to tackle the problem. Gervase owned warehouses and mills along the waterfront and knew first-hand how dangerous and precarious the crossing could be when the river was in spate. The way forward – just as it had been for the Romanising church-builders of earlier centuries – was to turn to quarries rather than woodlands. Work on a new stone bridge commenced in the 1190s. They built out from the Exeter side, each successive arch being either pointed and vaulted with chamfered ribs, or semi-circular on rectangular ribs. By the time the project reached deep water, the stone piers were resting on foundations of rubble and river gravel held in place by hundreds of oak stakes. Spanning the Exe in stone may have taken as long as twenty years, for the job was completed by Gervase's son, the town's mayor. It was a celebrated feat of engineering. The seventeen (possibly eighteen) arches carried a roadway over four metres wide, paved with flagstones. Roadside gutters carried rainwater through drain spouts into the river. The total distance between the abutments on each riverbank was a staggering 178 metres. (Today, the surviving eight and a half arches are marooned on an island around which swirls the modern obstacle to pedestrian mobility: motor traffic.)

The scale of engineering solutions was of course related to the size of the obstacles, and on the road linking London and Edinburgh one of the most formidable obstructions was the great river that divided the lowlands of southern Britain from the uplands of the north. After the Severn and Thames, the Trent was the longest river in Britain, but in terms of water-flow, it was the second biggest. With its enormous, 4,000-square-mile catchment basin, the Trent was capable of delivering so much water during episodes of high rainfall or snow-melt, that it would burst its banks and adopt new courses ('Trent' is a Celtic river-name that probably alludes to 'the trespasser', a river prone to flooding). One of the crossing-places carried the King's highway from Coventry to Derby across the river at a place where the floodplain narrowed; an ancient spot marked on an overlooking hill by a cluster of burial mounds. It was during the infrastructure era of the late fourteenth to early fifteenth century that this critical crossing was upgraded, the timber bridge being replaced by stone and a great causeway being built across the floodplain to connect the bridge to higher land. The combined, elevated structure ran for some three-quarters of a mile, arches in the causeway allowing flood-water to pass beneath. Walking or riding along the causeway after rains had flooded the valley must have been one of the most extraordinary aquatic experiences in England. Despite its Grade 1 listing – and reputation as England's 'longest inland bridge' and 'longest stone bridge' – the medieval causeway at Swarkestone is being battered to death by motor traffic using it as the main route between south Derbyshire and Derby.

Bridges – and causeways – were only as good as their caretakers. And the waters they spanned were a formidable adversary, continuously licking at foundations and piers. Some structures were better provided than others. In Exeter, the 'Sworn Officers' of the town included two Wardens of Exe-Bridge, whose duties included regular inspections of the structure and its adjacent banks (as well as the collecting of all 'rents, revenues, issues and profits' accruing to the crossing and the cleaning of 'dunghills' and 'heaps of dirt' from the roadway). The real challenge, however, was less in identifying a weakness or failure, than in finding the means to effect a repair. An entry in the Domesday Book refers to one man from every hide in Cheshire being required to contribute to the repair of the bridge and walls of Chester; any lord failing to provide the requisite hands would be fined a whopping £2 fine per man. Maintenance of the bridge over the Medway at Rochester was the responsibility of designated landholders, each of whom was charged with looking after one arch. Very often, bridges were built and maintained by individual benefactors, men like Gervase at Exeter, who had the social clout to raise enormous sums of money. Bequests could build a bridge, and so could the sale of indulgences. Many a bridge acquired a chapel; a place of devotion where the crossing of water could be accompanied by prayers – including to the benefactor who funded the span. Bridges were symbols of civic devotion; expensive to build and to maintain, and critical to peace-keeping (and war-making). They conferred a high-value sense of place. In many towns, they were the founding monument; the artificial structure around which the settlement fledged.

Complaints about the state of roads were no doubt a popular topic among travellers, and conditions certainly varied widely enough for practised grumblers to be well supplied with sloughs, floods and potholes. The system coped, and when it couldn't, it adapted. The principal means of adaptation was 'easement', the privilege held by road-users to take an alternative route across another's land if the primary route was unusable. Such rights of way led in places to the creation of braided roads and tracks as generations of successive travellers deviated in search of easier ground. Sections of road through boggy or steep terrain (where erosion could be an issue) could resemble long strands of frayed rope. Roads of this era were not the well-defined, surfaced-and-ditched thoroughfares surveyed and laid by the Romans of the second century, but earth-floored notions of passage where reliability was maintained through width and easement.

Among the lesser impediments to passage were encroaching road-side structures, seasonal sloughs and rutting so severe that carts and carriages struggled to make headway. Occasionally things got so bad that the government intervened: in 1285, King Edward I ordered the Prior of *Dunestaple* – now Dunstable – to sort out roads that had become 'so broken up and deep by the frequent passage of carts that dangerous injuries continuously threaten'. At the time, Dunstable was one of the most important road junctions in the south-east, the place where the Roman road from London to Anglesey (and therefore Ireland) crossed Icknield Way (indeed, the most prolific map-maker of the thirteenth century, Matthew Paris, got so carried away with the junction's significance that in one of his works he erroneously sketched the main Roman roads of Icknield Street, Fosse Way, Ermine Street and Watling Street all converging on Dunstable). The cross-roads was a royal stopover and there had been a royal residence here since the early 1100s, when the priory had been founded. It became a popular venue for tournaments. The life-threatening, chalky morass at Dunstable crossroads (now the tarred, fourteen-lane, light-controlled intersection of the A5 and A505) was both inconvenient and a threat to state security.

The greatest road-safety issue was the risk of robbery, or worse. Landscapes of danger had been polarising ever since the appearance in Britain of permanent settlements; once communities committed to remaining in a fixed place, the spaces 'beyond' had the capacity to become fearful voids. By the time landscapes were punctuated by towns, with their concentrations of anxious wealth, the intervening countryside presented itself to many as an existential threat. And crime – being a function of population density – increased as more people had reason to take to the roads. Back in the early 1100s, Henry I had decreed that main roads should be sufficiently wide for two wagons to pass or for sixteen knights to ride abreast, an attempt to maintain free physical passage but also a precaution against ambush; open sight-lines also made it less likely that travellers could be sur-prised. Over a century later, the problem had become more acute. An ambitious statute given at Winchester in October 1285 commanded that:

> highways leading from one market town to another shall be en-larged, whereas bushes, woods and dykes be, so that there be neither dyke nor bush whereby a man may lurk to do hurt, within 200 feet of the one side, and 200 feet on the other side of the way.

Oaks and other large trees were exempt from roadside felling provided that underbrush was cleared beneath them, and where the boundaries of parks were closer than 200 feet to the road, those boundaries had to be realigned or enclosed by a wall, ditch or hedge sufficiently im-passable 'that malefactors cannot get over or get back over to do evil'.

This diverse, far-reaching web of communications was recorded on the landscape in thousands of miles of muddy ruts and polished flag-stones. In abbreviated form, it also appeared on the earliest-known attempt to map connections between a number of Britain's 'centres'. The cartographer was not recorded, and neither was the map's date, but it recorded geopolitical interests current from the reign of Henry III through to Edward III. The period during which this spatial portrait was committed to a 116 × 55 cm sheet of vellum (or rather, two pieces sewn together) can be narrowed to between 1355, when the town walls it marks around Coventry began to be built, and 1366, when the island of Sheppey was renamed 'Queenborough', in honour of Edward III's wife, Queen Philippa. Compiled from hearsay, myth and empirical evidence, the map must have been a source of wonder. Along with 600 towns and villages were 2,940 miles of red lines, some of them appended with numerical values which appear to accord with old French miles. Six of these major 'trunk' routes radiated from a single point: London. Rivers dominated the vellum, great, fat veins that rose inland at lakes, feeding town and monastery as they made their way down to the encompassing sea. It was as if the cartographer began by plotting waterways and then attached settlements to their banks. In two fenland locations, circular rivers like moats depicted the Isle of Ely and the Isle of Axholme. Several other features sparkled with singularity. Cutting across the island, a line of red crenellations marked the *murus pictorum*, the 'Picts' Wall'; an island Droitwich was labelled in red *Hic fit sal* – 'Salt is made here'; right in the centre of Britain a navel-like double circle marked *Puteus Pek*, the Peak Cavern that Henry of Huntingdon had included in his twelfth-century 'wonders'. The only suggestion of wildlife was the depiction of a deer near Loch Tay ('excellent hunting here') and a wolf in Sutherland ('wolves live here'). In the realm of human geography, the places that shone like beacons were the towns, forty of them crowned with walls. London was granted the most lavish attention, its city symbol featuring a portcullis in silver leaf, crenellated towers, commanding windows and its name in gold lettering. Whoever put this map together knew well that the trick with cartography was in choosing what to leave out. The most pointed omission was any sense that Scotland and Wales were different to England. This was a Britain portrayed as a single monarchy with no internal borders; a Britain bound together by the red, arterial lines of its land-routes. The reality on the ground was far muddier.

By 1200, the town had become re-established as the most dynamic, influential landform on the British landscape. In Wales, there had been no towns at all when the Normans rode across the border, but between 1071 and 1310, seventy-seven towns were planted, with a surge occurring when Edward I contained Snowdonia within a belt of castles and new boroughs following his strike on Gwynedd. By 1300, there were over a hundred urban-type settlements in Wales, although perhaps only sixty passed the threshold of population, economy and administrative functions that would call them a town. Many were little more than a huddle of houses around an ancient church, or a small market hub for the surrounding countryside. Towards the end of the thirteenth century, there were signs of a changing tide. As the old military towns of kings and Marcher lords had grown and acquired economic functions, they earned associations with their local neighbourhoods and lost some of their alien identity. Places like Welshpool, founded by a Welsh lord, skipped the traditional military traumas and established themselves as commercial centres from the start.

In the north of Britain, kings and trade continued to stimulate urban growth. By the end of the thirteenth century, the original Scotti heartland between the firths of Moray and Forth had expanded under the rule of successive kings to create a greater Scotland which stretched from the far north to the Solway, including the Western Isles. The royal sheriffdoms that were formed to manage this vast and physically disparate territory underpinned the economies of emerging *burghs*. Under a royal charter of 1327, for example, all wool and skins within the sheriffdom of Forfar had to be sold to the burgesses of Dundee and all foreign merchants wishing to trade within the sheriffdom had to do so with Dundee merchants. The *burgh* had the shire on a tether. Some *burghs* had more productive hinterlands than others. Dundee controlled an agricultural catchment that included rich monasteries. Once Berwick was lost to the English, Edinburgh's hinterland reached the whole way from the Forth to the Tweed and beyond. Back in the first half of the thirteenth century, Aberdeen had been granted trading rights over a vast slice of eastern Scotland between Forfar and the Moray Firth. *Burghs* endured because they were propped up by charters which made each a minor, monopolistic fiefdom. And with the Crown skimming customs revenue on leading exports like hides, wool and woolfells (sheepskins), towns were increasingly important tools of state security. A significant devolution of power to towns can be traced back to at least 1319 when Aberdeen was granted the right to pay a fixed annual sum to the Crown with regard to *burgh* rents and other proceeds. Although the 'great customs' on exports were excluded from this arrangement, the partial detachment of town from Crown heralded a new era of urban autonomy. Other *burghs* in the kingdom of Scotland were soon the recipients of these royal grants in feu-ferme. It was within towns, too, that the clink of mints could be heard, the principal *burghs* being early centres of coinage. Between

1250 and 1251, the King of the Scots authorised no fewer than sixteen new mints in *burghs* (no better illustration is needed of the relative safety of *burghs* than the fate of this king, for Alaxandair mac Alaxandair was killed when he rode his horse over a cliff in the dark while trying to reach Kinghorn in Fife from Edinburgh). Economically, the most dominant *burghs* to emerge by 1300 were Edinburgh, Dundee and Aberdeen, with Perth a trailing fourth.

In England, the urban surge can be narrowed to the seventy years between 1180 and 1250 when the rural spaces between established towns became infilled with some 500 new boroughs and 2,500 new markets. By 1300, there were more than 120 'planted' towns in England. Adding further momentum to the cogs of market economics was the widespread popularity of trading fairs; an unprecedented 920 being licensed between 1199 and 1272. Some markets and fairs flopped, of course, and some of the licences were renewals rather than new events, but the net growth was enough to consolidate a wide-reaching economic landscape. In the urban constellation, small towns and country towns were dominated by the bright stars: York and Norwich, Lincoln, Bristol, Northampton, Canterbury, Dunwich, Exeter and Winchester, and the brightest of them all, London. One of the side-effects of phenomenal urban growth was the rebranding of suburbs. As towns grew ever bigger and spilled beyond their historic walls, the outer fringes of some had become refuges for urban subcultures, marginal zones beyond the control of civic authorities, a threat eventually addressed by the Second Statute of Winchester (1285), which commanded that:

> . . . in great towns which are walled the gates shall be closed from sunset to sunrise; and that no man shall be lodged in the suburbs, or in the outskirts of the town, except in the daytime – not even in the daytime unless his host will be responsible for him.

Town walls had graduated from being a last line of defence against marauding Vikings to being the first line of defence against the town's marginalised citizens. The statute also directed town bailiffs to inspect the suburbs every week or fortnight, and that anyone harbouring or lodging 'persons suspected of being in any respect violators of the peace' should be brought to justice.

London was a city on a different diet. By 1300, the city was five times wealthier than its nearest competitor among English towns and its population had grown to about 80,000. Decade by decade, the capital of England was leaving the rest of Britain behind. In the century up to 1300, riverside quays had been rebuilt a hundred metres or more out into the river so that ships with greater draught could berth; deeper water also meant that shipowners were less affected by tides. The land between the old shoreline and the quays was now occupied by a dense grid of warehouses, wine cellars, dwellings and workshops divided like the teeth of a comb by narrow lanes and alleys. The most hectic section of waterfront could be seen at the mouth of the Walbrook, just upstream of the bridge. The building innovations that had started to appear a hundred years earlier had now transformed the city. Timber-framed high-rises of two or three storeys were now commonplace, with cellars beneath, often of stone.

Feasting on the fruits that came and went with the Thames and with the roads radiating out from its walls, London swelled. Not only was the city disproportionately larger and wealthier than other towns in Britain, but it was also isolated from them. You had to ride out from London for sixty miles or so before finding a decent-sized town. Britain's second city, Norwich, had a population of 30,000 at the most. London was already a nation apart, connected to an urban network of production and exchange which reached the heart and peripheries of the adjacent continent, where it was creeping into the big league of cities. Though smaller than Paris, London was by now comparable in size to Ghent, Cologne and several Italian cities. In western Christendom, it was perhaps the fifth or sixth largest city. London had become part of a globalised web.

And so it was that, a thousand years or so after the urban boom and bust of Roman Britain, the island could be said to have been rebuilt to a state of 'peak town'. It was a 'peak' because – not for the first time – exogenous factors were about to put a brake on development. Population had been growing at an unprecedented rate and by the early 1300s, the number of people in Britain had probably doubled since the Norman invasion. England now accommodated nearly 6 million. Scotland had a population of between 500,000 and 1 million. Wales was probably at around 250,000. In Europe, the population had doubled between 1000 and 1300, from around 36 million to 79 million. Much of this growth had occurred since the end of the twelfth century. The vortices of economic energy surrounding towns had produced a dramatic tilt in the rural–urban balance. Maybe as much one-fifth of Britain's population now lived in towns, with half or more of these townspeople clustered in something like 700 towns of 2,000 people or less. But these towns were far from evenly spread. Virtually all of them were in England. In Wales, there were probably only three towns with populations over 2,000, and in Scotland just four.

The four-century ascendancy of *burh* and *burgh* saw field and pasture submit to the centripetal laws of market economics. Unlocked by writ and charter, agricultural landscapes and natural resources became the feed that fattened towns. The island's population was expanding faster than ever before, and so was the productivity of the land. Vast areas of woodland and marginal land were converted into fields. By 1300, the amount of land under cultivation had nearly doubled since the 1086 Domesday Survey, from 2.4 million hectares to over 4 million. The effect on rural settlement patterns was most striking in England, where three differing zones had developed: a central 'province', of mainly nucleated villages; a northern and western province, where settlement was mainly dispersed, and a south-eastern province, where there were both high-density dispersion and scattered nucleations. There were plenty of regional and local anomalies, of course, but the broad generalisation held true. The real point about rural settlements – whether nucleated or dispersed – was that there were so many of them: in England alone, there were between 10,000 and 12,000 large hamlets and villages, and between 3,000 and 4,000 *vills*, communities comprised of dispersed settlements.

But intensive land-use could kick back: decimation of woodland reduced a natural resource that was essential for fuel, building and toolmaking, while the loss of meadow and pasture caused shortages of livestock feed. In the drive to increase productivity, lords pushed the rural poor to the edge of revolt. In Wales, where marginal land had been put to the plough at the expense of stock-raising, population growth may actually have peaked by around 1300. With reduced livestock, there was less manure available to fertilise arable soils, and yields slumped. Outbreaks of disease among livestock caused episodic catastrophes. In some of England's open-field districts, pressure on land was beginning to cause a process of consolidation. Strips that had once been individually controlled began to change hands and accumulate with one individual. Blocks of land began to be enclosed and removed from communal use.

By 1300, country and town were vastly more populated and exploited than at any time in the island's past. The solar maximum and quiescent volcanoes that had basked Europe in a kindly climate for two or three centuries had provided ideal conditions for development. Then the seasons changed.

SEVENTEEN

Utopia

1300–1520

We knew it as 'the deserted village' and we never passed it without pausing at the fence beside the muddy track over the down. There wasn't a lot to see, really, but depending on the time of day and angle of sunlight, it was possible to sketch the edges of house platforms and the troughs of holloways that had once squelched to the footfall of weary field-workers. On the far side of the site (and you had to walk on along the track and double-back along the lane to see it) you could pick out the rectangular enclosure that used to wrap the village church. Usually, we'd find cattle grazing on the village. It was good grass for pasture.

Deserted villages have a poignancy that has fled many monuments and yet they're seldom identified by visitor-boards. The Ordnance Survey doesn't forget, however. The rumples of old walls and ways are usually pronounced enough to be picked up by large-scape maps. Not only is deserted Abbotstone marked on the 1:25,000 map of 'Winchester (North) & New Alresford', but its name is printed in the same font size as neighbouring villages of gentrified cottages. Abbotstone was assessed in the Domesday Book as having nine hides of land, with sufficient arable for five ploughs. In those days, the village land sloped from the down to the stream, where there was a mill and five acres of meadow. Abbotstone had every reason to flourish; it straddled one of the main routes through Hampshire and among other traffic its downland track carried pilgrims between Winchester and Canterbury. But something happened to Abbotstone in the early fourteenth century. By the late 1500s, the church was 'decayed and utterly collapsed' and Abbotstone had been consolidated with the neighbouring parish of Itchen Stoke. There are at least another 2,000 English villages like Abbotstone.

Let's start with climate. In terms of attribution, it's always the cloud beyond the horizon; the earth system that governs all and yet the force of change too often masked by more convenient causations: swerves of economics or religion, changes of ruler, epidemics, wars; the mood swings of humanity that traditionally engage historiographers. The fact is, that at around the time that Britain achieved both 'peak town' and 'peak population', the atmospheric forces that governed the planet's seasons were shifting in amplitude and location. Three times since Britain had lost its mantle of ice at the start of this book, solar minima had caused a wintertime high-pressure system to become clamped over central and northern Asia, disturbing climate patterns in the northern hemisphere. Each time this happened, westerly winds were interrupted, mid-latitudes were inflicted with intense droughts and northern latitudes were subjected to centuries-long periods of cold and storms. But climate doesn't change in smooth curves, and within the general trend of cooling there were variations. The two-century solar maximum that had brought warm climates to Europe from 1075 until 1275 came to an end when solar input dropped and then remained low until around 1370, a reduction that happened to overlap with a surge in volcanic eruptions from around 1100 until around 1350. The blast from one of these eruptions – likely to have occurred on Lombok Island, Indonesia, between May and October 1257 – reached an altitude of 43 kilometres and released more volcanic sulphur than any blast in the previous 6,000 years. The ways in which this new solar minimum interacted with the volcanic eruptions and with other earth systems such as atmospheric and oceanic circulations are all but impossible to disentangle with any certainty, but their effect on the ground was recorded by annalists. During the winter of 1309–10:

> such masses of encrusted ice were on the Thames that men took their way thereon from Queenhithe in Southwark, and from Westminster, into London: and it lasted so long that the people indulged in dancing in the midst of it near a certain fire made on the same, and hunted a hare with dogs in the middle of the Thames: London Bridge was in great peril and permanently damaged. And the bridge at Rochester and the other bridges standing in the current of the waters were wholly broken down.

Seasonal aberrations in the weather were always catastrophic. Those without wealth or power as a safeguard had limited resistance to prolonged droughts, rains and freeze-ups. The annals that recorded japes on the frozen Thames and the destruction (presumably by ice floes) of bridges, also related that 'poor people were oppressed by the severe frost, and bread wrapped in straw or other covering was frozen and could not be eaten unless it was warmed.' Cold was a killer. A few years after the freeze-up, the rains came. What followed was a catastrophe far beyond any that could be wreaked by kings, queens or churchmen. It was worse by far than any war the planet had hosted. So much water fell on the continent between 1315 and 1322 that crops failed for year after year and famine spread from land to land, killing perhaps one-tenth of northern Europe's population. England may have lost as many as one million as the population slumped from close to 6 million to 5 million. Cattle plague spread from the west, killing half of Europe's cattle. After that, through the 1320s and '30s, the weather problem became one of unpredictability: dry and warm summers interrupted by strong winds and storms. Then, in 1348–9, the 'universal pestilence' arrived on British shores.

Contemporary chronicles pointed the finger at Calais, or 'the east'. Landfall was recorded at a Dorset seaport. Characterised by rats, boils, black pustules, and the death of a 'great multitude', the pestilence raced through England and Wales and north beyond the Roman Wall where – according to the priest-historian John of Fordun – 'the mortality of men raged amazingly throughout the whole realm of Scotland.' For many at the time, the impression was one of apocalyptic devastation. A survivor in Rochester wrote that 'more than a third of the land throughout the whole kingdom remained uncultivated.' A clerk in Swinbrook, Oxfordshire, claimed that the plague 'completely emptied many rural settlements of human beings.' Deaths in areas with more dispersed, rural settlements were slightly less than in towns. Trade stuttered as the citizens of uninfected towns tried to protect themselves by denying entry to those from diseased districts. Fields were appropriated for mass burials. In London, a plot called No Man's Land, beyond the old city walls near Smooth Field, was set aside for plague burials and when this proved insufficient, a far larger plot was bought by Sir Walter Manny. According to the antiquary John Stow, some 50,000 were buried there during 1349. In other towns too – Worcester, Newark, Ripon, Beverley and York were just a few – walled grounds were consecrated for burial.

It was the most costly human tragedy ever to have befallen Europe. In three horrific years, the 'Black Death' killed around 25 million people – one-third of all Europeans. The death rate was comparable for Britain, where the population was also cut by a third. In Wales, around a quarter of the population died. Then, in 1361, the pestilence returned to Britain. With cruel timing, a damaging drought that year had led to a shortage of fruit and hay. In the context of the times, it was a small event, but a great storm in 1362 took the roofs of houses and churches, tore down bell-towers and uprooted trees. Britain reeled. Then the plague came for the third time, in 1369. And a fourth time between 1374 and 1379. Thomas Walsingham recorded in his *Historia Anglicana* that 'villages and towns, which had once been packed with warlike, provident and wealthy men, and with settlers, were emptied of their inhabitants and left desolate and abandoned.' Plague returned between 1390 and '93. From a peak population of 6 million in 1300, England had shrunk to around 3.4 million.

And that wasn't the end of it. A century after years of rains and storms had presaged a change in Britain's climate, the multi-century solar minimum exerted a second pulse of cold that lasted from 1400 to 1550, possibly accompanied by a slowdown in the thermohaline circulation, the great oceanic conveyor that distributed the energy absorbed by the planet from solar heating and brought the warm waters of the Gulf Stream to western Britain. It was a shut-down of the conveyor that had created the Britain of ice and tundra in Chapter 1; even a slowdown could cause far-reaching impacts to coastal ecologies and fisheries, and seasonal changes of temperature and rainfall. Why Britain's population continued to fall through the 1400s, while continental Europe's began to recover, is impossible to say with certainty, but the main levers on mortality were exposure to a deteriorating climate (through food shortages, inadequate housing and so on), recurrence of plague and the interconnected complex of economic and social knock-ons. Summers on Europe's largest island seem to have been generally warm, but winters were cold and stormy. Epidemics continued to strike, with bouts of pestilence from 1405–7, from 1433–9 and from the 1450s until the 1470s. By the 1440s, England's population may have slumped to a mere 2.4 million, remaining at this base level till the end of the century. The fourteenth and fifteenth centuries represented the single biggest depopulation event the island had known. The numbers of humans in Britain had more than halved. Such a massive reduction in the species most responsible for landscape modification had a dramatic effect on settlement density and land-use.

In Oxfordshire, Tusmore was already small and vulnerable before the pestilence struck. It occupied an unenviable, exposed site 400 feet above sea level on the band of oolitic limestone that ran like a broken spine from the Channel coast, diagonally through south-central England via the Cotswolds, to the Humber. Although the settlement had no river, there were streams and a small lake (Tusmore's name came down from the Old English *Thures-mere*, Thur's pool or possibly Pyrs-mere, a lake haunted by a giant or demon – apt, given the hamlet's fate.) Generations of Tusmore's peasants paid their rent, prayed in their little church and tilled the lumpy stonebrash. But it was marginal land and much of it had never been ploughed. The hamlet was one of the smallest – and poorest – manors in the Hundred of Ploughley. In 1279, Tusmore had just nineteen tenants, with its Rector the single freeholder. By 1341, it was showing symptoms of ill-health, two carucates of its arable land lying uncultivated, perhaps a result of the run of sodden years and famine. Within years, Tusmore had been wiped out, and in 1357, the lord of the manor, Sir Roger de Cotesford, obtained a licence to 'enclose his hamlet . . . and the highway . . . passing through it, the hamlet having been inhabited entirely by [his] bondmen but now void of inhabitants since their death in the pestilence.' Whether Sir Roger enclosed Tusmore's ruins so that he could create a park and grand house, or whether he wanted grazing land for sheep, is unclear. Today, Tusmore's lumps and dips are cocooned within a 3,000-acre private estate. There's no sign of the fourteenth-century church, but the hamlet is still marked on the grass in holloways and house platforms, just a stone's throw from the ornamental lake that records the old village mere, now the foreground for a neo-Georgian mansion built in 2000 as the centrepiece of Tusmore Park.

In many other hamlets and villages, successive bouts of plague diminished populations that were already shrinking. Places on poor soils or in marginal locations had their demise hastened. Cold Weston's name hints at its bleak location on a north-facing slope, 260 metres above sea level in south Shropshire. In earlier times, its enterprising residents had built a watermill and attended the small church of St Mary. In 1291, the parish had been assessed at £5 3s, but just fifty years later, in 1341, the assessment had plummeted to 4s 8d, less than one-twentieth of its thirteenth-century peak. Where there had 'once been an abundance of cattle', the assessors noted 'a waste place'. No parson could be persuaded to stay in Cold Weston, and the village's two surviving tenants were 'living by great labour and want'. Cold Weston exists today as house outlines and a deep, grassy holloway that used to be the village thoroughfare. St Mary's church is a private house.

To witness a hamlet or village emptied of people and abandoned to weeds and pasture was the cause to many of mourning and anxiety. It was a long, agonising cull of less-resilient settlements. Many factors could undermine the viability of a community, from loss of population through famine and plague, to marginal soil quality, poor location and the policies of the landlord. The determination of working-age inhabitants to stay on could sustain a settlement through crisis, but many villages 'died' because peasants took to the road. Any settlement with a productive component to its economy stood a better chance of making it. Britain was well endowed with exploitable resources, and plenty of villages survived the depopulation crash by turning to commodities like coal, lead, tin, iron, glass and tiles – products that could be traded at nearby markets. Very few settlements with productive potential were abandoned.

More than anything, the size of a settlement was the determinant of survival. Most of the half million or so houses that were deserted between the early 1300s and early 1500s were in parts of Britain where villages were thin on the ground; the most frequently abandoned settlements were isolated farmsteads and small hamlets, while those most likely to survive were the principal villages of parishes. The greatest damage to the rural economy was inflicted through the eradication of village markets. As much as 60 per cent of the markets that had been functioning before the plagues were not in existence by the early 1500s, while the survivors had to become more specialised. To fill the economic voids, town markets expanded their reach.

The culling of between 10 and 20 per cent of villages and hamlets prompted an astonishing spate of adaptation. As local economies adjusted to reduced labour-forces and new markets, the countryside became a test-bed of innovation, driven in part by a new agent of change. Towards the end of the fourteenth century, and in particular between 1380 and 1410, cash-strapped landlords released vast amounts of land for a fixed annual rent, for *ferme*, or *farm*. The 'farmers' who took on these leases proved to be dynamic innovators. As much as one-quarter of Britain's agricultural land was released, severing the direct relationship between the owners of the great estates and their land; direct management giving way to more flexible, semi-independent custodians. While many of the great estates became rent-collecting operations, a new breed of tenants experimented with practices that might turn their leased granges and demesnes into hard cash. On a couple of demesnes in the Vale of York, a farmer in Strensall – a district of heath and field – made use of his mixed holding by cultivating around 200 acres while also keeping 799 sheep, 92 horses and 198 cattle. With the market at York just seven miles along the gentle valley of the Foss, Thomas Vicars was strategically located for fattening and selling. (These days, Strensall is a 'convenient and sought-after location' in a 'thriving suburb of York', and a farmhouse without its land will cost a prospective commuter considerably more than half a million quid.) Saddled with fixed rents, farmers had to be adaptable and decisive; knowing which products earned effective returns at market, how to manage land and workers, and when to invest in livestock, employees, equipment and buildings. Many farmers were also flexible, with the freedom of movement to choose an estate that would repay their rent with profit.

In the bigger picture, there were two major modifications to the rural landscape. Firstly, in England and Wales, there was a progressive reduction of land under arable cultivation. As the countryside became denuded of its population, landowners turned from labour-intensive ploughing, planting, weeding and harvesting, to animal husbandry. Fields that had been tilled for generations were turned over to pasture. On the large estates, a long-term policy shift towards pasture could affect substantial blocks of agricultural land; on one of the manors belonging to Tavistock Abbey in Devon, the acreage of land under arable fell from 128 in 1298 to only 50 by 1420. Flocks of sheep and herds of cattle made more economic sense. A manor in Warwickshire saw its value rise threefold between 1386 and 1449 as it switched from mixed farming to pastures supporting twenty cattle, a productive rabbit warren and 1,643 sheep. To many at the time, it must have seemed that sheep were replacing people. Fresh meat provided an immediate return, and wool could be sold to clothiers or to dealers for export. In west Norfolk, the number of sheep grazing the Townshend estate astride the upper reaches of the Wensum increased from 7,000 in 1475 to 18,000 in 1516. In some areas, landowners retained arable land but worked it less intensively, leaving it under grass for years at a time and then ploughing it for cultivation and planting. Throughout the 'Great Depopulation', yields per acre fell; there were just not enough people to work the land, and without regular ploughing and assiduous weeding, harvests suffered. Norfolk had some of the best arable land in Britain, but yields fell from around 9–12 bushels per acre between 1250 and 1350, to 8–10 bushels between 1350 and 1450. The story north of the border was rather different. In Scotland, the hillier topography generally provided ample grazing close to settlements and the demand for arable land was held up by high oat prices and low prices for cattle.

The second great landscape change wrought by depopulation was the break-up of open fields. In many parts of England, these huge, communally worked fields, patterned with arable strips and worked for hundreds of years, were subdivided and enclosed as owners and tenants alike sought to convert their holdings into economically coherent blocks. Parts of England became a patchwork of small fields bounded by hedge, ditch, wall and bank. In many places, it was the peasants who led the way, creating enclosed blocks by consolidating the strips of open fields through exchange or by expanding onto marginal grazing. Peasant enclosures were the most versatile agricultural holdings of all, worked by people who knew their soil intimately, and who could balance cropping and grazing according to need. They were also the least controversial form of enclosure: locally enacted land reforms negotiated within the community and generally accommodated without dispute. In terms of acreage, they were the most common type of enclosure during the two centuries of whittling depopulation that followed the first plague.

Other forms of enclosure could be more troublesome. The gentle-men, the esquires and the knights with their medium-sized land-holdings – a couple of manors rising to as many as a dozen for a wealthy knight – were motivated agriculturalists. The proximity of their home to field and pasture gave the gentry a more intimate re-lationship with the source of their landed income than was possible for many of the big estate-owners. Gentry could keep close tabs on their acres, supervising rent collections, market prices and land-use. Repackaging their productive units through enclosure was one of their most effective economic levers. In places, entire settlements were seized and grassed over for livestock. Among those who took against enclosures was the scholar (and one of the two under-sheriffs of London), Thomas More. In *Utopia*, a fictional work delivered to the printer in 1516, More had one of his characters, Raphael, a Portuguese world traveller, telling John Morton, Archbishop of Canterbury and Chancellor of England, that there was a cause of theft 'peculiar to the English'.

'And what might that be?' asked Chancellor Morton.

'Your sheep,' replied the worldly Raphael.

More was making a political point (later he was beheaded, for reasons beyond a sheep gag) about the land-greed of noblemen and their 'great mob of useless retainers' who through their own greed were turning farmers into thieves. More's meek sheep were monsters 'so voracious and fierce that they swallow up people: they lay waste and depopulate fields, dwellings and towns.' A landscape of virtuous cultivation was being rubbed out and replaced by grassland. 'They leave nothing to the plough but enclose everything for pasture; they throw down homes and destroy communities, leaving just the church to function as a sheep fold.' And to ram home the seriousness of en-closures, More evoked the pestilence, the greatest terror known to England's political elite, accusing landowners of turning 'all habita-tions and arable fields into a wilderness . . . just so that one insatiable glutton, a grim plague to his native land, can merge fields and enclose thousands of acres within a single boundary.' At the time, there were perhaps 8 million sheep in England. They outnumbered human beings three-to-one.

So the countryside of 1500 was a very different place to its pre-plague state of 1300. A far smaller, more mobile rural population dwelled in fewer settlements. Driven by the market economy, land-holders specialised in activities that provided the greatest profit. After about 1470, many of the great estates returned in varying degrees to direct management, especially in Scotland and the north of England where lords found that they could make good money by controlling livestock. The gentry, whose principal income was usually derived from rent, looked for economies of scale by enlarging their operations as the market demanded. Farmers were well placed to respond to demand by enclosing land and diversifying into whatever the market needed. For the rural class hit hardest by the Great Depopulation, the catastrophes of the fourteenth and fifteenth centuries devastated then revitalised. The old free–unfree distinction of the peasantry was replaced by a ranking rooted in local economics, from labourers (waged workers who might hold a few acres), through husbandmen (a tenant farmer or small landowner) to yeomen (holders of 80 acres or more).

This leaner, adaptive countryside became dotted with buildings raised with new money. Neo-peasant architecture embraced a wide variety of modern styles that included traditional cruck-built houses and more elaborate 'wealden' houses with two storeys in their end-bays. A farmhouse near Mold in north Wales was built 20 feet wide and 50 feet long, with a spacious hall and separate rooms at each end, complete with lofts. Making a far greater impact on the landscape were buildings with communal functions. The century or more of serial catastrophes had seen a subtle, behind-the-scenes role-shift of family and community. As families were decimated by plague and dislocated through migration, the parish gradually assumed a more central role – an expression of social cohesiveness. Parish churches sprouted towers and grew aisles, clerestories, porches and chapels.

In general, the motives for devoting hard-won cash to ecclesiastical building projects were not – as a chapel-building cloth merchant of Long Melford put it – to 'win praise', but in order that 'the Spirit may be remembered'. Purgatory was a shitty prospect and the best way of winning remission was by encouraging fellow parishioners to pray for your soul by building extensions to the Lord's house, preferably *à la mode* perpendicular. In a further evolutionary leap, Gothic stretched for the heavens, emphasising the vertical and illuminating naves with huge windows subdivided by slender mullions into columns of light.

Towns emerged from the catastrophes leaner and more resilient. In Scotland, the Black Death, the Wars of Independence and the shrinking of customable trade contributed to a burghal shake-out that left four dominant centres. Between them, Aberdeen, Dundee, Perth and Edinburgh were paying 58 per cent of the customs returns by the 1370s. The most resilient of the Big Four, by far, was Edinburgh, paying a mighty 24 per cent, far ahead of its closest commercial rival, Aberdeen, at 15 per cent. Edinburgh's rise had been helped in part by the exposure of Berwick to English occupation; when it came to trade, Berwick's loss was Edinburgh's gain. Edinburgh grew in fits and starts along the spine of the ridge dominated by the fortress, and by 1365 a French visitor estimated that the town had 400 houses. In Jean Froissart's eyes, this urban eyrie was already 'the capital of Scotland, where the king chiefly resides when he is in that part of the country'. Important yet confined, implied Froissart, who pointed out that the Scottish capital was smaller than his own birthplace, Valenciennes. But Edinburgh carried on expanding and by the 1530s controlled 83 per cent of Scotland's wool trade and 68 per cent of hide exports. By this time the town's population may have crept beyond 10,000. Most of Scotland's towns had populations of less than 1,000, and it was only the dominant quartet of Perth, Dundee, Aberdeen and Edinburgh that topped 2,000.

Towns in Wales had a more torrid couple of centuries. Conflict with the English and plague nibbled at urban growth, and so in several cases did environmental misfortune. The rains that disastrously affected so much of Europe from 1315, had an exaggerated effect in several Welsh towns. Wrapped by the Wye and Monnow, the little town of Monmouth suffered three decades of flooding. And on the south coast of Wales, wind-blown sand and silting estuaries obstructed towns like Kenfig and Kidwelly, while a growing sand bar across the Twyi estuary prevented seagoing ships from reaching Carmarthen. In north Wales, silting on the Dee led to the decline of Overton, Flint and Holt.

In England, the embers of urban regeneration had been kept bright by the unquenchable heat of the marketplace. While village markets had been decimated, many market towns were strengthened as competition from smaller centres died off. There had been a few urban casualties. Bideford in Devon lost its regular weekly market, as did Chorley in Lancashire. In virtually every town, population dropped significantly and urban landscapes adjusted accordingly. The greatest losses occurred in the largest towns, episodically. Poll tax assessments of 1377 show that towns hit by the first plagues continued to be decimated through the fifteenth century; York's population plummeted from over 13,000 in 1377, to 6,000; Lincoln fell from 6,500 to 4,000, Boston and Beverley fell from around 5,000 to 2,000. The first effect of abrupt urban depopulation was invariably dereliction. Entire streets rotted and sagged. Burgage plots fell vacant. Parish churches fell silent. Civic authorities attempted to force landlords to rebuild their properties but it seldom worked. By 1473, the authorities in Coventry were urging landlords to pull down their derelict houses. By 1523, the town had 565 unoccupied houses. In Lincoln, the forty-six active parish churches that the town had filled before the plague shrank to nine by 1549. Norwich, the second largest city in the kingdom, had been particularly devastated by the plague. The claim that 57,474 died in a single year was an exaggeration, but there is no doubt that the great city was hollowed out by mortality. Two hundred years after plague first rampaged through East Anglia, a parliamentary Act recorded that Norwich was still afflicted by 'desolate and vacant groundes, many of theym nighe and adjoyninge to the highe stretes replenished with moche unclennes and filthe'.

With piecemeal dereliction came a gradual shrinking of urban footprint as suburbs and backstreets were abandoned in favour of central locations. A few small towns ceased to exist, either because their economic functions were too weak or through forces of nature. A port founded in 1247 to serve the Northumbrian stronghold at Bamburgh had failed by the mid 1500s, and Ravenserodd on the Humber was wiped out by 'merciless floods and tempests'. The towns that had a good plague were the specialists: the cloth producers of the Pennines like Leeds and Bradford, Halifax and Wakefield, and their East Anglian cousins Lavenham and Hadleigh. In Devon, the small towns of Totnes, Tiverton and Crediton rode out the crisis on cloth. Ports serving these towns did well, too. By the mid 1400s, more than 1,600 carts were passing through the gates of Southampton every year. Both Exeter and Ipswich contributed more to the Lay Subsidy of 1524–5 than they did before the plague. Towns adapted to survive. Birmingham became an ironmongery centre; Thaxted in Essex made cutlery; Sherburn-in-Elmet in Yorkshire went for pins. In Hertford-shire, Buntingford succeeded in swelling from a thirteenth-century hamlet to a modestly sized town of 350, largely because it fell between manors; in the absence of a single dominant lord, this enterprising little place was put on the map by an informal collective of motivated craftsmen and traders who took advantage of the settlement's location on Ermine Street. Meanwhile, just along the old Roman road, Standon and Chipping Camden went downhill.

In more resilient towns, the 'dereliction phase' was followed by a re-sumption of development, albeit on a far more selective basis than had been the case in the heady pre-plague decades. Bridges were repaired and built; guildhalls commissioned and parish churches extended or rebuilt. In some larger towns, Roman enterprise was emulated by the construction of conduits to bring clean fresh water to the heart of the community. The town authorities in Exeter instructed that a public la-trine be built over a mill-leat on the river. On the Solent, Portsmouth benefited through royal patronage when it was chosen as the best Channel location for the first, purpose-built dry dock constructed in Britain. It signalled the beginning of a more intrusive attitude to coastal landforms. For millennia, large vessels had been built and repaired in creeks or on beaches, supported by timber scaffolds which could be removed when the ship was ready to float. The dry dock was an excavated basin, lined with timber or stone, and fitted with a temporary, water-tight barrier where it opened to saltwater. Ships could be worked on with their hulls clear of water and mud. They were invaluable for warships, and expensive to build. Completed in 1495, the dry dock at Portsmouth was protected from tides by a double barrier of timber, infilled with shingle and clay. Construction was said to have cost a whopping £193, and removing the lock's 'gate' of shingle and clay to allow ships to pass in or out took twenty men twenty-four days.

No town was more resilient than London. Although the city probably lost half of its population during the demographic traumas of the fourteenth century, the breadth of its economic base meant that it suffered far less than smaller urban centres. Between 1400 and 1540, the value of goods crossing London's wharves doubled and the city's national share of imports and exports leapt from 45 per cent to 70 per cent. While towns the length and breadth of Britain stagnated or declined, London's wealthy merchants commissioned fabulous townhouses. Sponsored by private donors, the city spent nearly thirty years building a guildhall on the site of the Roman amphitheatre. Over 150 feet long and 50 feet wide, the Guildhall had space for a thousand guests and was illuminated at each end by spectacular Gothic windows. Eight years after the Guildhall was completed, a market, granary and schools were completed at Leadenhall. The prisons at Ludgate and Newgate were reconstructed, and water-piping and conduits were laid up Cornhill. The city gained its largest public 'place of easement' during this era, with the building by Mayor Richard Whittington of the Great House of Easement, an 84-seater flushed by the Walbrook.

Down on the river, the city's mercantile power was formalised by new buildings and by the stiffening in identity of the long, straight street that stretched all the way from the Tower to the Fleet. On the side of London with no town walls, Thames Street came to define the southern 'line', a key thoroughfare fringed on one side by short alleys and streets cutting down to the wharves of the river. Here were the warehouses on which the city depended, with the brewers and dyers, the Steelyard, Vintry and Queenhithe. Newcomers included a custom house just upstream of the Tower. At the eastern end of Thames Street, Blynesgate (or Byllynsgate) thrived as a general market for corn and wine, salt, pottery, coal and iron, fish and various other general commodities. Further up the slope from Thames Street, the parallel axis of Cheapside and its extensions was lined from the site of the old Roman forum to Newgate above the Fleet with the market zones of butchers, fishmongers, poultry-sellers, leather-workers, food suppliers, drapers and mercers, goldsmiths and grain merchants. The noxious tanners were outside the walls, down by the Fleet and cattle were marketed at the open ground of Smoothfield, or Smithfield.

London's momentum through two centuries of British depopulation was exemplified by the building of no fewer than forty company halls, the city's commercial pre-eminence advertised by the diversity of liveried sponsors: the Masons, Glaziers, Carpenters, Joiners, Ironmongers, Waxchandlers, Innholders, Brewers, Fishmongers, Salters, Vintners, Shearmen, Fullers, Weavers, Mercers, Drapers, Dyers, Haberdashers, Embroiderers, Skinners, Sadlers and Cordwainers, the Bowyers and Armourers, the Barbers, Goldsmiths and Girdlers. All of the Company Halls bar the Butchers and Cooks were set within the old Roman walls. Towards the end of this period, even raffish Southwark got a makeover when Henry VIII prohibited the riverside brothels; 'a continued raunge of dwelling houses' where 'ravening she-wolves catch hold of silly wretched men and plucke them into their hools'.

The river itself was modified for a new age. Four miles downstream from London Bridge, on the south bank of the loop taken by the

Thames around the tongue of marshland that would become the Isle of Dogs, a royal dockyard was founded at Woolwich in 1512, and the following year, another royal yard was founded right next door, in Deptford. As the closest naval yard to London, Deptford was in a strategically convenient location and it was here that a novel type of dock was built for warships. The wet dock was a large, watertight basin in which the water-level was maintained at the level of a high tide. Once admitted through the watertight gates, vessels inside the dock remained constantly afloat. Deptford Dockyard became the leading dockyard of the century.

The city had endured, with an embellished core, yet London was no Venice or Augsburg. In Thomas More's *Utopia* (written partly in Antwerp and printed in the university city of Louvain), London looked a very long way indeed from the author's Amaurot. In More's imaginary world, Utopians were so city-centred that they related more to their civic status than to their country. More's fictitious island had fifty-four cities set twenty-four miles apart (the distance a man can walk in a day) but only one of them was described. Amaurot, the capital, 'placed as it were at the navel of the country' on a great river called the Anyder, had more than a passing resemblance to London, an impression confirmed by marginal notes that referred to the Anyder's tidal characteristics being the 'same as the Thames', and to the great bridge that made London 'like Amaurot'. Not surprisingly, Amaurot was built to a higher standard than London. It was a planned city, with carriageways 20 feet wide, designed for the free flow of traffic and for the protection of citizens from the wind. Fresh water was conveyed across the city in earthenware ducts. The houses had three storeys and were not faced with timber but with flint, quarry-stone or brick. Cavity walls were filled with rubble and the flat roofs were sealed with 'a form of cement that costs next to nothing', which was both fire- and weather-resistant. Draughts were excluded from Amaurot's houses by glass windows or by sheets of linen treated with clear oil or gum. All the houses had 'matching terraces facing each other along the length of the street', and they all had gardens. In Utopia, each of the fifty-four cities was girdled by a green belt of agricultural land that extended outward for a minimum of twelve miles. More's *Utopia* was published in 1516 and by the end of the century it had been issued in eleven further editions and twelve translations from Latin into French, four into German, three into English and three into Italian. Amaurot was the city of dreams.

Utopia opened the door to a landscape century. Through a combination of geographical enquiry, humanism and printing, the great outdoors crossed the threshold and came indoors. Books and maps equipped a new generation of geographers with the tools to record Britain's extraordinary topographies.

EIGHTEEN

Heat Island

1520–1620

Smoke wrapped like a blanket around the roofs of Britain's largest towns. Woodsmoke and coal smoke. In London, where nearly twenty times more coal was being burned in 1500 than in 1300, sulphur dioxide levels above the city had risen by a factor of 11. In calm weather, coal-burning towns had sulphur dioxide concentrations 20–30 times greater than the outlying countryside. As early as 1512, atmospheric damage to furnishings was being noticed. The high-carbon lifestyle contributed to a specifically urban climate. Air flow was impeded by buildings, visibility was reduced and fogs were more common. Large towns were also warmer than the surrounding countryside. The burning of fossil fuels and the warmth generated by thousands of people living in close quarters, combined with the radiative characteristics of the built-up area; where vegetated surfaces and moisture-trapping soils used up much of their absorbed radiation through evapotranspiration and the release of water vapour, towns were built of non-reflective, water-resistant materials like stone, tiles and slates that were efficient at absorbing and storing heat. Temperatures in the streets of a city could be 4°C higher than the surrounding countryside. Imperceptibly, towns were warming. The halving of Britain's population during the catastrophes of the previous two centuries had led to an equilibrium of sorts and a period of relative prosperity. In terms of hands-on land, Britain was back where it had been after the Norman invasion. But at some point between 1470 and 1520, the population began to grow. Then, from the 1520s, it began to surge.

We have a witness to these rekindled decades. He read landscapes as stories. His roads were lines on pages that turned back through time. He rode and wrote until he went mad. John Leland had a royal commission to rescue bibliographic treasures from libraries which were being scattered and neglected following the suppression of the monasteries. He travelled for years, covering thousands of miles, from Northumberland to Cornwall, Kent to Caernarfonshire. But the king's librarian was also a private topographer; he looked at wood-block tomes and cracking vellum, but he saw rivers and ruins. He lost his mind but found Britain. When Leland died insane, his collected notes ran to hundreds of thousands of words. Some were jottings; others were worked descriptions. He knew he was compiling an unprecedented audit of a remarkable land, but he also knew that it could never be finished. The notes for East Anglia disappeared and politics precluded Scotland. The work that survived is the earliest detailed topographical description of England and Wales. John Leland was Britain's first fieldworker.

Leland's Britain was almost entirely green; a rumpled, rustic island sparsely dotted with tiny, intense hubs of urban activity. A Venetian nobleman visiting England in the early 1500s remarked that he found it 'very thinly inhabited' with 'scarcely any towns of importance'. In England, as much as 95 per cent of the population was rural. Wales was almost as rural as it had been in the days of Gerald. In Scotland, as much as 97.5 per cent of the population lived in settlements of less than 2,000. A recurring place-label in Leland's lexicon was 'cuntery'. It could be taken to mean a political unit or county (as in 'this province or cuntery', or 'the hole cuntery of Richemontshire'), or a geographical region ('the weste cuntery'). He also used it to describe topographical zones, for example 'wooddy cuntery'. And he used it as a label for that space outside a settlement. 'The circuite of the paroch by the cuntery adjacent' was a reference to the country – the 'countryside' – that lay within the parish and yet was not occupied by the settlement itself. In this context, 'country' was the green space beyond the houses.

Leland shuffled his countryside into broad categories. At one extreme were the 'mores', 'bogges' and 'craggi and stoni montaines'. Less repellent, and scattered across lowlands as well as highland, were the areas of 'waste' and of uncultivated 'hethe'. Forest and woodland had particular pertinence. Their depletion was both a symbol of agricultural progress and of the kingdom's diminishing resilience. Near Manchester, Leland noted that the 'lakke of woode' had caused the 'blow-shoppes' to decay. And in Huntingdonshire he learned that the whole county 'hath beene, as it is saide, forrest ground: but it is ful long sins it was deforestid' and that the deer have 'resortid to the fennes'. At the fertile – and in Leland's eyes, most attractive – end of the spectrum were the landscapes he frequently labelled as 'champayne', from the Latin *campus*, a word that had become attached to the open fields which still characterised so much of southern Britain. In Leland's lexicon, champayne was near perfect. The pleasures he enjoyed as he rode through this unfenced farmland chirruped from his pages: 'good pasture and corne al in champayne'; 'chaumpaine ground sumwhat plentiful of corne, but mostly layid to pasturage'; '10. good miles all by champayne, no wood, but excedynge good pasture and corne.' Crossing Monmouthshire, the 'contry is champain'. When he rode by farmland that was hedged, walled or fenced, it was cursorily labelled 'enclosyd grownde'.

Scattered across this tapestry were vignettes of particular beauty; the picture-perfect scenes that had quickened the pace of Gerald's quill when he gazed upon the Teifi's garden. Now and again, Leland's methodical notes bloomed with similar views. At Nether Stowey in Somerset, he found that Lord Audley's manor house was 'stonding exceding pleasauntly' amid pastureland, with separate parks for red and fallow deer, and a 'faire brooke' that provided for the domestic needs of the manor house, an image that placed the herds of rude antiquity in counterpoint to Tudor civilisation. In Herefordshire he climbed Dinmore Hill ('very stepe, highe, well woodyd') and enjoyed far-reaching views from the summit: 'a *specula* to se all the contry about'. And in Leicestershire he explored Belvoir Castle, rescued from decay by the Earl of Rutland, who had renovated the keep as a historic spectacle: 'a fair rounde tour now turnid to pleasure, as a place to walk yn, and to se al the countery aboute'. Visitors to Belvoir could enjoy the countryside views from behind a safety rail the Earl had erected around the old castle wall, and a garden now decorated the castle grounds. That this 1540s humanist should have been moved by the beauty of landscape was not so surprising. Only twenty years earlier, the word 'landscape' – *landschaft* – had been used for the first recorded time in German by Albrecht Dürer, when he referred to Joachim Patinir as '*der gut landschaft maler*' – the good landscape painter. Patinir, like Dürer, had broken with the fifteenth-century tradition of observing the natural world through the controlling frames of windows or arcades, by stepping outside and being absorbed by the countryside.

In Leland's England, ruins like Belvoir Castle were a recurring landmark. His kingdom was overlaid with an unmapped stratum which could be read in spilled masonry, mounds, banks and ditches. He was the earliest writer to have comprehensively logged these relic landforms and to have treated them as the physical echoes of a collective past. Joining him in the saddle, the reader was treated to a geographical mystery tour. Britain's narrative was written in the landscape, if you knew where to look. At Catterick, he came across ancient, squared stones which had been unearthed near the church. And ten miles further north, he was riding the slopes above the Tees when he came to several artificial mounds and numerous ditches, some flooded with water. These, he concluded, must have been the ruins of a military camp (on this occasion, it was the perimeter of a pre-Roman, Brigantine power-base). When he came to the four massive standing stones beside Dere Street, near Boroughbridge in Yorkshire, he assumed they must have been placed there by the Romans as waymarks on the long road north:

'as yn a place moste occupied yn yorneying, and so most yn sighte.'

Leland's politics may have been constrained, but his curiosity was uninhibited. On the downs near Avebury (was he perhaps on Windmill Hill?), he came across ditches, banks and barrows which he deciphered as the burial sites and camps of warriors. Some of the ruins he found were larger than most modern (Tudor) structures. Deep inside Cornwall, Launceston castle still appeared to crush the landscape with its sheer bulk, its motte 'of a terrible higth' and its massive, decayed keep surrounded by three wards which led Leland to wonder whether this wasn't the 'strongest' and 'biggist' fortification 'that ever I saw in any auncient worke in Englande'. Among these strewn quoins and felled arches there was one kind of ruin that Leland struggled to confront. It was his own monarch who had ordered the dissolution of the monasteries and there were many coy encounters with religious houses which had been 'suppressid': a 'late monasterie' outside Reading; 'a priory of blake monkes' in Wallingford, 'suppresssid by Thomas Woulsey cardinale', 'an house of Gray Freres' in Exeter, 'now a plain vacant ground'.

Towns glowed on Leland's landscapes like random embers. Their apparent rarity was due in part to the speed of travel. A traveller approaching the walls of a town in the mid-sixteenth century would be passing through a portal dividing two entirely different worlds. Compared to the countryside, towns were packed with people; they had their own specialised economies; a marketplace, tradesmen; they had complex social structures and political orders not found in villages and hamlets, and they exerted a powerful influence far beyond their physical boundary. Although he was a Londoner (and one who had studied at Cambridge, Oxford and Paris), Leland managed to approach provincial towns with a fieldworker's open mind, and it was this crud-on-the-boots attitude that allowed him to view each urban stopover as a fresh investigation. A town in Wiltshire serves as an example.

That day, Leland had been following the Fosse Way out of Cirencester. Over a thousand years after it had been surfaced by Roman engineers, the old straight road linking Lincoln and Exeter was still one of the main routes across eastern and southern Britain. For a man of Leland's temperament, it must have been monotonous riding. Ten miles from Cirencester, where the road began to dip towards a stream called Newnton Water, he left the highway and struck off to the left and rode across 'champayne grounde, fruteful of corne and grasse'. After a couple of miles the country road sank to cross the Avon by way of a stone bridge and Leland was able to lift his eyes and observe the 'toune of Malmesbyri . . . on the very toppe of a greate slaty rok, . . . wonderfully defendid by nature'. Riding up the slope he found that the town's four gates (they're named after the cardinal points of the compass) were all in ruins. In places, the town's walls still stood 'ful up', but were now 'very feble'. Set loose in a town, Henry VIII's librarian-turned-topographer became an urban explorer, a wide-eyed Vespucci tracking uncharted alleys, interviewing natives, recording landmarks; adding coordinates to his infinite and uncontainable mental map of Britain. He concluded that Malmesbury started as a castle, and that it probably wasn't a town at the time of the Saxons but that it was named after 'Maildulphus a Scotte', who had also founded the abbey.

Malmesbury was humming. There was a 'good quik' market every Saturday and the square was furnished with an octagonal market cross, 'a costely peace of worke . . . made al of stone' that had a vaulted roof 'for poore market folkes to stande dry when rayne cummith'. The security was good, too: when the annual fair was held at the feast of St Aldhelm, a body of armed men were deputised to keep the peace. Intriguingly, Malmesbury also seemed to be gripped by redevelopment. Politesse – and an understandable attachment to his own head – prevented Leland from explaining why the abbey church had recently been put up for sale, but he did reveal that it had been bought by the town and was now the parish church. Behind the purchase was 'an exceding riche clothiar' called Stumpe who had filled 'every corner of the vaste houses' of the dissolved abbey with his cloth-weaving looms. Stumpe was Malmesbury's Master Bigg, acquiring land and buildings for commercial redevelopment. His next plan was to build streets for clothmakers in the open ground at the back of the abbey. Leland didn't mention whether Stumpe was also behind the conversion of the old parish church into Malmesbury's town hall. Or whether the conversion of the church's west tower into a residential dwelling was also a Stumpe project. Another small church on the abbey site had been occupied by weavers and their looms.

Burning brightest in Britain's urban constellation were towns which specialised in products derived from locally available raw materials. Many of these specialist centres were barely more than villages. On the low watershed between the Avon, the Trent and the Severn, John Leland had come across small communities making the most of their proximity to coal and iron from Staffordshire and Warwickshire. In a largely rural land, these tiny hotspots of production were deeply fascinating. One of a great many Leland encountered was set in a pleasant landscape of woodland, pasture and 'meatly good corne', north of Alvechurch, where he found a hamlet called *Dyrtey* whose 'praty strete' was lined with smiths and cutlers. Just across a brook, he entered *Dyrtey's* slightly larger neighbour, enchantingly arrayed each side of a single street which rose gently up the valley-side: 'The bewty of Bremischam, a good market towne . . .' began Leland, before elaborating on the 'many smithes in the towne that use to make knives and all maner of cuttynge tooles, and many lorimars that make byts, and a greate many naylors'. He judged that most of Bremischam's economy was based on its smiths. Bremischam was doing well; small was beautiful. Bremischam would become Birmingham.

The most active towns had social and political structures that included labourers, traders, craftspeople, property developers, churchmen and civic enforcers. Towns were familiar with rolling redevelopment, both brownfield and greenfield. Not all could boast fancy octagonal market crosses, but it was the market function that mattered. Market towns exerted a considerable reach beyond the strictly physical limits of their medieval walls. These were complex, dynamic, influential communities. The kingdom's innumerable, dissolving, ecclesiastical foci were being replaced by a new economy that revolved around secular, urban centres; gilded crosses replaced by market crosses.

One other type of Tudor town merits a mention. In a kingdom beginning to redefine its identity through naval supremacy, Portsmouth was unique. Although the town itself had – according to Leland – just 'one fair streate' and 'but one paroche chirch', the defences and naval base were impressive. Two round towers protected the narrow entrance of the huge natural harbour and to prevent unwanted maritime guests a massive iron chain could be drawn between the towers. On its landward side, Portsmouth was protected from attack by a ditch and earthen wall revetted with timber, massively gated and topped with large cannon of iron and brass. But the sight that seems to have intrigued him most was the 'great dok for shippes' built fifty years earlier by Henry VII. When the curious topographer peered into its depths, he could see the rib-timbers of *Henry Grace de Dieu*, one of the largest ships to have been built within living memory.

The contrast between the many towns Leland encountered in England and the few he found in Wales was striking. The populations of the four largest Welsh towns – Wrexham, Carmarthen, Brecon and Haverfordwest – were just 2,000 or so and there were around another eight settlements that may have supported 1,000 people. Leland's description of Wrexham included the observation that it was 'the onely market towne of Walsch Maylor', although it was commended for its 'goodly chirch collegiate . . . one of the fairest of all North Wales'. Carmarthen had been the site of a Roman fort, joined in 1094 by a Norman castle and a settlement. By the early 1200s, Carmarthen's port was licensed to trade in wool, pelts, leather, lead and tin. But Leland recorded little beyond the note that the town 'hath incresid' with the decline of neighbouring Kidwelly. Haverfordwest – like many Welsh settlements – was walled and its three parish churches ('one of them withowt the toune in suburbe') suggested that the most westerly settlement in Wales, sited on the superb anchorage of Milford Haven, had benefited from its trading links to Ireland and the continent. Brecon came across as one of the more memorable towns on Leland's Welsh itinerary; walled, with four gates, and 'a mighti great chapel (S. Mariae), with a large tour for belles of harde ston costely squared with the expences of a thousand poundes'. Brecon had expanded and had a suburb called 'Porthene'. Throughout his Welsh tour there was a noticeable absence of eye-catching civic structures such as stone-built houses, water conduits, market crosses and so on. The harbour of Tenby, one of the seven larger settlements along the south coast of Wales, had 'a peere made for shyppes' and 'is very welthe by marchaundyce: but yt is not very bygge having one paroche chyrche'. Leland was astonished to discover that Tenby had no well, the population being 'forced to fech theyr water at S. John's withowt the towne'.

If a theme emerged at all from Leland's descriptions of the larger Welsh settlements, it was one of stasis or decay. In many cases, development had been hampered by several centuries of being an intermittent war-zone. Raglan was 'bare'; Arberth was 'a poore village'; Criccieth 'hath beene a franchisid toune, now clene decayith'; Mold 'was ons a market toune' but 'the wekely market is decayed . . . and a greate numbre of houses be withowt token almost destroyed'. Mold was barely a village, with just two streets 'byside other little lanes' and 'in al be scant 40. houses'. A reader with Leland in Wales would never encounter an urban hubbub comparable to Malmesbury. One of the few places to emerge positively was Welshpool, or 'Walsche Poole', which Leland praises for being 'wel buildid after the Walsch fascion'.

The urban absentee in Leland's notes was London. By 1550, the population had reached 75,000. Set against European cities, London was not numerically remarkable: Naples was nearly three times larger, at 212,000, Venice had a population of 158,000, Paris 130,000, Lisbon 98,000 and Antwerp 90,000. The biggest city in the world, Beijing, had already passed the half-million milestone. London was creeping into the continental big league, but at home, it was already the kingdom's primate city, the 'metropolis'. At the time, towns in Britain seldom mustered populations greater than a thousand, and most country towns in southern and central England had populations of, at most, five or six hundred.

Nourished by its role as a centre of commerce and trade, government and court, and by a constant influx of migrants from rural districts and from abroad, the city on the Thames had extended into a loose collective of riparian settlements. The three most defined urban blocks were still the City of London within its perimeter of medieval walls; Westminster, just over a mile upstream, and Southwark on the south bank of the river, linked to the City by London's only bridge. Between the City and Westminster, a ribbon of up-market housing had developed, with gardens that ran down to the river. New suburbs were spreading. North of the medieval walls, the 'great marsh' or 'moor' described by Fitz Stephen nearly 400 years earlier was being drained and colonised by houses. More sought-after were the estates beyond Westminster where there was a scattering of villages favoured by royal officials and courtiers keen to avoid the overcrowded streets of Westminster. One of the most desirable was the old Anglo-Saxon riverside village of Chelshithe, also known also as Chelsey or Chelsea, which was close to the Roman roads heading west and south-west, and convenient for London. A decade or so after *Utopia* was published, it was here that Thomas More built himself a red-brick house with bays, a porch and flanking casement windows, set in a 14-hectare estate with river access. It was a good address. There were views across the water to the woods and fields of Battersey and from a rise in his garden, More could see the distant roofs of London and St Paul's. Upwind of the city, Chelsea was famed for its clear airs.

By the mid 1500s, London was also extending its reach far down its enormous estuary. A map of 1588 showed a string of suburbs extending downstream from London Bridge: On the south bank there was Rotherhithe, and in sight of it on the north bank, the sprawling houses of Ratcliffe; its name derived from the 'Red Cliff' of gravel undercut by the river current sweeping around the long bend in front of Wapping Marsh (Ratcliffe is one of London's lost villages, remembered on the modern A–Z as Ratcliff Cross Stairs, and in the names of a couple of nearby streets: the village market used to be held in what is now Ratcliffe Cross Street). Further downstream, the 1588 map showed Deptford strung out like a rope along the south bank, and Greenwich with its spires and tower, then the fortified barrage at Blackwall and distant village of Woolwich with its hilltop church. The Thames was usurping the Solent. Naval ships had to sail to the Tower of London to take on board their ordnance and the Thames was well provided with potential dockland. From 1550, the royal fleet was based in the deep-water channels of the Medway, where Chatham was founded as a royal dockyard. A mast pond and additional storehouses were added in 1570 and a dry dock in 1581.

The geographical feature that brought so many to these crowded streets was celebrated in 1586 by one of England's leading scholars, William Camden:

> For many Cities have drawn their names from Ships, as Naupactus, Naustathmos, Nauplia, Navalia Augusti &. But of these none hath better right to assume unto it the name of a ship-Rode or Haven than our London. For in regard of both Elements most blessed and happy it is, as being situate in a rich and fertile soile, abounding with plentifull store of all things and on the gentle ascent and rising of an hill, hard by the Tamis side, the most milde Merchant, as one would say, of all things that the world doth yeeld: which swelling at certaine set houres with the Ocean-tides, by his safe and deepe chanell able to entertaine the greatest ships that be, daily bringeth in so great riches from all parts that it striveth at this day with the Mart-townes of Christendome for the second prise, and affordeth a most sure and beautiful road for shipping. A man would say that seeth the shipping there, that it is, as it were, a very wood of trees disbranched to make glades and let in light, so shaded it is with masts and sailes.

William Camden's county-by-county chorography, *Britannia*, was a patriotic fusion of geographical fieldwork and historical research and the places he described were understandably the most glorious in Christendom. But a contemporary visitor from Germany felt pretty much the same way about London. Paul Hentzner had included London in a three-year tour of Europe which began in 1597. London's size and its 'extensive suburbs' had impressed him, and so had the Tower and the Thames, 'every where spread with nets for taking salmon and shad'. He found London's bridge a 'wonderful work ... supported upon twenty piers of square stone, sixty feet high, and thirty broad, joined by arches of about twenty feet diameter.' Houses lined both sides of the bridge so that the entire structure had 'the appearance of a continued street'. Looking upward to the parapet of the tower on the bridge, the German was able to count thirty severed human heads rammed onto iron spikes.

And London continued to swell, despite recurrences of plague. The city lost 20,000 inhabitants in 1563, 18,000 in 1593 and 30,000 in 1603. The poor were most afflicted. By 1600, London's population was thirteen times greater than the next largest city in Britain, Norwich, which had lost nearly a third of its population to an outbreak of plague in 1579. No other country in Europe had a capital that exceeded the size of other towns by such a margin.

London was top-centre on the first page of a stunning book of maps that began circulating several years before the publication of Camden's *Britannia*. The first county atlas of England and Wales was inscribed by engravers working for a cartographer called Christopher Saxton, whose pioneering set of thirty-four maps revealed the two countries in such great detail that a browser could scrutinise landscapes without saddling a horse. England and Wales were portrayed as a paradise of rivers and gentle hills dotted with woods, fenced parks and Christian settlements wrapped by a coast of sheltered havens. Counties could be compared for density of settlement, for physical features, for size and location. The first page was composed so that the whole of Kent, Sussex, Surrey and Middlesex could be absorbed at a glance, together with the Thames from its estuary to Windsor. On a map speckled with tiny, solitary churches, London was a forest of spires and towers occupying the central reach of the Thames.

At the other end of the cartographic spectrum were 'single spire' places like Hutherffeld. Just one among thousands in the atlas, Hutherffeld was an ordinary village set among the outlying mole-hills of the eastern Pennines. It was a part of Yorkshire's West Riding where isolated communities bolstered the meagre returns of subsistence agriculture with small-scale, cottage-based cloth production. Hutherffeld's place-name is clouded in moorland mist. It may have derived from *Hudraed* and *feld*, 'the open land of a man called Hudraed'. Or perhaps 'huther' had evolved from 'huder', the Old English word for 'a shelter'. In the Domesday Book the settlement was recorded as *Odresfeld*. One generation after the pestilence, the poll tax of 1379 recorded just 14 individuals and 35 married couples – a total population of perhaps 250 – distributed in half a dozen dispersed settlements. By Saxton's day, it's unlikely that it amounted to more than a handful of dwellings along a single street which turned at its eastern end to run down to a bridge over the Colne. Like so many sixteenth-century villages, it had suffered neglect. In 1537, the bridge had been so dilapidated that William Broke had left 6s 8d in his will for its repair. And in 1571, a few years before the publication of Saxton's atlas, the chancel of the parish church was reported to have been in such a dire state that 'the rayne raineth into the churche, and it fell down vii yeres sence and slewe the parishe clerk'. Much of the dilapidation was due to the effects of the Reformation and abandonment of chancels for taking the sacrament, but Hutherffeld seems to have been more neglected than many parishes. Complaints were still being made in the 1590s about the state of the church.

Anybody gazing for long enough at the Yorkshire page in Saxton's atlas would eventually have noticed that the towns with the biggest symbols and their names in capitals were all located on rivers. *BRAD-FORTHE*, *LEDES*, *HALYFAX*, *WAKEFELDE*, *BARNESLEY* and *SHEAFELD* were all prominent and all were on rivers. Hutherffeld (upper and lower case) was also on a river, but for whatever reasons, it had failed to graduate to capitals.

*

Geographers like Leland and Camden and Saxton knew their land through road trips. They recorded from the saddle; they rested and researched at roadside way-stations, from inns to accommodating landowners. And they shared the road with packhorse trains, carts and lumbering freight wagons.

Some counties were slower to move through than others. During the reign of Henry VIII, the common ways of Kent had become 'so deep and noyous, by wearing and course of water' that carriages, pedestrians and horses could only pass at 'great pains, peril and jeopardy'. Mud and flood were the main impediments, both of them a result of undrained land. Floodplains such as the Severn could be notoriously difficult after rain, as rivers and brooks spilled their load across the surrounding levels and roads. By the mid-sixteenth century the first General Highways Act had to be passed 'for amending highways . . . now both very noisome and tedious to travel in, and dangerous to all passengers and carriages'. Leland's obsession with bridges and causeways was a measure of the critical role they played in keeping traffic on the move. On horseback, it could take an entire day to travel between two towns.

Rivers were morbidly fascinating. In Leland's pages, they flooded and thundered and they gagged on mining spoil. They were obstacles to be subverted, the only hindrance likely to impede a horseman's progress. There was an alternative view of rivers, but it was little recorded because its practitioners did not belong to the world of books and maps, and because – with a few exceptions – Britain's rivers were ill-suited to navigation by the kinds of boats that made economic sense.

When Paul Hentzner observed that in England there were 'but few rivers', he wasn't casting aspersions on the country's drainage but noting that navigable waterways were rare compared to the continent; the Brandenburger's home rivers included the huge – by English standards – Oder, Elbe, Spree and Havel. Britain was a long, thin island with a hilly interior of mixed geologies that had produced relatively short, swift, variable streams; these were productive natural habitats and ideal sources of power for watermills, but were unhelpful to heavily laden boats. Most rivers were too fast, too short and too shallow. In their lower reaches, larger rivers like the Severn, Thames and Tay tended to meander uneconomically. Not only that, but many of Britain's regions were defined by their drainage basins and were not connected to each other except at their outflow into the sea, and Britain's coastal waters were too tortured by tides and storms for flat-bottomed river-vessels. Rivers were also prone to reductions of depth during drought and dangerous spates during rainy episodes. Tree trunks, sandbars and mudbanks could cause blockages. Although there were sections of some wider rivers that could be sailed, most inland vessels were propelled by oars or poles, or hauled by teams of men or horses from the riverbanks, where towpaths were all too often obstructed by field boundaries and tributaries. Banks tended to erode and slump into the stream. Particularly troublesome to traffic were the innumerable weirs directing water-flow into artificial channels to feed mills. In the early 1600s, weirs on the Medway in Kent stood six to eight feet high and formed walls across the river.

Attempts to improve Britain's temperamental watercourses predated the Romans. In their own way, the forager-hunters of the Carrs and West Country fens had modified their watery habitats with lakeside platforms, raised timber tracks, causeways, bridges and weirs. For thousands of years, modifications to rivers had been undertaken without serious impediment to river travel (although watermen shooting the rapids between the piers of London's bridge might have disagreed). But the multiplication of mills and weirs collided with rising volumes of river freight to produce a quick-fix which inconvenienced all parties. Many weirs were modified with a simple timber gate or 'lock', which allowed freight vessels to pass through the obstruction. Flash-locks were an inconvenience born of necessity. Boats heading downstream would wait for the lock to be opened then flash through the gap on the released current; boats heading upstream would have to wait below the weir while the lock was opened and sufficient water drained through the gap for the water levels to equalise.

Britain was – as ever – lagging on the technology curve. On the continent, flash locks had already been used to achieve the impossible: the crossing of a watershed by a navigable waterway. Moving freight on water between two drainage basins was the economic *graal*; summit canals opened the way for low-cost, inter-regional trade. Not surprisingly, the Hanseatic League was involved. Opened in around 1398, the ground-breaking canal crossed a 17-metre watershed on the salt-route to the Baltic port and principal Hanseatic city, Lübeck. But flash locks were already old technology. A far more versatile type of lock was already spreading. The pound lock was the most important hydraulic device of the age. It was quick and safe, and could be installed on a river or a canal. It took the form of a pair of gates each side of a basin – the pound. Having entered the basin, vessels waited between the two closed gates while water levels were equalised through hand-operated sluices. In effect, a pound lock was a mechanical engine for lifting or lowering, and once installed, it formed a vertical step in the watercourse where previously there had been an angled gradient. The mechanisation of rivers was an entirely new level of interference. Pound locks were operating in the Low Countries from the fourteenth century, and there was a further breakthrough when they were modified to scale greater heights by installing them in sequence, like a hydraulic staircase, the upper gates of one lock functioning as the lower gates of the next. By 1604, work had started in France on what would be Europe's first summit-level canal to use pound locks, a mighty enterprise intended to cross the watershed between the drainage basins of the Seine and Loire.

Britain of the late 1500s was a jigsaw of regional economies interacting by road, by sea lane and by river. Moving goods by road could cost up to twelve times more than transportation by waterway, but most rivers deep enough for vessels were impeded by weirs, and canals had not been excavated since the days of the Romans. A rare exception was a cut excavated in the mid 1560s by Exeter traders to bypass the Countess of Devon's weirs on the Exe. It was only three feet deep and ran for less than two miles, which proved too short a distance to compete with road haulage. In general, efforts were directed towards maintaining the status quo: keeping rivers and roads open rather than building canals.

Glimmers of innovation were, however, apparent at mines. At a local level, the challenge of shifting coal from the pithead to the nearest wharf or point of distribution had been resolved by the construction of purpose-built wagonways. Engineered to follow gentle curves and gradients with a laid surface to reduce rolling resistance, wagonways bypassed the sharp ascents and descents, bumps and sloughs, common on ordinary roads. More coal could be moved with less ox- and horse-power. Wooden wagonways had been laid in Britain since at least the late 1500s, when Austrian miners brought the idea to copper mines in the Lake District, where they set planks end-to-end and hand-pushed their ore-trucks along them. By the early 1600s, the challenge of shifting coal efficiently had motivated a speculator outside Nottingham to lay a two-mile timber wagonway between his mine and the distribution point. Between the planks was a slot which guided a pin protruding downward from the trucks. Huntingdon Beaumont may or may not have been the very first to have laid a surface wagonway in Britain, but he was the earliest to invest so much money in the idea. Like so many pioneers, he came unstuck, and after a failed attempt (again using wagonways) to improve the flow of coal at Blyth, near Newcastle, he ran into debt and died in prison.

Britain's communications were hopelessly inadequate. Demand far outstripped the capacity of road and river to deliver. Trade within regions was increasing and so was trade through ports: in the final three decades of the sixteenth century, the tonnage of shipping sailing up the Thames to London increased by 50 per cent. East coast ports like Newcastle, Lynn and Yarmouth were growing on trade with the Baltic. Britain's economic potential was summarised by that observant visitor from Brandenburg, Paul Hentzner:

The soil is fruitful, and abounds with cattle, which inclines the inhabitants rather to feeding than ploughing, so that a near third part of the land is left uncultivated for grazing. The climate is most temperate at all times, and the air never heavy, consequently maladies are scarcer, and less physic is used there than any where else ... There are many hills without one tree, or any spring, which produce a very short and tender grass, and supply plenty of food to sheep; upon these wander numerous flocks, extremely white, and whether from the temperature of the air, or goodness of the earth, bearing softer and finer fleeces than those of any other country: this is the true Golden Fleece, in which consist the chief riches of the inhabitants, great sums of money being brought into the island by merchants, chiefly for that article of trade ... It has mines of gold, silver, and tin (of which all manner of table utensils are made, in brightness equal to silver, and used all over Europe), of lead, and of iron ... Glasshouses are in plenty here.

Once again, population growth was becoming a lever of landscape change. By 1600, the numbers of people living in England had virtually doubled in only 100 years and now stood at 4.1 million. In Wales too, the population had nearly doubled since 1500, from around 210,000 to 380,000 in 1600. Scotland's population was probably about one million and it too experienced a surge from the late 1500s through to the early 1600s. The numbers deficit caused by the Black Death had been corrected and the population had climbed back to its 1300 high point. In England at least, the economy had recovered some of the balance and scale that had been current during the Roman occupation. But it had been a thousand-year dip. Some kind of equilibrium had been reached between population and resources; a provisional harmonisation too between economics and politics. The Britain that emerged from the sixteenth century was a relatively healthy habitat; an island recovered at last from Roman occupation.

The population surge tamed the last vestiges of wild Britain. It was a tipping-point captured on the maps of Timothy Pont, a young graduate of the university at St Andrews. Between 1583 and 1595, Pont and his associates embarked upon a series of journeys throughout

Scotland, drawing and redrawing a series of detailed sketch maps covering much of the mainland between the Solway and Pentland Firths. Although Pont's primary interest was human settlement and its related features, he sketched a wealth of contextual information. River systems, lochs and coastlines were used as natural meridians on which to plot settlements, while features like mountains and woods were added as residual landmarks. It was an unprecedented endeavour. Glasgow appeared as a cross of streets lined with buildings; Linlithgow as a many-windowed 'Palace' gazing across a loch to 'The Park' (the view has scarcely changed). In his sketch of Dundee, the long, thin burgage plots were clearly shown, running at right-angles to the High Street. Pont's most detailed bridge appeared beside the houses and soaring spire of St John's Kirk at Perth. This was the lowest crossing point on the mighty Tay, and Perth was one of the most strategically placed towns in eastern Scotland. Pont's bridge was buttressed against the Tay with four enormous piers. Annotations on some of the maps alluded to the productivity of rivers and lochs. Loch Tay had 'Fair salmonds, trouts, eels and pearles'; deep in the mountains north of Dundee, Loch Lee could be fished for 'salmond trowts, trouts, pycks, pearches'. Over in the western Highlands the vicinity of another remote, mountain loch – Loch na Sealga – was noted as an 'excellent hunting place wher are deir to be found all the year long as in a mechtie parck of nature'. (There are still deer in the glen, and although Pont's extensive 'fyrs' have gone, their black roots protrude from the peat of the river bank.) Pont's woods revealed how much had changed in Scotland since the ice had melted. In the Highlands, the wildwood had shrunk to a small proportion of the land area and was generally removed from human settlements; in the Lowlands the woods were smaller, fragmented and often in the proximity of a great house.

Pont caught Britain on the cusp. The Scotland he recorded with his pictograms, sketch-maps and notes was the last part of mainland Britain to retain a memory of the wildwood and its wildest predator. Pont's is one of the final glimpses we have of wolves in Britain. In

Wales and England they'd been hunted to extinction by the fifteenth century. The 'excessively old' wolf that Paul Hentzner saw in Queen Elizabeth's zoo at the Tower of London may have been the lone survivor of a Scottish pack. The last record of a British wolf is dated 1621, when a huge bounty of £6 13s was paid in Sutherland.

Among Pont's remarkable maps, the most poignant by far is the sheet covering the tract of mountain, moor and loch that finds its apex at Cape Wrath. This north-western extremity of Britain holds greater title to 'Land's End' than the peninsula in Cornwall, for this was the scene of the island's final unwilding. It's clear from the map that this remote district was familiar to Scotland's coastal settlers. At a place that could only be reached by boat, or by a long walk over exposed moors, Pont drew a settlement at 'Sandwatt beg' and at an even less accessible spot a few miles inland, he showed a pair of peaks divided by a pass labelled 'Bhellach maddy', adding for the benefit of non-Gaelic speakers: 'or Woolfs Way'. But the most moving label on this particular sheet is also the largest. In enormous, cursive lettering, Pont wrote across the entire Cape Wrath hinterland: 'Extreem Wildernes'. The eradication of wolf-run wilderness marked the completion of an 11,000-year process that had begun with tentative footsteps onto a remote peninsula and ended with an island habitat entirely under the dominion of humans. Pont's was a parting glimpse at a lost landscape.

BOOK THREE

PART SIX

Conurbia

NINETEEN

The Clowd

1620–1725

Sealand – or Zeeland – is the part of Doggerland that rose again. Maps of the early 1600s displayed the great, tripartite delta of the Schelde, the Maas and the Rhine as an intricate maze of channels separating islands speckled with towns and villages. It was the golden dawn of the Netherlands, when the surveyors and marine engineers of this water-land were world leaders. The Dutch were able to make fields and cities from the sea. Walk the ruler-straight wharves and dykes of Zierikzee today and it's impossible not to be struck by the town's precarious deliberation; it appears to float on the edge of the Oosterschelde, an urban pontoon engineered hundreds of years ago by Zeelanders who understood hydraulics. One of them was Cornelius Wasterdyk Vermuyden. His mother had come from Zierikzee, the capital of Schouwen, one of the larger islands in the delta. His uncle, a specialist in impoldering and embanking, had been instrumental in creating Zierikzee's harbour. Cornelius was a teenager when Jan Adriaanszoon began creating the world's first polder by draining away the sea using windmills. The Beemster polder was ready for cultivation only three years after work on its ring-dyke was completed and profits from the crops quickly covered the cost of the reclamation. (Flushed with hydraulic success, Adriaanszoon added 'Leeghwater',

Low Water, to his name.) In 1610, while Beemster was rising from the sea, engineers built a dam between Schouwen and its neighbouring island, Duiveland. While Britain was shrinking before rising seas, the Netherlands was being made larger.

Cornelius Vermuyden sailed to England in 1621. Later, he claimed that he'd been responding to an invitation to drain the Great Level, the enormous wetland around the Wash, but soon after landing he was diverted to a series of more urgent projects. A 'violent tide' in September 1621 had taken out a section of the Thames bank at the outfall of Dagenham Creek, flooding much of Dagenham Marsh and Hornchurch Marsh. Vermuyden repaired the bank, fitted a sluice and received a parcel of reclaimed land by way of payment. This project overlapped with another urgent commission further downstream, on a reach of the Thames described in the 1610 edition of Camden's *Britannia* as 'flat and unholsome'. Stretching for five miles or so along the north shore of the estuary were half a dozen low-lying marshlands divided by winding tidal creeks. On his county map of 1576, Saxton had labelled them 'CANVE INSULE'. Unlike the adjacent mainland, which Saxton had thickly populated with villages and towns, this Essex Zeeland appeared to be devoid of human habitation. In *Britannia*, Camden filled the cartographic void by writing about the human geography of these marsh-islands. 'Canvey' was dotted with 'dairy sheddes' known locally as *wiches*, which were used by 'young lads' for cheese-making. Grazing the islands were some 400 sheep, prized for their 'sweet and delicate taste' and for the quality of their milk. The snag, according to Camden, was that Canvey was so exposed to flooding from the sea that 'often times it is quite overflowen, all save hillocks cast uppe, upon which the sheepe have a place of safe refuge.' But stemming the tides that swept up the narrowing Thames was a challenge that far exceeded the financial resources and technical skills of Canvey's landowners. The Dutch came to the rescue. One of Canvey's leading landowners, Sir Henry Appleton, was familiar with a wealthy Cheapside haberdasher called Joos Croppenburgh. The Dutch haberdasher struck a deal with the Essex landowner: in return for financing and facilitating the drainage of flooded lands on Canvey, Sir Henry and his fellow landowners would grant Croppenburgh one-third of the island.

Canvey was converted into a Dutch polder. Land that been inundated annually was permanently drained and made available for development. Many of the three hundred or so Dutch labourers and specialists who worked on the project settled on the new island and built themselves Dutch-style houses and cottages. Vermuyden's role is unclear, but in 1626 he was named alongside Croppenburgh in litigation related to the Canvey project, so he appears to have been a significant contributor. The immigrant legacy is remembered in Canvey's Dutch street-names (Zealand Drive, Tilburg Road, Korndyk Avenue, Zider Pass, Rembrandt Close) and in the Cornelius Vermuyden School, rebuilt in 2012. The shepherds and their *wiches* ceased to be Canvey's defining culture. They faded from the landscape as wordlessly as the foragers and hunters who had roamed Doggerland's marshlands. The inhabitants of the next wetland Vermuyden drained did not go so quietly.

It was the single greatest exercise in landscape modification Britain had ever seen. Fifty miles north of the Wash, beside the lower Trent and Don, lay 30,000 hectares of glistening fen, much of it Crown land. This was not a rescue mission, but an attempt at landscape 'improvement'; a conversion of wetland to arable. The intended area surrounded the Isle of Axholme, a ridge of ground rising to some 40 metres above sea-level. The isle's four parishes were home to over 2,000 people who had used the surrounding wetland for generations, grazing their herds of cattle, sheep and pigs each summer on common pastures which were enriched each winter by the 'thick fatt water' of the winter floods. Much of this common land lay under water from November through till May. Axholme's aquatic commons were a multi-functional resource: a source of firewood, timber and turves for fuel and building; a place for fishing and catching wildfowl; a source of hay for animal feed. On higher, drier ground, fields provided a wide range of crops, mainly barley, but also rye, oats and wheat, peas and beans. Hemp and flax were cropped for a thriving cottage industry of spinning and weaving. John Leland had observed of Axholme that it was 'meatly high ground, fertile of pasture, and corne'. So fertile that the farmers of two of Axholme's manors were keeping no fewer than 12,000 cattle through the winter by grazing them on a combination of the unflooded levels, on arable fields and on enclosed pastures. The people of Axholme lived well. And then a Dutchman arrived with plans to drain their wetland.

Rivers would have to be moved. In the first phase of the plan, Vermuyden diverted the River Don through a new channel to the Aire. The rivers Idle and Torne were diverted into the Trent. Nothing on this scale had been attempted in Britain before. The people of the wetland objected to the loss of their traditional lands, and to Vermuyden's use of Flemish labour. There were riots and deaths. Embankments were torn down. And in 1629, Vermuyden was knighted 'in recognition of the skill and energy that he had displayed in adding so large a tract to the cultivable lands of England'. Sir Cornelius moved on to the Cambridgeshire fens and the 160,000-hectare Great Level. As cuts and sluices dried out the wetland, the fen 'slodgers' took umbrage at the loss of habitats which had provided them with fish and wildfowl, withies and reeds, and peat for their fires. There were riots, but the penalties for unrest were too severe for local objectors to stand before an army of drainers backed by the Crown. In the course of a few years, two of England's greatest wetlands were lost to grids of drains and fields.

Wetlands had always been fragile habitats, their aquatic topography entirely dependent on abundance of rain, rates of silting and effects of human interference. Many of the lakes, ponds and marshlands that had succoured so many for so long had already disappeared from the British lowlands, starved of water by ditching for fields. In Norfolk, the peat diggings that had flooded to form lakes after the fourteenth century had shrunk dramatically as the 'broads' became infilled with saw sedge and reeds. The broads however pulsed with wildfowl like mallard and teal, widgeon and pintail. Inspired it seems by the Dutch, local landlords had begun to cover long, tapering channels with nets to trap ducks. One of the first was at Waxham on the coast, where Sir William Wodehouse erected such a trap 'known by the foreign name of a koye'. Decoys quickly caught on.

Greater by far than any of the inundated peat-diggings of Norfolk was Whittlesey Mere, a vast inland lake set in the shrinking wetland between Ely and Peterborough. Despite fenland draining and the inexorable creep of cornfields and orchards, the mere had endured as the largest surviving body of freshwater in southern Britain, and unlike Norfolk's lakes, Whittlesey was clearly visible from a major highway, the Great North Road. A description of this wetland relic was preserved in the journal of Celia Fiennes, the first person to leave a record of visiting every English county. Passing by in 1697, en route from London to Nottingham, Fiennes

> . . . came in sight of a great water on the right hand about a mile off which looked like some Sea it being so high and of a great length, this is in part of the fenny country and is called Whitlsome Mer, is 3 mile broad and six mile long, in the midst is a little island where a great store of Wildfowle breeds, there is no coming near it in a mile or two, the ground is all wett and marshy, but there are severall little Channells runs into it which by boats people go up to this place; when you enter the mouth of the Mer it looks formidable and its often very dangerous by reason of sudden winds that will rise like Hurricanes in the Mer, but at other tymes people boat it round the Mer with pleasure; there is abundance of good fish in it. This was thought to have been Sea some tyme agoe, and choak'd up, and so remaines all about it for some miles a fenny marshy ground for those little Rivers that runns into the Sea, some distance of miles . . .

Fiennes probably borrowed the dimensions of the mere from William Camden's *Britannia* of 1586 (a book she kept in her Wiltshire library), and perhaps her 'Hurricanes' were an embellishment of the 'violent water-quakes' claimed by Camden to threaten Whittlesey's 'poore fishermen', although she sensibly didn't repeat his theory that they were caused by 'evaporations breaking violently out of the bowels of the earth'. Writing a century apart, both Camden and Fiennes saw Whittlesey as both wondrous and dangerous. In southern Britain, it was the last of its kind.

Islands communicate through their coasts. At saltwater portals around the rim of Britain, cargoes were handled and stored; ships built and repaired; voyages conceived and concluded; information traded. Ports and dockyards had architectures of their own: quays and warehouses, ponds where masts and spars could be soaked in water to preserve their suppleness; graving slips where ships could be hauled out of the water with capstans so that their hulls could be scraped clean and treated with tar; docks that could be drained so that shipwrights could replace rotten timbers. The modification of coasts was being led by the Thames.

In the early 1600s, the royal dockyards of Woolwich, Deptford and Chatham were joined by a new commercial yard. The East India Company had been granted a licence to trade in 1600, and by 1620 they had more than ten trading stations on Asian coasts and were delivering average returns to their investors of 100 per cent. Having started by converting men-of-war to armed merchantmen, and by commissioning ships from East Anglian yards, they moved in 1608 to the Thames at Deptford, and in 1614 to Blackwall, where they developed a greenfield site just downstream of Blackwall Stairs, long established as a convenient place on the river for travellers to embark and disembark. It was a practical location for a new dock. Blackwall had available marshland, and being some way downstream of Deptford, it had a greater depth of water. By 1620, the marsh at Blackwall was a teeming dockyard, with dry docks, slips, a tar house, a saw pit for forming masts, a spinning house for turning hemp into cordage, a forge and a smith's shop for making anchors, nails, tools, and cables. There were storehouses for provisions, canvas and timber, a slaughter house and salting-house for butchery and preservation of meat, and a barber-surgeon's room. A cooper was kept busy making oak casks. Workers were housed in a 70-foot, two-and-a-half-storey accommodation block and the exposed, western side of the yard was protected by a four-metre brick wall. Within the wall was a tap house and victualling house, built to deter workers from wasting time 'in going up to the Towne'. In its first two decades, between 200 and 400 workers toiled at the Blackwall yard.

'London' could no longer be thought of as a settlement with defined limits. Much of the city's economic energy was being tapped oceans away and one of the city's primary organs of operation lay six miles downstream of the bridge. Ships stopping at Blackwall avoided having to negotiate the Thames's serpentine meanders, which were often fraught with fickle winds. The Blackwall yard was the largest employer of men in the greater London area. London and the Thames were heart and artery, and for the first time, the river was becoming clogged. Around the British coast, roadsteads, rivers and quays were becoming overcrowded with ships whose turnaround times were still dependent upon tides and winds. In estuaries and rivers with extreme tidal ranges like the Thames (and the Severn), difficulties were particularly acute. Vessels had three options for loading and unloading goods: they could beach on a high tide and refloat when the tide returned; they could moor alongside a deep-water structure such as a quay or jetty; or they could drop anchor, or moor to a buoy in deep water and then rely on smaller, shallow-draught lighters to handle their cargoes. A wet dock avoided all of these inconveniences. Sealed from the falling tide by a watertight gate, a ship could remain afloat. Wet docks were much more convenient for fitting out ships and raising masts, and for loading and unloading cargoes, especially shipments of livestock. The commissioning in 1659 of a large new commercial wet dock at Blackwall marked the beginning of a new age for the Thames. With a basin of nearly four hectares, it was the largest wet dock in the kingdom, capable of taking the enormous ships of the East India Company. For centuries, the tidal margins of the Thames had been embanked to protect farmland and realigned to create wharves, jetties and dockyards. But the commercial wet dock stopped the tide and made it possible for the largest, fastest ships on the oceans to come and go at their convenience. The Thames had been tamed.

London was on a trajectory of its own. Fed by rural–urban drift and immigration from overseas, the city and its suburbs were straining to accommodate the numbers. By 1660, the population had swollen to at least 500,000, up from 150,000 in 1580. The plagues of 1563, 1593 and 1603 were symptoms of overcrowding, and the death rate – particularly in the poorer districts outside the walls – was horrendous. Around one-fifth of the population died during the 1603 plague. The plague of 1665 was far worse, killing around 100,000. And that was followed within months by the most destructive event ever to strike Britain's built landscape. In the early hours of Sunday, 2 October 1666, an unquenched oven in Thomas Farriner's bakery on Pudding Lane set fire to his premises and quickly spread to the adjoining buildings. Pudding Lane was (and still is) one of the short, steep thoroughfares dropping down the ancient riverbank from Eastcheap towards the waterside. The gradient helped the flames leap up the hill. The Roman core of London turned into an inferno. It had been a long, dry summer, and the timber buildings were as dry as kindling. Riverside warehouses were packed with mountains of combustible materials. Hemp, oil and pitch ignited in blossoms of red heat which spread before a strong east wind. By Monday evening, the flames had nearly reached the Fleet Ditch and were tearing at buildings on Cheapside. It was probably on the Tuesday that the stones of St Paul's exploded and lead from the roof ran like streams in the streets. An immense cloud darkened the city, spreading west. Ash gathered in drifts. For four days the flames raged, north to the old walls and west across the Fleet and up Ludgate Hill to Fetter Lane. A subsequent map by Wenceslaus Hollar described the damage. It was a startling document. 'The Blank Part whereof represent the Ruins' covered virtually all of the Roman city. It was as if fifteen hundred years of urban evolution had been erased. All that remained was the street plan.

Shocking though it was, the 'Great Fire' demonstrated in the most spectacular fashion that this city was more than a place of timber and tile. London was an idea; an economy and a community so powerful that it was self-healing. No more than ten or so people died in the conflagration. Around 100,000 were made homeless, and four-fifths of the buildings within the old city walls were destroyed, among them 13,200 houses, 87 churches, the magnificent Guildhall and 52 livery company halls. Plans were drawn to reinvent London along the lines of Rome or Paris, with circular piazzas and a grid of rectilinear avenues, but they remained talking points as the city's feverish army of landlords and tenants reconstructed their properties on the original foundations. A few modifications to the street plan were implemented: some streets were widened; the important riverbank axis of Thames Street was raised in level and a new axis – King and Queen Street – was cut through the ruins connecting the Guildhall site with the quay just downstream of Queenhithe, an idea that had appeared on an abandoned plan drawn up by diarist and writer John Evelyn. The River Fleet, the last natural landmark within the greater city, was straightened and turned into a brick-vaulted, artificial waterway. Within ten years of the fire, London was booming again.

'Greater' London didn't just refer to the scale of its physical sprawl, but to its capacity to find enhancement through catastrophe. Many of the warehouses destroyed in the first two days of the fire were rebuilt with far greater internal volumes. Throughout the city, timber was used widely in the rebuild, but brick (and to a lesser extent, stone) became far more apparent and London's architects used the opportunity to impose their own interpretation of French, Dutch and Italian baroque styles on their gleaming new creations. The Royal Exchange and Guildhall, livery company halls, several market buildings and fifty-two churches rose again in stone.The grandest building took shape more slowly. On the crown of Ludgate Hill, an architect called Christopher Wren started work in 1675 on a new cathedral.

Beyond the city of London, the big story was being told by towns and industry. The post-Roman population crash and the plagues of the fourteenth and fifteenth centuries had granted Britain's diminishing natural habitats a couple of extended recovery-spells. Norman colonisation had brought changes but these had been relatively superficial – more concerned with style than enduring transformation. During the early seventeenth century, hyperactive regional economies became a major topographic force, expressed most visibly through urbanisation. By the middle decades of the century, perhaps as much as a quarter of Scotland's population were living in towns. In Wales, it was about a third. And England was around 40 per cent urban – a figure skewed by the enormous size of London (without London, England was no more urbanised than Wales). In the century since Leland had taken to the road, the proportion of people living in English towns had increased eightfold.

The key to urban growth was a market and a speciality. The little West Riding place Saxton had labelled *Hutherffeld* serves as an example. The village had experienced a steadily climbing population and by the middle of the seventeenth century there were around 600 people living in the parish. *Hutherffeld*'s fortunes changed in November 1671 when letters patent granted the local landowner, John Ramsden, the right to hold a weekly market for cattle and merchandise. Overnight, *Hutherffeld* became a different place: a marketplace. Successful markets turned into towns. Of the 800 or so market towns in England and Wales by the mid 1600s, some 300 had developed specialisations in marketing a single, defining commodity. Newcastle was the main source of coal for London. Northampton was supported by shoes, Oxford by gloves. Walsall specialised in horse harnesses and Wigan in pewter. Sheffield concentrated on 'edge tools' and, west of the Pennines, Manchester had become the commercial hub for textiles produced in outlying Lancashire towns.

The other transformational force gathering momentum was the pillaging of landscapes for natural resources. Nearly six thousand years had elapsed since pioneering agriculturalists had bored deep into the South Downs in search of seams of flint. Since then, Britain had become perforated with pits and shafts reaching for tin, lead, coal, copper, silver, fluorspar (used by doctors to treat colic). Some mining sites had so many adjacent pits that extractive landscapes had developed. In terms of tonnage, nothing came close to coal. Since the mid sixteenth century, output had been rising to meet demand for smelting furnaces and domestic fuel, the latter provoked by the long, cold winters that had struck episodically since the 'Little Ice Age' had begun to bite in the early 1300s. In Scotland, where mean annual air temperatures had fallen by perhaps 1.5°C, a small glacier had reappeared in Coire an Lochain in the Cairngorm mountains. In London, the Thames froze. 'Frost,' wrote diarist John Evelyn in January 1684, 'more & more severe, the Thames before London was planted with bothes in formal streets, as in a Citty.' Most dwellings in Britain were not built to cope with months of sub-zero weather and in towns beyond the reach of cheap firewood, coal was the only solution. Evelyn found himself caught between a cold and a cough: coal was being burned in London in such quantities that, for the first time, air pollution was becoming a hazard. So thick was the smoke at Court in White-Hall that people 'could hardly discern one another for the Clowd'. It was, Evelyn warned, dangerous to health, causing '*Catharrs, Phthisicks, Coughs and Consumptions* [to] rage more in this one City, than in the whole Earth besides'. The swelling population and dependence on coal had made Britain a world leader in aerial pollution.

Huge reserves of coal were buried in measures each side of the Pennines, across south Wales, along the coast of Northumberland and Durham and across the central belt of Scotland. Smaller coal measures occurred in several other regions, too, from Somerset to Cumberland and Kent. According to a survey of 1649, the best place in late-seventeenth-century England to appreciate a high-carbon landscape was the Tyne:

> Many thousand people are employed in this trade of coals; many live by the working of them in the pits; many live by conveying them in wagons and wains to the river Tyne; many men are employed in

conveying the coals in keels, from the staithes, aboard the ships: one coal merchant employeth five hundred, or a thousand, in his works of coal, yet, for all his labour, care, and cost, can scarce live of his trade . . .

The French cartographer Albert Jouvin de Rochefort (travelling 1672–6) also wrote in astonished terms of the mines outside Newcastle, 'which are here so plenty, that it may justly be called the magazine whence all Europe is furnished with that commodity'.

And two decades after Rochefort's visit, the pits on the Tyne were even more extensive. When the indefatigable horsewomen Celia Fiennes rode north from Wiltshire in 1698 to see the pits on the Tyne, she paused on a hilltop viewpoint two miles outside Newcastle: 'I could see all about the country which was full of coale pitts.' Fiennes never missed a good coal-pit.

Riding the shires of England between 1685 and 1698, Celia Fiennes took particular interest in what she referred to as 'industrious' places. From a viewpoint beyond St Austell in Cornwall, she recorded 'at least 20 mines all in sight'. These were tin mines, and the physical scale of the operation included stamping mills, coal-fired furnaces and complicated devices required to keep the pits drained of water: mills turned by horses and new 'water engines that are turned by the water, which is convey'd on frames of timber and truncks to hold the water, which falls down on the wheeles, as an over shott mill'. Fiennes reckoned there were over one thousand men at these mines, which were working 'almost night and day, but constantly all and every day including the Lords day which they are forced to, to prevent their mines being overflowed with water'. Riding on westward towards Tregony, she passed 'by 100 mines', some working, others abandoned to flooding. Further west still, near Redruth, she rode 'over heath and downs which was very bleake and full of mines'. Copper, this time, and these works were less extensive because the ore was shipped off to Bristol for smelting, the ships returning with coal to be sold in Cornwall 'at easier rates'.

Joining the creeping expansion of mining landscapes were the places Fiennes usually referred to as 'works'. She was fascinated by them. They were novel, thrilling spectacles; places where furnaces roared and vats bubbled; where ingenious 'machines' could be inspected and wondrous objects produced. In Canterbury, she found paper mills powered by water-wheels and houses packed with silk-weaving looms. She wandered through Northwich, 'full of Salt works the brine pitts being all here and about and so they make all things convenient to follow the makeing the salt, so that the town is full of smoak from the salterns on all sides'. Few of her era knew England better; Fiennes's episodic quest to ride through every English county provided her with a unique comparitive overview. And leading her from shire to town was the prospect of discovering English folk making things. Cloth had turned Leeds into 'the wealthyest town of its bigness in the Country'. And it was workshops that made the city of Norwich look 'what it is, a rich thriveing industrious place'. Places were defined by what they produced.

Despite the smoke and clamour associated with pits and quarries and works, these 'industrious places' were still seen to harmonise with the natural world. Extraction sites were another form of farming. In Poole harbour, iron sulphate was being collected and processed for use as a mordant or 'fixer' in the dyeing industry, and for tanning and ink manufacture. One of these 'copperas' works was on Brownsea Island and so Fiennes took a boat trip so that she could see the raw stones being 'gathered' and placed

on the ground raised like the beds in gardens, rows one above the other, and are all shelving so that the raine disolves the stones and it draines down into trenches and pipes made to receive and convey it to the house; that is fitted with iron panns foursquare and of a pretty depth at least 12 yards over, and they place iron spikes in the panns full of branches and so as the liquor boyles to a candy it hangs on those branches: I saw some taken up it look't like a vast bunch of grapes . . .

Fiennes labelled the site as 'the Copperice workes' and yet she coloured her description with allusions to gardens, planting frames, rain, candy and grapes. The dissolving beds, the trenches, pipes and the building with its 'great furnaces' and heating pans would have occupied a conspicuous section of the island's shoreline and been smoky, noisy and alien to Brownsea's insular ecosystems. To Fiennes, the existence of vast reserves of ores and coal in Britain was evidence that the island had been blessed with unusual diversity. Mining was God's compensation in regions of poor agriculture. In Derbyshire, for example,

> you see neither hedge nor tree but only low drye stone walls round some ground, else its only hills and dales as thick as you can im- agine, but tho' the surface of the earth looks barren yet those hills are impregnated with rich Marbles Stones Metals Iron and Copper and Coale mines in their bowells, from whence we may see the wisdom and benignitye of our greate Creator to make up the defficiency of a place by an equivolent as also the diversity of the Creation which encreaseth its Beauty.

A couple of decades earlier, de Rocheford had taken the same view as he gazed across the uncultivable moorland outside Newcastle: 'The country hereabouts would be the worst and most steril that I have seen in England, were it not for its mines of sea-coal.' To de Roche- ford's eye, too, coal-mining compensated for unproductive soil. Pits enhanced the view.

Britain was too dependent on roads. The introduction in 1564 of the long-wheelbase wagon had increased payloads, and the swivelling front axle had improved turning circles. But road travel could be as demanding as a sea voyage. Some parts of Britain were so badly served that a 'stranger' would have to adopt an expeditionary attitude to survive. When Thomas Browne and his brother decided to leave the flatlands of Norfolk and venture north and west to the Pennines, the account of their 'Tour in Derbyshire' in the summer of 1662 read like a winter expedition in the Alps. Even before they reached the foothills, Browne was complaining that he'd 'never traveled before in such a lamentable day both for weather and way'. Ahead of them lay 'strange mountainous, misty, moorish, rocky, wild, country' (East Moor, between Chesterfield and Bakewell), which delivered in due course floods, mud and bogs and a road of 'rocky uneveness', 'high peaks' and 'almost perpendicular descents'.

When Celia Fiennes came this way in 1697, she too commented on the 'steepness and hazard of the Wayes' and on the need to hire guides in order to travel safely. Guides were frequently essential for travellers outside their home region. Maps were expensive and printed at too small a scale to help with detailed navigation. People encountered along the way would be unlikely to know what lay beyond the nearest market. Most 'locals' were just that; people who had spent their entire life within the radius of a one-hour walk. Fiennes complained at one point of 'common people' who 'know not above 2 or 3 mile from their home'. In most counties, signposts were so rare that in the County Palatine of Lancaster, they stood out like beacons. It was here that Fiennes was delighted to find 'Posts with Hands' erected at all cross-ways. In Cheshire, she was accosted by highwaymen.

Road surfaces were inconsistent. There were sections of pitched ways like the ones Fiennes enthused about in the County Palatine of Lancaster, but the state of most roads was patchy. Fiennes' journals are spattered with so many sloughs, floods and potholes that her rides read at times like an obstacle course. The 'very deep bad roads' between Uppingham and Leicester (now the A47) were 'full of sloughs, clay deep way' and delayed her so much that it took eleven hours to make thirty-three miles, a pace so slow that 'a footman could have gone much faster'. In Northumberland her horse floundered through bogs on the way to Haltwhistle and in Cornwall she was almost thrown from the saddle when her horse fell into a deep, flooded hole. The paved highways of the Romans had long gone. On Watling Street at a place Fiennes called 'Hockley in the Hole' (now Hockliffe), the 'sad road' was so 'full of deep slows', she decided that 'in the winter it must be impassable'. It was a notorious blackspot: in 1633, landowner Sir Edward Duncombe had to respond to complaints that he'd failed to maintain the road from Hockliffe to Woburn with a commitment to lay 400 loads of gravel and stone on the road, annually.

The roads issue had, as it happened, reached a tipping point. Poor drainage and increasing volumes of traffic, and in particular heavily loaded wagons and carts, and herds of cattle, were mashing surfaces into flooded ruts, potholes and morasses. In various parts of the country, attempts had been made to reduce the damage, charging premiums on cartloads weighing more than a ton, and then, through the Highways Act of 1662, prohibiting wagons with wheels of a width less than four inches, and those being hauled by more than a seven-horse team. The following year, local justices were granted the power to levy tolls on a section of the Great North Road, the tolls to be used for improvements. Toll roads got off to a slow start. A few more were created in East Anglia and Gloucestershire, but it wasn't until authority for them was switched from local Justices to independent bodies of trustees that the practice of using tolls to maintain roads took off. The first of the new turnpikes was established in 1706 and covered the section of Watling Street that Celia Fiennes had complained about, between Hockliffe and Stony Stratford.

One of the most spirited supporters of turnpikes was the author of *A Tour Through the Whole Island of Great Britain*. Daniel Defoe had travelled Britain widely, and like Celia Fiennes thirty or so years earlier, he'd kept notes. Unlike Fiennes, he included Scotland and Wales in his itineraries. He too found that roads veered between acceptable and lethal. In common with most travellers of the age, he had many narrow escapes. Crossing the Pennines by Blackstone Edge, he'd been beset by a blizzard and been 'blinded' by snow, and in Wales he'd found the going so difficult that 'Hannibal himself would have found it impossible to have marched his army.' He knew first-hand how turnpikes could transform road travel. Encountering the notorious sloughs of Hockley in the Hole, he was delighted to 'now see the most dismal piece of ground for travelling, that was ever in England, handsomely repaired'. Soon, an average of eight turnpike Acts were being passed every year. In an appendix to the second volume of his travelogue, Defoe set out his masterplan for improving Britain's road system. In a nutshell, he wanted to rebuild the Roman network:

> I have seen the bottom of them dug up in several places . . . a laying of clay of a solid binding quality, then flint-stones, then chalk, then upon the chalk rough ballast or gravel, 'till the whole work has been raised six or eight foot from the bottom; then it has been covered with a crown or rising ridge in the middle, gently sloping to the sides, that the rain might run off every way, and not soak into the work.

And he supported the principle that the slave labour and limitless access to raw materials enjoyed by the Roman 'lords of the world' could be substituted in the modern age by 'this new method of re-pairing the highways at the expense of the turn-pikes'. From what he'd seen, Defoe believed that great progress had been made by the turnpikes and tolls of recent years. By way of example, he cited 'that great county of Essex' and the road from London to Ipswich, a road 'formerly deep, in times of floods dangerous, and at other times, in winter, scarce passable; they are now so firm, so safe, so easy to trav-ellers, and carriages as well as cattle, that no road in England can be said to equal them'.

Turnpikes and wagonways were opening the way to new interven-tions in the landscape. Wheels rolled faster on constant gradients and gentle curves. Embankments, cuttings and bridges were becoming more common. Defoe wrote excitedly of the stretch of Watling Street from St Albans to South Mimms (a place whose identity has been inherited by a service station on the M25) where 'the bottom is not only repaired, but the narrow places are widened, hills levelled, bot-toms raised, and the ascents and descents made easy'. Defoe's dream was a web of roads reaching out from London 'completely sound and firm, as Watling-street was in its most ancient and flourishing state'.

Equally significant were the engineering advances being made in wagonways. One of the pioneers had been Huntingdon Beaumont. Dying in Nottingham prison was not the send-off he deserved. His wagonways were the answer, but he was killed by his creditors. By the 1660s, some forty years after he died, there may have been as many as nine wagonways in Tyneside. From there, the idea spread to the coalfields of Yorkshire and Cumberland, and to Scotland. Unlike those on the continent, British wagonways were horse-drawn, pin-guided, above ground, and big. The key to the mechanical efficiency of the system lay in minimising gradients and curves in the wagonway. A visitor to the Tyne in the 1720s described how one Colonel Liddell had built a five-mile wagonway from his pits at Tanfield 'over valleys filled up with earth, 100 foot high, 300 foot broad at the bottom; other valleys as large have a stone bridge built across, and in other places hills are cut through for half a mile together'. Gravity carried the loaded wagons downhill to the Tyne, then horses towed the empties back up on a parallel way. One of Liddell's bridges carried the Tanfield wagonway over the ravine of Causey Burn. It was a courageous construction, with a single arch of some 105 feet, standing 80 feet above the river. Completed in 1726, Causey Arch had the largest span of any bridge in Britain. The architect, a local mason called Ralph Wood, used squared sandstone, anchoring his gigantic span on buttresses built out from the sheer banks of the ravine. Some 450,000 tons of coal was run down the Tanfield Way in 1727 alone. Assuming the wagonway was used on working days only, a bystander would have seen, on average, one wagon passing every 45 seconds. The celebrated arch was Wood's second attempt to span Causey Burn, an earlier bridge of timber having collapsed. The mason was so anxious about the structural integrity of his new stone bridge that he committed suicide by stepping from its apex. The bridge still stands.

The engineering that was draining wetland, filling wet docks, levelling turnpikes and raising wagonways was not being applied to river navigations, where vessels were still impeded by weirs and flash locks.

In his book *The Natural History of Oxfordshire* (1677), antiquarian Robert Plot described shooting a lock with the current as being 'not without violent precipitation', while the upstream passage frequently required the help of 'a Capstain at Land': a capstan on the bank that could be used to haul a vessel against the flow of water. Neither direction could be achieved 'without imminent danger'. Flash locks were hazardous and time-consuming and weir owners could refuse to open a gate because they needed the water for a mill. Boats on the Thames could wait on the mud for a month or more until a lock was opened. When three pound locks were installed on the Thames between Burcot and Oxford, Robert Plot hailed them for their 'ease and safety'. But inserting pound locks like the ones installed in the 1560s on the ship canal at Exeter was difficult. Building them required both an Act of Parliament and the support of local landowners who frequently took a not-in-my-back-plot attitude. Towns objected to improved rivers on the grounds that their merchants and seamen would suffer if trade was diverted to other rivers and towns at the heads of navigation; their fear was that a more navigable river would carry trade straight past their own quays without stopping. Other vested interests included agricultural landowners who argued that bargemen would trample crops and steal livestock, and horses would strip meadows. There were also fears that locks would raise river levels and cause flooding, and that an enlarged river system would distort market prices: in Leicestershire, it was argued that people benefited from cheap corn because there was 'no Navigable River near to carry it away'. Road hauliers also objected to river improvements because bargemen threatened their custom.

With so many objections to improving river navigation, canals remained an unrealisable dream. The most-discussed scheme was a link that would connect the North Sea to the Atlantic by way of the Thames, Bristol Avon and Severn. The idea had cropped up in the reign of Elizabeth, and in a pamphlet of 1610 titled *A Profitable Worke to this whole Kingdome*, Thomas Proctor advocated bringing 'one river to or neare unto another, as Thames to or neare Severne, or Severne neare or to Thames', so that coal from the Forest of Dean could be conveyed by water all the way to the capital. The idea that Britain's waist could be cut through from sea to sea was eagerly seized upon in 1655 by Francis Mathew, who mooted a variation which depended upon the upper reaches of the Bristol Avon being made navigable to Malmesbury, where a canal would be excavated through to the Isis at Lechlade or Cricklade. Mathew envisaged a fleet of 300 billanders – Dutch-style sailing barges – operating in squadrons, each commanded by an Admiral and Rear-Admiral, 'carrying their Flags of proper Colour'. Every billander would carry 30 London chaldrons – nearly 40 tons – so the entire fleet would be shifting over 10,000 tons at a time to London. To protect the Newcastle coal trade, Mathew's colourful squadrons would sail only in winter. There were many other attempts to link London and Bristol by canal, including a Bill introduced to the Lords in 1668 by the Earl of Bridgewater, who was disappointed that 'some foolish Discourse at Coffee-houses laid asleep that design as being a thing impossible and impracticable'. Futurists like Proctor, Mathew and Bridgewater had intractable foes ranging from landowners along the routes of their proposed canals to the proprietors of river navigations who foresaw the end of lucrative regional monopolies. Countless canals were contemplated, yet none built. Meanwhile, between 1660 and 1724, the length of navigable rivers nearly doubled from 685 to 1,160 miles.

While canals were blocked by self-interest, hydraulic engineering leapt ahead on the coast. By the end of the century, the four-hectare Blackwall Yard wet dock was looking like a village pond. Upstream towards the city, a 25-hectare commercial basin was excavated at Rotherhithe. The Bill relating to its construction was given Royal Assent in 1696, and it was completed several years later. The Howland Great Wet Dock was a game-changing project; the first large commercial wet dock in Britain, it was nearly 1,000 metres long, with an entrance lock 13 metres wide. It could take no fewer than 120 ships, afloat. The Howland Great Wet Dock was a laying-up and fitting-out facility rather than a port complex with wharves and warehouses, but it proved that basins on this scale could be built and operated. It took an even more congested port than London to convert the concept of the commercial wet dock into a full-blown artificial trading port. That port was Liverpool, where traffic had raced ahead of available wharves.

Behind Liverpool's new dock was one of the most experienced engineers in the kingdom. Thomas Steers probably grew up at Rotherhithe on the Thames and had served as an infantry officer among the polders and dykes of the Netherlands. Married and settled back at Rotherhithe, he'd watched the construction of the Howland Great Wet Dock, and in 1707 Elizabeth Howland had commissioned him to prepare a hydraulic survey. By 1710 his reputation was sufficient for him to have been called to Liverpool to draw up plans for the port's first wet dock. Covering nearly nine hectares, it was somewhat smaller than the Rotherhithe dock, but it was part of a complex that turned Liverpool into the most important port on Britain's west coast with a basin large enough to take a hundred ships. These gigantic facilities were a significant leap towards fabricated coasts. The slips, quays and docks of old were relatively tiny modifications to natural shorelines; wet docks on the scale of Rotherhithe and Liverpool were of a new order: artificial, controlled inlets that excluded tides.

Not for the first time, British landscapes were on the brink. The last wilderness had been edged off the map. The archipelago had been cleared and settled. Fens had been drained and coasts realigned. London had ballooned into the biggest city in western Europe. Very little of this was innovative. Out on the edge of Europe, removed by seawater from direct interaction, Britain was a land of topographic mimicry. Big ideas had usually been imports. But that changed as Britons finally turned to water for inland transport and discovered the power of coal. On an island abundant in both, their exploitation would be driven by engineers, apprentices, farmers, dilettantes – a disparate, virtual community of creative thinkers whose schemes ranged from crackpot to practical. The invention revolution would change the face of Britain.

TWENTY

Unnatural Geographies

1725–1811

Canals finally crossed the water, but not to England, Wales or Scotland. The birthplace of David Riccardo has not been established, but his mother, Rachel Burgos, was born in Bombay and his father was an English-born Jew who for a while was Director of Munitions and Mines to Friedrich Augustus, Elector of Saxony and King of Poland. Riccardo had lived in Dresden, although he married Rachel in Amsterdam. Their son, David, is thought to have been an officer in a regiment of engineers and to have spent time in Germany, France and Holland, where he developed an interest in fortifications and canals. By the 1720s, David Riccardo had become Richard Castle and was in England. In 1728, he was in Ireland, working for Edward Lovett Pearce, a well-connected ex-captain of Dragoons who had become a Member of Parliament and then been selected as architect of the new Parliament House in Dublin. Pearce appears to have needed Castle's architectural expertise and in the subsequent wrangle about credit for the building, an anonymous contributor to the *Freeman's Journal* claimed that one of them was 'a Gentleman of Rank and liberal Education, [who] seems to have possessed a classic and polished Taste. The other a Foreigner of great Experience and Skill in building [who] does not appear to have possessed much Elegance of Fancy.' *Grands Projets* of this scale required a dreamer and a doer.

With his eye on the role of Ireland's Surveyor General, Pearce became a trustee of the Navan Turnpike and of Ireland's first canal, a freight route that would facilitate the shipping of coal from the deposits of East Tyrone to the markets of Dublin, removing with the strike of a pick a troublesome dependence on English coal. It was a conceptually bold and technically challenging project: a cut eighteen miles long that climbed over the watershed between the largest inland lake in Ireland and the open sea. Digging began under Richard Castle in 1730, but it was a long, fraught project. Castle seems to have been out of his depth and in 1737 the engineer who had turned Liverpool into a world-class port was called across the water. Thomas Steers completed the Newry Canal, which finally opened in 1742. *The Dublin News-Letter* of 30 March 1742 described how the first vessel loaded with Tyrone coal passed through the canal and arrived at Ireland's capital with a flag at her topmast and guns firing. England looked on.

Thirteen years after Irish labourers completed the first summit canal in Britain and Ireland (and more than 350 years after the Hanseatic League had built their summit canal to Lübeck), England tentatively dipped its toe in canal water. It wasn't even a premeditated act of canalisation. The first parliamentary authorisation relating to the Sankey Brook Navigation in south Lancashire was a commonplace 'Act for making navigable' an existing waterway, in this case some streams and a river – the Sankey Brook – flowing from the St Helens coalfield to the River Mersey. Like Newry, Sankey had a broker and an engineer: John Ashton was the money-man and Henry Berry understood water. In his will, Ashton referred to himself as a 'Merchant and Cheesemonger', a disarmingly modest title for a speculator whose bank balance had been built on the African slave trade and on ownership of the Dungeon Salt Works on the north bank of the Mersey estuary. On paper, serious money could be made if a means could be contrived to ship coal in greater volume, at less cost, from the relatively close collieries at Haydock and Par into Liverpool. Henry Berry had been clerk to the great Thomas Steers, who – it was later reported

– had given the youngster 'the finest instruction'. Berry knew about hydraulics, and he'd grown up beside Sankey Brook.

Liverpool's Common Council authorised Berry to undertake a preliminary survey of the Brook and its streams. Berry knew that it was a navigational non-starter: the river was too narrow, too variable in flow and, in wet weather, it burst its banks. But there was little chance of getting Parliament to agree to the construction of a completely new 'cut', a canal. What happened next was described somewhat coyly in Berry's obituary. After conducting his survey and confirming that the Brook would be an 'impractical' navigation, Berry 'communicated his sentiments to one of the proprietors', who learned that 'the object they had in view could be answered by a canal'. That proprietor was presumably the wily merchant John Ashton, who approved a plan to extend the scope of the waterworks. At the time, improving a river navigation was understood to include dredging and widening, making cuts to eliminate difficult bends and installing locks to maintain a constant water level. Berry and Ashton went a bit further, bypassing the awkward Sankey with a new ten-mile cut, using the adjacent Brook as a top-up reservoir and as an overflow. Work stated on 5 September 1755, 'but the project was carefully concealed from the other proprietors, it being apprehended that so novel an undertaking would have met with their opposition'. In November 1757, the *Liverpool Chronicle* ran an ad stating that 'Sankey Brook Navigation' was 'open for the passage of flats to the Haydock and Par collierys'. Britain, at last, had joined the Canal Age.

The Sankey Brook Navigation was so successful that it destabilised the local economy. Within two years of the canal opening, the Duke of Bridgewater found that his Worsley coal was being undercut on the Manchester market by cheaper coal coming along the Sankey from St Helens. Packhorses couldn't compete with barges; roads couldn't compete with canals. And this was a surging market: Manchester needed fuel for its rapidly expanding textile industry. As it happened, the Duke of Bridgewater was already a canal convert. His father had dreamed of a coal-carrying waterway to link the family's mines at

Worsley with the needy hearths of Manchester, a swelling city dependent upon packhorses for its fuel. And the young Duke's education had included a Grand Tour of the continent, under the tutelage of a brilliant classical scholar called Robert Wood whose archaeological works included books on the ruins of Palmyra and Balbec. Wood did his best to excite his charge with classical sites, but the sickly boy was keener on canals. At the time, the Canal Royal en Languedoc was regarded by Grand Tourists as one the the most spectacular sites in Europe.

The Duke's ally was his land agent, John Gilbert, a hands-on engineer who had begun adulthood as a twelve-year-old apprentice in a factory making small metal parts. Gilbert saw that a canal tunnel that was bored from the foot of a cliff horizontally into the mineworkings would serve more than one purpose: it would drain his flooding mines, it would avoid the difficult and expensive process of lifting coal up to the surface and it would avoid the laborious six-mile overland route to central Manchester; one horse would be able to tow a barge loaded with 30 tons of coal, more than ten times the volume that could be hauled on the road. In 1759, the Duke obtained an Act to construct a canal from the mine to the River Irwell, where the coal could be transferred to river craft for the final few miles into Manchester. This was not a good solution, and the Duke knew it. The Mersey and Irwell navigation was notoriously unreliable and was once described as 'tedious, expensive, and liable to great interruption'. True to form, the navigation company tried to charge the Duke the highest possible tolls. That, and difficulties with choosing a convenient route for the canal, prompted Gilbert to introduce to the Duke a brilliant 43-year-old engineer called James Brindley.

Several years older than Gilbert, Brindley had begun his working life apprenticed to a millwright and wheelwright in a village outside Macclesfield and then set up on his own in Leek, where his reputation spread as a tinkerer and improver of machinery, albeit one who was better at relating to his own imagination than to his fellow human beings:

In appearance and manners, as well as in acquirements, Mr. Brindley was a mere peasant. Unlettered and rude of speech, it was easier
for him to devise means for executing a design, than to communicate his ideas concerning it to others.

By 1750 the 'mere peasant' had a second workshop, leased from the
Wedgwood family in Burslem. In Brindley's world, there was no mechanical device that could not be improved. He worked on water- and
windmills, on atmospheric engines and won acclaim for a drainage
scheme. His nickname was 'the Schemer'. He pumped out a flooded
colliery near Manchester by using an overshot wheel powered by
water conveyed through an 800-metre-long tunnel from the River
Irwell. He'd also taken out a patent on 'fire-engines'.

It was at this point – summer of 1759 – that the Schemer and the Duke
were united by land agent John Gilbert. Brindley conducted a forty-
six-day 'ochilor servey or a ricconitoring' of a ten-mile canal route
which would link the Duke's coal mine and the Manchester markets.
It broke the rules of hydraulic engineering. Unlike the Newry and the
Sankey, the new canal would pay no attention to natural waterways.
It would be an entirely artificial cut, taking its cue from the contours
of the landscape rather than from rivers. Brindley adjusted the route
so that it could accommodate a subsequent branch connecting with
Liverpool. Being level, there would be no locks. The main obstacles
on the route were the low-lying floodplain of the Irwell and the river
itself. Brindley proposed to float his canal across the floodplain on a
lofty embankment 900 yards long and 17 feet high, and to hurdle the
river with a huge aqueduct, a plan so outrageous that it provoked the
engineer John Smeaton (who had recently overseen the building of
Eddystone lighthouse) to comment: 'I have often heard of castles in
the air, but never before saw where any of them were to be erected.'

Less than two years after Brindley started work, coal barges were
gliding above the treetops of the Irwell valley on a 200-metre aqueduct raised above the River Irwell on three great stone arches. A
bargee guiding 50 tons of coal at the walking pace of his horse along

the canal's still waters could gaze down upon knots of boatmen struggling to shift far smaller loads against the river current (the Barton aqueduct became known as the 'Castle in the Air'). By 1761, the canal was open through to Manchester where the average price of coal halved from 7d per hundredweight to 3½d. Of equal importance to the town's businesses and homes, coal now arrived regularly rather than intermittently. At the Worsley end of the canal, a basin was excavated at the foot of the cliff, now pierced by a tunnel to the coal seams. Coal was loaded directly onto barges deep underground and taken by water all the way to Manchester. The following year, the Duke was seeking parliamentary authority to extend his canal to the Mersey. If successful, he'd be able to ship his coal to Liverpool as well as Manchester, without using the despised rivers of the Mersey and Irwell navigation.

Before the Bridgewater Canal had even reached the Mersey, plans for a national network were being sketched by Brindley and other canal engineers. Back in early 1758, Brindley had been commissioned to survey the route of a proposed canal linking the Mersey and the Trent, by way of the Staffordshire potteries. This was a 'scheme' far greater than anything he'd been invited to join in the past: an artificial waterway that would link the seaports of Liverpool and Hull and connect Britain's western and eastern coasts. The route had been investigated three years earlier by a couple of surveyors from Liverpool under the pay of the town's Corporation, but now the lord lieutenant of Staffordshire, Lord Gower (who also happened to be the Duke of Bridgewater's brother-in-law), wanted Brindley to have a go. Brindley's journal entries for February 1758 refer to days investigating the canal that would create this coast-to-coast 'novocion' or 'novogation'.

Although the Trent-to-Mersey did not have the political momentum to develop, Brindley had been introduced to the idea of the 'Grand Trunk': a main stem to which 'branches' could be attached, forming an inland navigation system connecting every part of England and all of its major seaports. Nothing on this scale had been built since the Romans had rolled out their road network across Britain.

The Grand Trunk could create a parallel web of waterways without currents, shallows, weirs and obstructed banks. In the imaginations of engineers, the Grand Trunk became the Great Cross, a system of artificial waterways which would finally overcome the deficiencies of rivers. Centred on the coal and iron around Birmingham, hundreds of miles of canals would be excavated linking the four points of the cross: London, Liverpool, Bristol and Hull. Low-cost freightways would connect east with west, port with port. There would be no hinterland; places that mattered would be conjoined. The northern segment would connect the Trent with the Mersey; the North Sea with the Irish Sea. Other canals would link the Trent and Mersey with the Severn and Thames. It would be a triumph of engineering over nature. Engineer Brindley never did think much of rivers, anyway. During a parliamentary exchange which became canal-lore, the engineer from Derbyshire was asked what he thought rivers were for: 'To feed navigable canals.'

The success of Brindley's link from Manchester towards Liverpool stiffened political wills and loosened purses. By 1767, the Bridgewater Canal had crept twenty-four miles along the contours all the way from Stretford to Runcorn, again without the use of any locks. The Mersey was in sight. Another innovation for Britain – a flight of ten locks – finally connected the Duke's coal mines and Manchester with the sea. If the Sankey Brook Navigation could be said to have begun the process of regional destabilisation, the Duke's Cut took it a stage further: land carriage between Liverpool and Manchester was priced at 40 shillings per ton; carriage on the tiresome currents of the Mersey and Irwell was priced at 12 shillings per ton. The Act of Parliament relating to the Duke of Bridgewater's Canal restricted him to charging 6 shillings per ton. The Duke could undercut road hauliers by a factor of seven, and still make a huge profit. Brindley's contouring line and flight of locks were the final break with natural watercourses. He'd created an entirely artificial waterway. There was virtually nowhere in industrialising Britain that a canal could not reach. Cheap coal unleashed a clamour for canals. Conceived as local solutions to inefficient and expensive transportation, canals came back to the agenda as a systematic network. Britain's entrepreneurial sluices opened and released a wave of canal building.

The first sods of the Grand Trunk were cut by a lame master-potter from Staffordshire in July 1766. The barrow had to be wheeled by James Brindley. Josiah Wedgwood had been crippled in his teens by smallpox and grown up in Burslem on the packhorse trail that kinked and climbed through hilly country between the Peak District and the Liverpool road. Burslem and the neighbouring villages were remote and badly served by appalling roads: holloways that became torrents in wet weather and rutted mires so deep and sticky that horses and donkeys would break their legs. Carts and wagons were rare. But for hundreds of years, these cut-off communities had augmented their income by working the superb local clays. The youngest of thirteen children, Josiah used to walk the seven-mile round trip to school at Newcastle-under-Lyme and by the age of fourteen was apprenticed to his brother in the family pottery, learning the art of 'throwing and handleing'.

'I myself began at the lowest round of the ladder,' remembered a man who would become famous for resilience and vision. It was vision that prompted him to buy a 140-hectare estate on the route of a canal before it existed. After apprenticeship, Wedgwood had set up as an independent potter in Burslem and developed his passion for experimental glazes. His £10 a year rental lent him an ivy-smothered cottage, a few tile-covered sheds and a pair of kilns, but by 1765 he'd opened his own showroom off London's Grosvenor Square and the year after that, he bought the Ridgehouse estate in the valley a couple of miles south-west of Burslem. The Grand Trunk was due to make its way along the valley as it sought the line of least resistance around the southern end of the Pennines. The man who had grown up isolated by mires knew the value of location. The Grand Trunk would bring in his essential raw materials like the white clays of Dorset, Devon and Cornwall and enable his fragile wares to be conveyed on smooth water away from the pottery to markets in Britain and America.

The Ridgehouse estate became an Etruscan town. On this greenfield site in a Staffordshire valley, Wedgwood built a large manufactory and laid out new grounds and accommodation for his workers and their families. Commemorating the source of his artistic inspiration, he called the place Etruria. More than many at the time, Wedgwood understood that the art of industry was the coordination of production and organisation of labour. In one of his earlier manufactories, he'd erected a belfry so that he could summon his workers at the start of each day, the old practice of blowing a cow's horn having failed to deliver the required enthusiasm (the pottery became known as the 'Bell Works'). He was keen on order and discipline. And he came to learn that good housing, close to the works, was fundamental to industriousness. He wasn't the first owner of a manufactory to buy the motivation of his workers through the provision of accommodation but he was ahead of his peers in providing housing for so many. Around 300 could dwell in the neat little terraced houses he built at Etruria, with their two rooms and scullery on the ground floor and a couple more rooms upstairs. And these were not back-to-backs, but spaced rows of houses with gardens. A few had three storeys. Inhabitants were known as 'Etruscans'. The Etruria pottery was inaugurated three years after Wedgwood had bought the undeveloped site.

The Grand Trunk – the Trent and Mersey missing-link – began to grow branches before the main stem was completed. In the same year that the backers of the Grand Trunk obtained their Act of Parliament, another bunch of backers got an Act to build a canal linking the yet-to-be-constructed Grand Trunk with the Severn navigation. The promoters of the Staffordshire and Worcestershire also employed James Brindley to oversee the engineering. Having shown that he had the courage to experiment, Brindley now displayed fearlessness in the face of a scheme the scale of which far exceeded anything he'd tried before. The Staffordshire and Worcestershire required no less than 4 aqueducts, 43 locks and a tunnel; the 93-mile Trent and Mersey was more than twice the length of the Duke's Cut to the Mersey and required 76 locks and a tunnel measured at 2,919 yards (2,669 metres).

The Staffordshire and Worcestershire was completed in 1772, the year Brindley died, and the Grand Trunk was completed in 1777.

Canals and industry were part of the same story. They depended upon each other for expansion. And the element common to both was water. Water filled canals and water powered industry. Thirty miles east of Etruria, the River Derwent had been turned into an axis of invention by the son of a Lancashire tailor. Young Richard Arkwright was Britain's most celebrated apprentice. His parents had been too poor to educate him at school and after serving his apprenticeship with a barber, he'd set up his own shop in Bolton, where he'd invented a waterproof dye for periwigs. By the time he reached his thirties, he was looking into one of the greatest technological challenges of the age: how to mechanise the domestic spinning wheels that whirred through the calloused fingers of cottagers the length and breadth of Britain. Collaboration with a watch- and clockmaker from Warrington (John Kay had the technical nous; Arkwright the money) produced a cotton-spinning machine that substituted fingers with wooden and metal cylinders. The machine could spin 128 threads at a time, threads furthermore that were stronger than those twisted by hand on a wheel.

The implications for human geography were immense: a single technical advance had replaced the ancient practice of home spinning with mass-production in a manufactory – a factory. Arkwright's ratchet of breakthroughs continued with a move in 1771 from his horse-powered mill in Nottingham to the valley of the River Derwent on the southern flanks of the Pennines, where he built a mill powered by water flowing from Cromford Sough, an old lead-mine channel which provided a constant supply of warm water. Arkwright's was the first successful water-powered cotton mill in the world. Five-storey Cromford Mill ran day and night and employed 200 people, far more than this stretch of the Derwent could provide. Like Wedgwood, Arkwright provided housing for his workers. By 1777, Arkwright had built another mill at Cromford. The Derwent, its worker settlement and Sir Richard – he had been knighted in 1786 – were being talked about far beyond the borders of Derbyshire. One of those who rode this way was Sir John Byng, an army officer turned tourist who, like Celia Fiennes and Daniel Defoe, was 'seized with this journalizing frenzy'. Byng could not believe his eyes:

> Below Matlock, a new creation of Sir Richard Arkwright's is started up, which has crowded the village of Cromford with cottages, supported by his three magnificent cotton mills. There is so much water, so much rock, so much population and so much wood that it looks like a Chinese town.

There were many other 'Chinese towns'. After a visit to the Duke of Bridgewater's coal basin at Worsley, Josiah Wedgwood commented that it had 'the appearance of a considerable Seaport town. His Grace has built some hundreds of houses, and is every year adding considerably to their number.' Standards were often higher in these industrial dormitories than in Britain's ancient towns. At Etruria, Wedgwood had sunk three wells fitted with pumps so that every worker had access to crystal-clear water. John Byng had found industrial Cromford 'so clean, and so gay, as to quite revive me, after the dirt and dullness of Bakewell'.

Another industrial hotspot could be seen on a stretch of the River Severn, where it cut through a gorge upstream of Bridgnorth. Geology had provided the makings of a fortune, for all in the same place were coal measures, ironstone and plentiful wood and water. It was here, in a side-valley called Coalbrookdale in 1709, that Abraham Darby built a coke-fired blast furnace to manufacture cast iron. A fairly small-scale operation abruptly changed gear in the 1750s when Darby's son, also Abraham, started to supply his coke-blast iron to the leading forges in the Midlands. A clear-watered, wooded dale became an industrial gulch packed with furnaces and forges fed by dammed ponds. Where mighty broadleaves had swayed in autumn gales there were now spoil heaps, brickyards and wagonways rumbling with horse-drawn loads of coal, timber and clay. By the late 1700s, Coalbrookdale was one of the most manipulated landscapes in Britain; a complex that oozed like a river of molten metal along the tight valley upstream of Bridgnorth. When the agricultural writer Arthur Young visited Abraham Darby's coke-blast iron works in the summer of 1776, he found the surrounding countryside 'too beautiful to be much in union with the variety of horrors spread at the bottom; the noises of forges, mills, with their vast machinery, the flames bursting from the furnaces with the burning of coal and the smoke of the lime kilns.' By the 1790s, Coalbrookdale was on the tourist trail, drawing folk like James Plumptre, a 21-year-old Cambridge graduate and prospective Reverend, who had been told to visit the dale in the dark when the spectacle would be 'far greater; the gloomy woods, sounding waters, and gleamy fires, giving it an effect, like that poets tell us of in their descriptions of the Shades below'. (Unfortunately, he couldn't hire a chaise, so missed the flames of Hell.)

Common to these concentrations of volcanic energy were spectacular structures. In the case of Worsley, it was the subterranean labyrinth of coal canals and loading quays, and the fabled 'Castle in the Air'; in the case of Etruria, it was the Grand Trunk; Coalbrookdale had a great bridge built by Abraham Darby III over the River Severn in 1779. It was the world's first bridge to be constructed of iron. For Britain, an island that had taken its cue from the continent for so long, the bridge was rich in symbolism. Henceforth, innovation would cross the gulf in both directions. Darby's builders used traditional wood-working joints to fit their metal edifice together, but the tensile strength of iron meant that the bridge's many members were more slender than would have been possible with timber. Viewed from the riverbank, Darby's ferrous creation appeared to be suspended over the void on a spider's web of slender filaments. Britain's ancient forests were going up in smoke, but industrialists were replacing them with works that filled people with awe. Plumptre was impressed:

The Iron bridge is an amazing work of art, but is calculated rather
to surprize, than please, and seems more for curiosity than use. I

think this will be known only by fame, when a stone bridge of the same expence would not have been the worse for the time it had stood: but it serves to shew the power of human art.

Coalbrookdale was celebrated for its scale and for its spectacle. Stretching for a couple of livid miles through the countryside, it was the largest site of its kind in Britain. Combining extraction, manufacture and invention, it represented the first generation of landscapes that could be called 'industrial'.

While industrial landscapes seethed and clattered with increasing urgency, Britain's historic towns and cities were being propelled into new identities. The small, out-of-the-way West Riding parish of Hutherffeld underwent a spasm of water-powered growth. Since being granted market rights in 1671, this tiny cog on the Colne had been expanding and developing its radius of influence. By the 1720s, Hutherffeld was appearing on maps as Huthersfield, with its single street proudly labelled 'The Town Street'. (The upper end of the street was known as 'Top o' th' town' and the lower end 'Bottom of Town'; it was a practical kind of place.) On ground floors, families laboured over looms, turning wool into cloth. In some of the smaller cottages, looms were squeezed amid the furniture of kitchen or bedchamber while larger establishments might have dedicated workshops and additional outbuildings for dyeing or finishing. On the valley slope where streams tumbled to the Colne there were fulling mills powered by water. Further down Town Street, an open space close to the church had become the Corn Market and a larger space where a pair of narrow lanes entered town from the north and north-east had been turned into the Beast Market. For want of their own place, local clothiers draped their wares over the churchyard wall and tombstones. Market status had yet to create a genuine town from the Hutherffeld Saxton had mapped as one of many villages hosting a parish church. But by the later 1700s, it was the nucleus of half-a-dozen or so surrounding settlements whose combined population may have been around five thousand and whose primary source of income was the manufacture of cloth. Aside from its marketplaces, there was little to separate Huthersfield from its neighbours, but in November 1766, it acquired a definitively urban landmark.

On open land at 'Top o' th' town', Sir John Ramsden built a gigantic trading hall for the neighbourhood's clothiers. The Cloth Hall dwarfed every other building in town. There was nothing remotely vernacular about Ramsden's monumental trading house. Elliptical in plan, it was built of brick, rather than local stone. Walking around the 805-metre, curved perimeter of the building took several minutes, an experience that must have amazed the dale's cottage clothiers. The exterior wall was over eight metres high. Ramsden's brick Coliseum was finished with a stone-quoined, triptych entrance dignified with a pediment and a clock tower. On top of the lot was a columned Renaissance lantern (complete with capitals) and a cupola crowned with a decorative orb; a Florentine flourish deep in the West Riding. Superficially, the exterior of the Cloth Hall was a spectacular demonstration of Ramsden's wealth (and questionable taste), but the scale of the thing demonstrated an extraordinary belief in the commercial potential of this out-of-the-way town. Inside the Hall were 116 trading stalls regularly ordered around the interior of the wall, and in two aisles that cut at right-angles to each other across the long and short axes of the ellipsis. Ramsden referred to these aisles as 'transepts'.

Huthersfield's clothiers moved indoors from the churchyard. Quantitatively, the Cloth Hall was just a big building in a small town, but its sudden, colossal presence in this West Riding backwater modified Huthersfield's identity. In a pattern common to emerging manufacturing hubs, the appearance of a landmark building unshackled a surge of developmental energy. The Cloth Hall was quickly followed in 1771 by the building of a new 'shambles' to accommodate the town's forty or so butchers. Seven years later, a map revealed a manufacturing town that had far outgrown its one-street status. The Street was still the central axis for scores of outlying cloth producers but suburbs now stretched like tentacles from both ends. The more desirable section of

The Street was the upper, cleaner end adjacent to the new Cloth Hall, and it was here that the town's principal businesses had gathered: a smithy, joinery, bakery, cutlery, stables, warehouses, wool-shops and inns. The Market Place was midway along The Street, opposite the George Inn. On an area of open space at Bottom of Town, cloth that had completed the fulling process was dried and stretched on ten-terframes. Beside the Tenter Ground ran the leat – locally known as a 'goit' – which brought water from the Colne, across the common to Huthersfield's corn mill. On Tuesdays, market came to town and from dawn till dusk various open spaces were turned into trading zones. The Market Place became a pop-up emporium packed with stalls, wagons, carts and milling traders. By the end of the century, Huthersfield was appearing on maps as Huddersfield.

Despite being wedged into a deep V-shaped, Pennine valley, Huddersfield responded to the call of canals. In 1776, Sir John Ramsden connected his town to the outside world by excavating a four-mile canal along the bed of the Colne valley to connect Huddersfield with the Calder and Hebble Navigation – a canal completed in 1770 with the intention of connecting the River Calder navigation at Wakefield with Sowerby Bridge in the West Riding. Huddersfield gained an entrepôt, with a wharf, yard, warehouse and access road on the fields beyond Bottom of Town. Both the Cloth Hall and canal facilities were slightly south of the town centre and by 1778 Sir John was contemplating his own 'new town', a huge development on the fields to the south of the old main street. Ramsden wanted to use the Cloth Hall as the focal point for a new street plan: a long, broad, straight street would run down the slope from the Cloth Hall towards the canal. New side-streets would form the basis of a rectilinear grid. Sir John's new town would marginalise Huddersfield's narrow, kinking town street and create a new axis of enterprise on the green fields to the south. It was a hectic age.

If Huddersfield could be characterised as one of the brash, urban fast-breeders, Norwich still saw itself as the provincial paterfamilias, defined by its medieval town walls and unable to accept that it was no longer England's second city. Between 1751 and 1774, local architect Thomas Ivory uprated the city with his own brand of classically inspired exteriors: a Methodist Meeting House, an octagonal chapel, the Assembly Rooms and a theatre, numerous houses and the Artillery Barracks. It wasn't that Norwich was shrinking, but that other cities were growing faster. For much of the eighteenth century, the population of Norwich had fluctuated fairly erratically, and by 1800 it had crept to 37,000 or so. But by then it had slipped to ninth in the provincial rankings, behind Newcastle/Gateshead, Plymouth, Sheffield, Leeds, Bristol, Birmingham, Liverpool and Manchester. The new generation of ports and industrial cities were expanding at unprecedented rates. Liverpool, a waterside parish of just 6,000 in 1700, mushroomed by 1800 to a city of 80,000. Its neighbour Manchester blossomed from 17,101 in 1758 to 70,409 by 1801.

Manchester was emerging as a special industrial case: a town with a unique set of attributes. Across the Pennines from Huddersfield, Manchester had been one of Lancashire's main towns back in the sixteenth century when Saxton published his county maps. Well connected and straddling a confluence of rivers, it was at the centre of a district that had specialised in the home production of textiles for so long that 'proto-industrialisation' was part of the culture. People were used to innovating and networking, and Lancashire had developed a solid middle class of entrepreneurs with access to capital for machinery. There was also a tradition of engineering and of education in scientific societies and technical academies. In the south of the county between the Pennines and the Mersey estuary, Manchester had unparalleled industrial potential.

Among west-coast ports, Liverpool had gained most overall from the increase in transatlantic trade, but Glasgow was making up for lost time. A beneficiary of the Union of 1707, Glasgow was able to capture and then dominate the tobacco trade and to set up manufactories to fulfil rising demand from America and Europe. Industries sprang up on the Clyde, from cotton to chemicals, metal goods and machinery. By the end of the century, Glasgow had become the only large port on the west coast of Britain that was also an industrial city. Its problem was shoals: the fourteen miles of river between the city and Dumbarton where the Clyde widened into an estuary were partially obstructed by no less than ten submerged banks which reduced the depth of water to a metre or so at low tide. Erosion, exacerbated by the prevailing south-westerly winds, had been widening the Clyde and reducing its depth so much that 30-ton lighters using Broomylaw Quay could be trapped in dry seasons for weeks on end. There had been various attempts to resolve the problem and in 1768, the city authorities sought the advice of an engineer called John Golborne, who had won acclaim for his work on the New Cut Canal outside Chester – at the time, the greatest work of its kind in Britain. Golborne's radical solution for the Clyde was to change the profile of the river 'by removing the stones and hard gravel from the bottom of the river where it is shallow, and by contracting the channel where it is worn too wide'. The engineer was confident that he could increase the low-tide depth over the shoals to two metres. In July 1770, Golborne's men began work and by the end of 1772, they'd built over a hundred 500-foot jetties to reduce the width of the river and were scraping away the shoals using dredging ploughs attached to punts on the river bank. One of many visitors to the operation was a friend of Golborne's from Wales, an accomplished traveller and Fellow of the Royal Society, Thomas Pennant, who called for breakfast with the engineer in the summer of 1772, then joined the spectators: 'After breakfast,' noted Pennant, 'survey the machines for deepening the river.' What he saw were huge dredge-buckets made of cast-iron and timber being drawn to and fro across the Clyde by capstans. He was told that each bucket could bring up half a ton of gravel; 1,200 tons a day. Golborne succeeded in removing so much gravel from the river in front of Broomylaw Quay that it was possible for lighters of 70 tons to berth. The city authorities were so pleased that they presented Golborne with silver cup and a £1,500 gratuity.

Back from the waterfront, Glasgow itself was also in the throes of reinvention. A new bridge over the Clyde – the first for 400 years

– was completed in 1772, the year Pennant visited and although he used the word 'disgrace' in describing the bridge's design, he found the rest of the city generally pleasing. Glasgow's houses were of stone and 'generally well built, and many in a good taste, plain and unaffected'. The main street was 'unfortunately not straight' but the view from the cross at the principal intersection had 'an air of vast magnificence'. The tollbooth was 'large and handsome' and there was an imposing statue of King William in front of the Exchange. But it was Glasgow's market buildings – so often the symbol of civic progress – that really fired Pennant's admiration. The days of scummy streets and outdoor stalls were over:

> The market-places are great ornaments to the city, the fronts being done in very fine taste, and the gates adorned with columns of one or other of the orders. Some of these markets are for meal, greens, fish or flesh: there are two for the last which have conduits of water out of several of the pillars, so that they are constantly kept sweet and neat. Before these buildings were constructed, most of those articles were sold in public streets; and even after the market-places were built, the magistrates with great difficulty compelled the people to take advantage of such cleanly innovations.

Glasgow, decided Pennant, was 'the best built of any second-rate city I ever saw'.

Glasgow's population ballooned by an astounding 75 per cent in the twenty years to 1801; 77,000 people dwelled in a city of two towns: the old town gathered around the original medieval 'cross' of streets, and the neat, rectilinear grid of the 'New Town' on the greenfield site framed on two sides by the historic axes of the High Street and Trongate.

Glasgow and Liverpool shared similar, west-coast, maritime models, with urban topographies evolved to support expanding ports. Over on the east coast, Edinburgh's story was entirely different. By the 1750s, Scotland's most important urban centre had changed little since Daniel Defoe tempered his excitement at the city's spectacular location with revulsion at its overcrowded squalor. Edinburgh's problems had a lot to do with its mountainous geography. The town was confined to the crest of a long, steep-sided ridge that descended from the castle at its western end to Holyrood Palace at its eastern end. Between the two ran Edinburgh's main street, one mile long and lined on both sides – as if they were teeth on a two-sided comb – by narrow alleys, or *wynds*. A second, parallel street ran along the lower, southern levels of the ridge.

Edinburgh was a dump. And in 1752, the Convention of Royal Burghs went public with the insanitary truth. Confined to its ridge, it had only 'one good street' that was 'tolerably accessible only from one quarter'. Each side of this single street were narrow lanes whose 'steepness, narrowness, and dirtiness, can only be considered as so many unavoidable nussances'. The city's houses 'stand more crowded than in any other in *Europe*, and are built to a height that is almost incredible'. Internally, these houses had 'a great want of free air, light, cleanliness, and every other comfortable accommodation'. According to the Convention's pamphlet, some of these *lands* had as many as ten or a dozen families stacked 'overhead of each other in the same building', all of them using a common staircase 'which is no other in effect than an upright street, constantly dark and dirty'. The list went on. And on. The city's principal street was 'incumbered' with markets of many kinds, and the shambles down on the side of the noxious North Loch was 'rendering what was originally an ornament to a town, a most insufferable nussance'. There was a 'great deficiency' of public buildings, there was no exchange for Edinburgh's merchants, so safe repository for the town's public and private records, no meeting place for magistrates and town council, or for the Convention of Royal Burghs.

But there was more to Edinburgh's future than physical regeneration. The Convention wanted to build a metropolitan beacon. In its current state, Edinburgh was an embarrassment; instead of setting 'the first example of industry and improvement', it was 'the last of our trading cities that has shook off the unaccountable supineness which has so long and so fatally depressed the spirit of this nation.' On this island of Great Britain, there could be only one centre for an emerging Scottish identity. They held back from associating the words 'capital' and 'Scotland', instead referring to 'the chief city of North Britain'. But their meaning was clear. The ideal subscribed to was London:

> Upon the most superficial view, we cannot fail to remark its health-ful, unconfined situation, upon a large plain, gently shelving towards the *Thames*; its neighbourhood to that river; its proper distance from the sea; and by consequence, the great facility with which it is supplied with all the necessaries, and even luxuries of life. No less obvious are the neatness and accommodation of its private houses; the beauty and conveniency of its numerous streets and open squares, of its buildings and bridges, its large parks and extensive walks. When to these advantages we add its trade and navigation; the business of the exchange, of the two houses of parliament, and of the courts of justice; the magnificence of the court; the pleasures of the theatre, and other public entertainments: in a word, when we survey this mighty concourse of people, whom business, ambition, curiosity, or the love of pleasure, has assembled within so narrow a compass, we need no longer be astonished at that spirit of industry and improvement, which, taking its rise in the city of LONDON, has at length spread over the greatest part of SOUTH BRITAIN, animating every art and profession, and inspiring the whole people with the greatest ardour and emulation.

The Convention's vision was a London-of-the-North. On the site of tenement ruins in the High Street, they wanted an exchange, 'with proper accommodation for our merchants'. On ruins in Parliament Close, they wanted accommodation for the courts of justice, offices, apartments and a new library which would be 'The great charter-room of the nation'. Most significantly in terms of ambition, they wanted an Act of Parliament so that the royal burgh could be extended to the south and north, 'removing the markets and shambles, and turning the North-Loch into a canal, with walks and terraces on each side'.

The task of converting a toxic medieval town on an unpromising site into a world-class city was helped by the fact that parts of Edinburgh were already falling down. Indeed, the Convention's proposals had been triggered by the sudden collapse in 1751 of one of the high-rise *lands* which were clustered like urban brochs around the castle. The Town Council pounced, undertaking a survey which found so many properties to be 'insufficient', that 'several of the principal parts of the town were laid in ruins'. There was no shortage of brownfield sites. The foundation stone for the new Exchange was laid in 1753.

In the grand scheme proposed by the Convention, the city's success would depend as much on lateral expansion as it would on building admirable civic institutions. Edinburgh needed to grow. A couple of small developments moved forward, but the really huge prize lay on the far side of the valley containing North Loch, the fetid cesspit that acted as a vast semi-moat separating the city from the level farmland to the north. If the city could hurdle North Loch, the way would be open for development on a spectacular scale, with the bonus of a direct link between the city and its port of Leith. This was the construction project that would reinvent Edinburgh as a modern city: the new unconstrained metropolis that would break free from the ancient confines of the east–west axis of the High Street.

Expansion of the town's footprint, the Royalty of Edinburgh, had been mooted back in the 1680s by James II, but the Revolution of 1688 had intervened and it wasn't until the 1720s that the Town Council acquired land to the north. The North Loch Estate was a superb site for a planned town: a broad whaleback with dramatic views north to the Firth of Forth and south across North Loch to the rocky spine of teetering Edinburgh. Although the new site was only a quarter of a mile from the old town, access was a problem because the two were separated by North Loch itself; it wasn't until the 1750s that sufficient pressures were generated for progress to be made.

The draining of North Loch began in 1759 and in 1763 the first stone was laid of a huge bridge which would cross the valley from the old town to the new. Above three arches 72 feet wide would run a roadway 40 feet wide and 1,134 feet long. It was only 89 feet shorter and 7 feet narrower than Charles Labelye's new bridge over the Thames at Westminster. But Edinburgh's new bridge was not built to span water; this was Edinburgh's new axis, linking the past and the future, the old and new, land and sea. It was a symbol of civic belief.

Three years after work on Edinburgh's bridge began, a young architect, James Craig, was presented with a gold medal and the freedom of the city in a silver box for winning the competition to find the way of exploiting the land beyond North Loch. The sheer scale of Craig's vision was graphically revealed in a stunning copper-engraved map of 1766 by the Edinburgh cartographer John Laurie. There was old Edinburgh, crookedly strung along its mountain spine; and there beside it was new Edinburgh, a rectilinear grid of broad streets pinned to the crest of the whaleback by two huge squares, one at each end. As striking as the contrast in street plans was the similarity of scale: the new town was as big as the old.

While the likes of Drummond, Craig and Laurie were working up ambitious plans for the New Town to the north, a speculative developer called James Brown had already begun building the first major new development to the south. Brown had bought his 11-hectare greenfield site for the bargain price of £1,200 and had set his sights on Edinburgh's urban elite: the lawyers, nobles and politicians who had the money and aspiration to live in modern, spacious, low-rise houses set around the four sides of an Italianate square. One twentieth of the rent from tenants was to be spent on lighting and cleaning the square. When Thomas Pennant came to Edinburgh in July 1769, at the beginning of a tour of Scotland, one of the places he wanted to see was Brown's new square. Pennant found 'a small portion is at present built, consisting of small but commodious houses, in the *English* fashion. Such is the spirit of improvement, that within these three years sixty thousand pounds have been expended in houses of the modern taste, and twenty thousand in the old.' By now, Edinburgh was changing every year. In 1772, Pennant was passing by on another tour, and this time he was able to refer to the land beyond North Loch as 'The New Town'. And the vital umbilical between old and new was at last in place: The 'very beautiful' bridge had been completed since his last visit and North Loch had been drained and was now 'a deep glen'. The 'Castle in the Air' had been honoured.

Edinburgh had recast its own geography. In bridging and draining the great divide of North Loch, the city had opened the way to almost unlimited northward expansion. The dream expressed in the opening lines of the Convention's pamphlet of 1752 was on the road to realisation: 'Among the several causes to which the prosperity of a nation may be ascribed, the situation, conveniency, and beauty of its capital are surely not the least considerable.' Edinburgh's makeover was probably the largest and most ambitious exercise in identity-change taking place at the time. Here's Plumptre walking into Edinburgh in 1799:

> The entrance to Edinburgh this way is very striking. Immediately in my foreground I had a party of gentlemen playing at golf. Beyond, on my right, the rock on which the castle stands, a most stupendous mass of itself, and with a great mass of buildings upon it. This however is greatly disfigured by the new barracks, which are built in a heavy formal manner, and far out-top the other parts of the castle. The old town, with many fine and many new buildings in it, stretching to the right, the new Town in front, the finest series of regular and handsome stone buildings perhaps in the world; the Firth of Forth appearing on each side beyond it, and the whole view terminated by lofty blue mountains.

Britain was a two-speed archipelago. The fast landscapes – industrial sites, ports, towns and cities – were developing so quickly that twelve months could render a place unrecognisable. In stark contrast – for they were always adjacent – the agricultural landscape appeared to evolve at the pace of a weary haymaker plodding home at sundown. A battle was being fought between country and town. There had been a time when towns had been barely visible dots on the freckled countenance of Britain; a time when the economy was driven by people working the land and living in crofts and cottages. But by the end of the seventeenth century, the contribution of agriculture to the total national income had fallen to less than half. By the end of the eighteenth century, it had probably dropped to less than a third.

Perhaps the countryside should be seen as 'less fast' rather than 'slow'. Out in the fields, a quiet revolution was rippling through the shires. While new agricultural practices such as four-field crop rotation, selective breeding and adoption of the Dutch plough were raising crop yields, the land itself was being modified by agriculturalists as never before. During the eighteenth century, the amount of land in England that was being used for arable meadow and pasture expanded by 38 per cent as more fens were drained and the process of enclosure gathered momentum.

The enclosing of open fields into smaller fields had been going on for centuries, but from the 1750s it accelerated dramatically in a swathe of English counties stretching from Yorkshire to Dorset, with the Midlands being most affected. The greater part of the increase was being driven by Acts of Parliament. If owners of at least three-quarters of the village land agreed to enclosure, a petition was drawn up requesting that Parliament pass an Enclosure Act for that village. Unless there were sufficient objections, the Act was passed and commissioners appointed to oversee the enclosure of the land, a process that began with the drawing of a map showing all the individual strips and their respective owners. A second map was produced allocating new fields to landowners, who were then required to demarcate their plots with hedges, walls or fences. They also had to build new farmhouses and construct access roads.

The pay-off for parishes that opted for enclosed, compact farms was a more efficient use of land. There were fewer uncultivated borders between strips; fewer weeds that spread from a neighbour's ill-tended strip; fewer diseases spread between communally grazed animals; fewer hands needed to cultivate crops and tend beasts. On the downside, the new compact farms could be cruelly exclusive: existing farmers who failed to prove ownership of strips, faced eviction, even if they'd worked the land for generations. And poorer farmers who were allocated the smallest plots frequently found that they couldn't compete with the larger landowners. Evictions and departures were common, with the dispossessed making their way to towns and cities where they joined the industrial underclass. Rights of access to common lands were rubbed out as tracts of woodland and grazing and meadow that had been shared since time immemorial were ruled off-limits. The impact of parliamentary enclosure legislation was regionally limited and concentrated into a fairly short and frenetic episode during the latter part of the eighteenth century. Per decade, the number of Enclosure Acts rose from 36 between 1740 and 1750, to 660 between 1770 and 1780.

In the regions affected, enclosure created new landscapes. The imposed fields, with their ruler-straight borders, were quite unlike anything seen in the countryside since the days of Rome. A parish that had lived for centuries with an open landscape striated with hundreds of strips arranged in extensive blocks was transformed into a neat patchwork of rectilinear fields demarcated with physical borders. Chequerboards of drystone walls crept across the Cotswolds, Yorkshire Dales and Midlands. New farmhouses punctuated new field patterns. Desire-paths were blocked and replaced by paths and roads oriented on towns and markets. Counterintuitive right-angles were introduced where paths or roads were forced to follow the hedged or walled perimeter of a newly bordered field. James Plumptre found the new fruitfulness intoxicating. Riding in a chaise on the new turnpike between Cambridge and Barton, he enjoyed seeing so many ditching gangs at work on each side of the road: 'In a few years,' he wrote, 'when the country is drained and the fences get up the improvement will be wonderful.' (Today, the enclosures Plumptre saw being created are a pattern of neat, rectilinear arable fields with a regular geometry unaltered but for the M11, which cuts like a gigantic wheel-rut across the enclosed land between Coton and Grantchester.)

Enclosure was a regional measure and if time is stilled for a moment, this England of the late 1700s had the appearance – superficially at least – of a land where town and country still harmonised. To a visiting Prussian clergyman, Karl Philipp Moritz, England had it all:

> The circumstances that renders these English prospects so enchantingly beautiful, is a concurrence and union of the *tout ensemble*. Every thing coincides and conspires to render them fine, moving, pictures. It is impossible to name, or find a spot, on which the eye would not delight to dwell. Any of the least beautiful of any of these views that I have seen in England, would, any where in Germany, be deemed a paradise.

Nobody visiting the highlands and islands of Scotland would have compared them to paradise. Beyond the eyes of Edinburgh, a convulsion was occurring. From Lewis to Kintyre in the west, from Strath Naver to Strath Earn in the east, townships were being abandoned as Gaeldom was overcome by the commercial and demographic forces charging across Europe. So thirsty for raw materials was the siphon of southern industrialisation that no part of Britain lay beyond reach. Demand for wool, mutton, cattle, kelp and labour was encouraging Scotland's landed classes to extract more from their estates. And there were more hungry mouths. On islands and in glens, the population had expanded beyond the ability of traditional farming to provide enough food. Between 1755 and 1811, the numbers of people living on the islands of the Outer Hebrides ballooned from 13,000 to 24,500. Over the same period, the population of Mull and the Inner Hebrides nearly doubled from 10,000 to 18,000. Small islands were particularly exposed to crisis. When Thomas Pennant's ship anchored off Canna in 1772, he found the islanders 'in such want, that numbers for a long time had neither bread nor meal for their babes: fish and milk was their whole subsistence at this time'. Canna's last harvest had failed and the islanders had nearly run out of fish-hooks. On Skye, Pennant found islanders 'sunk beneath poverty, or in despair'. Beyond the ruling walls of castles like Dunvegan were people who 'prowl like other animals along the shores to pick up limpets and other shellfish, the casual repasts of hundreds during part of the year in these unhappy islands'. He reckoned that there were no more than two or three slated houses on the entire island of Skye and that all of the rest were roofed with 'fern, root and stalk'.

The old ways of the highlands – and much of the lowlands – had sufficed, more or less, for centuries. The traditional *fermtoun* of fifteen or twenty homes was intimately related to its lands: the higher-quality, intensively worked 'infields' and the more extensive, lower-quality 'outfields'. Much of the labouring was communal. There was always peat to be dug and stacked, homes to be repaired or built, plots to be ploughed. The long 'rigs' or ridges that formed raised beds for cultivation needed continual tending if their valuable soils were not to be washed away in the adjacent drainage furrows. The social subdivision of communities into the elite rent-paying 'tacksmen', sub-tenants and cottars had worked in a world of self-contained clans, but was undermined by outside markets and rapid population growth. The stratified social order was imposed against an open landscape on which there were few walls or hedges, ditches, dykes or roads.

Travelling a decade or so after the Clearances started, Pennant's account does not record forcible evictions, and neither do those of Samuel Johnson and James Boswell, who came to the Highlands the following year. But both parties of southerners recorded the momentous cultural shift being written in the landscape. Johnson wrote of 'the mischiefs of emigration' and wondered whether there was not some way 'to stop this epidemick desire of wandering, which spreads its contagion from valley to valley'. With little appetite among outsiders from 'more fruitful countries' to replace emigrant losses by moving to such an impoverished part of the world, Johnson foresaw 'a lasting vacuity' in the Hebrides and predicted that 'an island once depopulated will remain a desert'. Moorland was enclosed for sheep-grazing, tacksmen laid off and tenants turned out of their crofts. The clan chief had been replaced by 'a trafficker in land'. The most brutal evictions occurred in the early years of the nineteenth century. At short notice, entire townships were emptied and torched. In Sutherland, Donald Macleod climbed to a high point and counted 250 blazing houses, a conflagration that lasted for six days 'till the whole of the dwellings were reduced to ashes or smoking ruins'. Evicted tenants emigrated or were resettled in planned settlements where they were expected to conform to the economic theories of Adam Smith, adapting to their changed circumstances by learning to fish, or to weave or to burn kelp for iodine. Over one quarter of the population of Sutherland was cleared in twelve months. Within two or three generations, many townships became uninhabited imprints on the landscape: congregations of slumped house-walls attended by neat striations of grassed-over run-rig.

Britain's green and pleasant landscapes had absorbed industrial sites, an entirely artificial system of waterways, teeming towns and cities and a staggering surge in geological profiteering: the annual total of coal being extracted from British seams had leapt from 3 million tons in 1700 to 10 million tons by 1800. The use of coal in London alone had increased seven-fold since 1650 and in 1800 the Little Ice Age showed no signs of releasing its chilly grip on the archipelago. Bubbles trapped in the ice of Greenland were already recording that the amount of CO_2 in the atmosphere had crept up from around 275 parts per million in 1750 to around 285 parts per million by 1800. Britain's population was growing at a faster rate than at any time in its past. In 1600, the population of Britain had been around 5 million. By 1700, the total had increased to around 6.5 million and by 1750, it had crept up to 7.5 million, around twice the population during the Roman occupation and roughly level with the subsequent peaks of circa 1300 and 1650. But the population peak of 1750 wasn't followed by the normal decline or stagnation. Instead of being capped by the land's ability to produce food, the population graph steepened. By 1800, Britain was home to 10.5 million.

It wasn't just a matter of volume. The pace of life was quickening, too. Turnpikes finally came of age. When Defoe had been riding the shires, the average number of turnpike Acts being passed each year was a little over eight, but by the 1750s the annual average had climbed to over forty. In a booming Britain, parishes simply could not keep pace with the demands being made by traffic. More active turnpike trusts smoothed corners and reduced gradients by making cuttings. Continuing legislation gave trustees increasing control of the expanding turnpike system. In 1741, new legislation gave trustees the right to install weigh-bridges. Any load over three tons carried a surcharge of 12 shillings per hundredweight. A decade later, all turnpike trusts within thirty miles of London were compelled to install weigh-bridges, and to charge an additional one pound per hundredweight to any wagon drawn by six horses. In 1753, nine-inch tyres became obligatory. From 1765, penalties had to be paid on wagons whose axles were equal lengths (axles of different lengths meant that wheels made four lesser ruts rather than two deeper ones).

Legislation mended old roads and built new ones, but at a heavy cost to freedom of movement. Trustees of turnpikes were entitled to erect toll gates so that the tariffs could be collected and on some turnpikes, gates were erected across side-roads to prevent traffic seeking free alternatives. The penalty for pulling down or destroying 'any turnpike-gate, post, rail, wall, chain, bar, or other fence, set up to prevent passengers from passing without paying toll' was transportation for seven years or three years in prison. Turnpike trusts were neither popular nor universally honest. In 1752, the Kensington trust was found to have generated £10,000 income in three years, yet incurred a debt of £3,300; their stretch of turnpike amounted to fifteen miles, which could be maintained for around £100 a mile.

While long-distance travel gathered pace, London slowed down. Streets were clogged. Herds of cattle and flocks of sheep arriving in the city from the north and north-west were driven along Oxford Street and High Holborn to the livestock market at Smithfield. The solution was to build the first orbital bypass in Britain, a wide, straight road skirting London's built-up fringe, all the way from the Edgware

Road through the leafy suburban villages of Paddington and Islington to Smithfield Market. 'The New Road' was like no other in London: 40 feet wide, it carved a swathe through the countryside, with a clause in the Act prohibiting any buildings to be erected within 50 feet of the road's edges. The bypass created a barrier between London and its leafy northern greenbelt and attracted a surge of building development. The old walk out of the city towards Primrose Hill, along The Green Lane and Love Lane was lost beneath the expanding urban grid. By 1800, the New Road was traversed twice a day by the Paddington stage-coach, which could make it to the city in two and a half hours in the morning, while the return trip in the evening would take three hours, 'considering the necessity for precaution against the accidents of night travelling'. The New Road was infamous for its sticky clay and 'miry ruts'; these days we know it as Marylebone Road and Euston Road.

By the end of the century, a new word had sidled into popular usage. It first appeared in *The Times* on Thursday, 17 October 1799, when alert industrialists would have spotted a tempting proposition on the Property page. A freehold estate was to be auctioned at the Angel Inn, Cardiff. Nant Dyrys was a picture of old Wales situated in a 'delightful picturesque little valley, surrounded by a chain of hills', with a stone-and-tile house, convenient out-houses, and about 30 acres of enclosed land, 11 acres of productive oak coppice, 95 acres of upland and a 55-acre sheep-walk. According to the advertisement, the estate was 'supposed to contain copious veins of coal and iron ore'. It was, continued the ad, a 'situation such that an Iron Manufactory may be established thereon to great advantage, being distant only 8 miles from the Neath Canal, and 10 from the Merthyr Canal'. True to property ads, it was creatively misleading: the eight miles to the Neath Canal involved a rough, serpentine road over a mountain pass, while the longer route downstream to the Merthyr Canal was by way of a narrow horse-track threading the valley of the Rhondda Fawr. Prospective developers of Nant Dyrys were reassured that communications between their new iron foundries and the outside world could be facilitated by the construction along Rhondda Fawr of a canal or a 'rail way', the latter a landform new to *The Times*, and new to most of its readers. The 'rail way' was, of course, an old idea, but it was about to be reinvented.

Running through Merthyr Tydfil east of the High Street is a long, thin road with gentle bends known as Tramroadside. In other parts of Britain, it might have been a 'wagon way' or a 'rail way' or a 'rail road', but in Merthyr it was a 'tramroad'. The Merthyr tramroad was quite was a feat of engineering. Dropping some 100 metres over a distance of nine miles, it linked the ironworks of the upper Taff with the Glamorganshire Canal downstream at Abercynon. The entire route had been carefully surveyed, bridges built and tunnels bored. Retaining walls protected the tramway from falling rock and prevented it from slumping down slopes. Iron rails cast in an L-shape were nailed into oak plugs set into stone sleepers. As an exercise in terraforming, tramroads like the one on the Taff were comparable to canals in that they supplanted the undulations of a natural landscape with a controlled gradient capable of shifting far more freight than could be moved on conventional roads.

The agreement to build the Merthyr tramroad was signed in 1799, and the route was in operation by 1803. One horse could haul a load of ten tons down the valley, then drag the empties back, completing the nineteen-mile round trip in one day. Teams of horses increased the size of loads. The Merthyr tramroad was one of many in Britain and it would have remained in relative anonymity had it not been used the February after it opened to prove the viability of a mobile high-pressure steam engine – a steam railway locomotive. The test was masterminded by a Cornishman, Richard Trevithick, who was convinced that such devices could be used for hauling both people and freight. Trevithick's gasping contraption dragged five wagons loaded with ten tons of iron and seventy men down the valley to Abercynon. A sheared bolt prevented a return trip till the following day, but Trevithick was able to boast that his machine had travelled at nearly five miles an hour and consumed only two hundredweight of coal. Like many an inventor before him, Trevithick was ahead of his time and the test run at Merthyr was neither mentioned in national newspapers nor immediately followed by widespread conversion to steam-haulage on tramways. But it was a world first. Britain, for so long a net importer of inventiveness, had begun to produce big ideas. So big, that the island would lead the world into a new kind of landscape.

TWENTY-ONE

Chained Earth

1811–1920

By the early 1800s, London was the world's largest city. More than that, it was the largest city that had *ever* existed. London had become an uncontrolled experiment. There was no pre-existing model. This unprecedented concentration of population came with multiform novelties: never before had one million humans milled in one place; never had one settlement consumed so much food, produced so much excrement, needed so many roofs, nor burned so much coal. Behind the polished glass of the symmetrical villas on the broad rectilinear streets in west London, there was a real fear that the impoverished hordes further east would erupt from their squalid maze of over-crowded lanes and alleys.

The only way of comprehending such a vast urban sprawl was by abbreviating it onto a map. Edward Mogg's *London in Miniature* (1809) revealed that the main built-up zone was still concentrated north of the river, where expansion had been greatest to the east and west and now reached all the way from the suburban fields of Bethnal Green to Hyde Park. From this elongated core, ribbon development oozed outward to the suburbs of Bow, Hackney, Islington and Camden Town. South of the river, the built-up zone stretched downstream from Southwark to Rotherhithe and upstream to Lambeth.

It was out at the suburban rim that the city was at its most dynamic, relentlessly rolling outward over field and pasture in a breaking wave of scaffold and brick. In the sixty years or so since the publication of John Rocque's 1746 *Exact Survey of the Citys of London, Westminster, ye Borough of Southwark*, the metropolis had spilled and filled. The countryside bordering Oxford Street had gone. Where Rocque had shown an ancient pattern of hedgerows and field-paths either side of Tottenham Court Road, the entire area west to 'Edgeware Road' was now infilled with rectilinear grids of broad streets. Somehow, the country lane that used to dogleg north from Oxford Street past 'Marybone Gardens' had survived in outline as an anomalously crooked city street (now known as Marylebone High Street). Mogg's map included prospective estates, traced in grids of faint lines north of Russell Square and west of Baker Street. In the open country between Primrose Hill and the hamlet of Kilburn on Edgware Road, a developer had his eyes on the fields of St John's Wood Farm, where Mogg depicted a proposed 'pleasure ground'. On a huge 42-acre footprint, the 'BRITISH CIRCUS' would be over one mile in circumference. (These days, St John's Wood tube station marks the spot.) Into this rampaging agglomeration rode an architect.

Lambeth-born John Nash once described himself as a 'thick, squat figure with round head, snub nose, and little eyes'. Ugliness and beauty were recurring extremes of reference during the troubled life of this brilliant architect-turned-planner. Maritally challenged and self-exiled to the heartland of the Picturesque – Wales – for the first part of his professional career, Nash fell in with the likes of Richard Payne Knight, author of *The Landscape, a Didactic Poem*, Sir Uvedale Price (*An Essay on the Picturesque, As Compared with the Sublime and The Beautiful*) and Humphry Repton (*Sketches and Hints on Landscape Gardening*). Nash was inspired and his private life of bankruptcy and betrayal at last found an uneasy balance with a professional career built on designs for exquisite country villas and prisons. Back in London by 1797, the squat architect built himself a handsome house on Dover Street and established a reputation for Picturesque country houses in styles varying from classical and Italianate to castellated and Gothic. He was better at context than detail, a characteristic that opened the door to a second career.

Nash was in the right city in the right reign. London's long record of resisting planners was briefly interrupted by the madness of George III, whose maladies obliged Parliament in February 1811 to transfer sovereign power to the Prince of Wales. Aged forty-eight, the Prince Regent had grown from a tall, handsome, charming young man known for the 'irresistible sweetness of his smile' into a corpulent wreck characterised on his fiftieth birthday as 'a man who has just closed half a century without a single claim on the gratitude of his country or the respect of posterity.' As it happened, the Prince Regent was already making his pitch for posterity with John Nash.

Between the New Road and Primrose Hill was an increasingly valuable area of farmland known as Marylebone Park. It still lay – just – beyond the reaches of London's urban sprawl and in 1811 the leases on its farms came to an end. A competition was held for the most suitable plan to develop the park. Nash won with a scheme that changed the shape of London. Instead of streets and houses, he imagined a new park – the Regent's Park – framed with grand terraces and decorated with monumental villas, groves of trees and a lake. This aristocratic garden suburb would be framed on its northern edge by a new canal – the Regent's Canal – linking the Grand Junction at Paddington with the Thames at Limehouse. (Nash was one of the canal's directors.) The Regent's Park development promised the best of both worlds: it was located just a few minutes by carriage from the heart of the city, yet it would offer the Picturesque amenities of a rustic landscape. To restrict access for London's underclasses, Nash took the precaution of minimising the number of bridges over the canal.

The Regent's Park was only one element of a truly grandiose scheme. Nash also planned to gouge a regal boulevard right across London from his new park to Pall Mall. To achieve this feat of reconstruction would require the hacking of a clear corridor through the jumble of cramped houses and businesses blocking the route south of Portland

Place. It would be like taking an adze to stubble. Thus would Regent's Park be connected by Regent's Street to the Regent's Carlton House on Pall Mall. Lined with buildings in the latest architectural fashions, the boulevard would form a new axis across London and separate the modern from the antiquated: the 'West End' from the rest. An extension of this grand avenue would turn a right-angle at Pall Mall and head east to Charing Cross, where further demolition and street-making would create a huge urban *place* dominated on its northern edge by two new blocks for the King's Stables.

It was a scheme worthy of Hooke or Wren; a visionary plan in a city that had resisted all attempts to rationalise its chaotic form. Astonishingly, it actually happened. The route you walk today down Portland Place, past the BBC, over Oxford Circus, down Regent Street to Piccadilly Circus, Pall Mall and Carlton House Terrace, is the great divide conceived by Nash. And if you return along Pall Mall eastwards, you'll find yourself in the *grand place* Nash imagined, now known as Trafalgar Square. The 'Metropolitan Improvements' masterminded by Nash and the Prince Regent were the first great planning breakthrough in London since Roman surveyors had pegged the alignments of the river-crossing and road junctions nearly 2,000 years earlier. But it was an abnormal alliance during an era of chaotic urban expansion.

Land was being consumed as if it was an infinite resource. The unsentimental hand of economic advancement swept the countryside, imposing parliamentary enclosures, levelling hills, cutting drainage, putting common land to the plough. Trees were often in the way. In his protest poem, *To a Fallen Elm*, John Clare – one time hedge-setter and day labourer – grieved for the tree felled 'with wrong's illusions'. This was the tree that had 'murmured in our chimney top' when Clare was a boy; the tree that had played the 'sweetest anthem autumn ever made'. Clare shared an attachment to his elm that might also have been felt by the folk who roamed the wildwood eight thousand years earlier; the believers who upturned the splayed oak stump of 'Seahenge' on the water-margin of Norfolk. The elm stood for Clare's homeland, his parish, the diverse ecosystem of youth. The clearing and felling of common land drove rabbits from their hills to 'nibble on the road'. His hedges were ripped out and replaced by dykes; his parish gates thrown off their hooks: 'The bees flye round in feeble rings / And find no blossom bye'. Clare's parish of Helpston was enclosed between 1809 and 1820. He was there when it happened. The destruction and the denial of access to an ancient, shared landscape was an existential loss:

There once were lanes in nature's freedom dropt,
There once was paths that every valley wound,—
Inclosure came, and every path was stopt;
Each tyrant fix'd his sign where paths were found,
To hint a trespass now who cross'd the ground:
Justice is made to speak as they command;
The high road now must be each stinted bound:
—Inclosure, thou'rt a curse upon the land,
And tasteless was the wretch who thy existence plann'd.

Clare's landscapes were being reworked by labourers with a new armoury of tools. The rapidly developing implements industry was producing seed drills, harrows and ploughs, and new devices like horse hoes and clod crushers. Through the 1840s and '50s, threshing machines progressively took over from hand flails, and the long-handled scythe began to replace the traditional sickle. Methods of cultivation that had been practised since the first farmers stepped ashore in the south-east some time around 4000 BC were being dropped overnight. The very soil beneath Clare's feet was changing, too. The availability of manure had been rising with the increase in flock and herd sizes, and by the 1830s, farmers were also fertilising soils with chalk and lime, clay, marl, saltpetre, salt, soot, hoofs and *shoddy* – chopped woollen rag. The new idea, however, was bone.

Human bone dust made its newspaper debut in 1822. A report placed in the *Manchester Guardian* in November claimed that 'more than a million bushels of human and inhuman bones' had been shipped during the previous year to the port of Hull, where they'd been 'forwarded to the Yorkshire bone-grinders, who have erected steam engines and powerful machinery, for the purpose of reducing them to a granulary state'. Most of the pulverised bones had been sold through the agricultural market at Doncaster to the farmers of Yorkshire who found the product 'a more substantial manure than almost any other substance'. It might have been regarded as a fairly humdrum item of trade news had the report not also included the information that the imported bones included those of soldiers who had fallen in the recent battles at Leipzig, Austerlitz and Waterloo. The *Guardian* report was reprinted verbatim in the *Gentleman's Magazine* of November 1822, and in *The New Annual Register, or General Repository of History, Politics, Arts, Sciences, and Literature, For the Year 1822*. Human bones were still being traded in 1829, when *The Spectator* reported that a cargo of skeletal parts from the battlefield at Leipzig had been imported through Lossiemouth, for 'an agriculturalist of Morayshire . . . intended for manure'.

Bones were the means by which the people of Britain would be 'rendered independent of *foreign produce*'. To Sir John Sinclair, founder of the Board of Agriculture, the efficacy of bone as a fertiliser was 'perhaps the most important discovery, connected with the cultivation of the soil, that has been made in the course of a great number of years.' It had been believed since the 1760s that granulated bone made excellent manure, but it wasn't until 1829, when the Doncaster Agricultural Association published a report on the subject, that its use became more widespread. By the time Sinclair published *The Code of Agriculture* in 1832, bone-mills were 'very common' in northern England and could be erected for between £100 and £200. Most were driven by steam engines, but some were water- or horse-powered. Once reduced to dust, 25 bushels of bone would fertilise one acre. If the report in the *Manchester Guardian* was correct, the bone imported during 1821 would have been sufficient to fertilise 40,000 acres. The value of bone being imported for fertiliser rose from £14,395 in 1823, to £254,000 in 1837. Even allowing for a steep rise in the price of bone dust, the land area being fertilised by bone by the late 1830s must have been hundreds of thousands of acres. Unfortunately for the future of osteologic fertilisers, there weren't enough bones for Britain's fields. Sinclair recommended that cattle skeletons be shipped as ballast from Brazil and that Britain's coasts be exploited for 'the shells of sea-fish, coral, and shell-marl', all of which were 'equally useful as a manure'. He directed industrialists to Caithness, Forfar and 'other districts of Scotland, where shell-marl abounds'. Sinclair's enthusiasm may have contributed to the disappearance of prehistoric middens on Scottish coasts. He also advised that banks of coral could be stripped from the shores of the Western Isles of Scotland, Argyllshire and Loch Broom. (Together with the desecration of battlefields, the loss to archaeology and marine diversity doesn't bear thinking about.)

Fields would never be the same again. By 1835, nitrate of soda was being used as a fertiliser and in the same year, a Liverpool merchant traded the first cargo of Peruvian guano to arrive in the country. Between 1841 and 1847, guano imports soared from 1,700 tons to 220,000 tons. The 'Chemical Age' exploded across field and fen with the publication in 1840 of *Organic Chemistry in its Application to Agriculture and Physiology*. Baron Justus von Liebig's book argued that chemistry could increase yields and reduce costs. For readers of the translated work in a Britain faced with runaway population growth and finite acreage, Liebig had come up with the agricultural grail. By 1842, *The Farmer's Magazine* was writing about a new world in which the 'heavy expences in agriculture' would be obviated by 'chemical manufactories'. At his home in Hertfordshire, an agriculturalist called John Lawes worked out how to manufacture superphosphates by adding sulphuric acid to phosphate rock. By 1843, he had a fertiliser factory up and running on Deptford Creek, close to sugar refineries whose waste product – bone charcoal – was a convenient source of phosphate. Lawes and Sir Henry Gilbert – who had been a pupil of Liebig – established Britain's first agricultural research station at Rothamsted, just outside Harpenden.

In concert with fertiliser, marginal and difficult land was being improved for cultivation and grazing by underground drainage. For centuries, much of the water falling on the old open fields had been channelled along the parallel, ploughed furrows between the ridges, a process that had never been of benefit to soil quality, since topsoil and nutrients were flushed to the field-edges. In a few parts of the country, more effective drains had been buried beneath the surface. Farmers in Essex and Suffolk appear to have had the most advanced system, lining deep trenches with alder, heather or thorn and then replacing the soil. In other counties, peat or stone were sometimes used to line the trenches. In Leicestershire, farmers cut lines of V-shaped sods, which they truncated by cutting-off their lower portion. Replaced in their sockets, the sods provided a cavity through which underground water could flow. In Hertfordshire, stone-lined soakaway pits were cut at the foot of sloping fields. A Warwickshire farmer called Joseph Elkington won fame in the late 1700s for his 'rod of Moses', a crowbar he used for deflecting springs that were flooding fields.

'Underdrainage' attracted more adherents than heroes, but James Smith of Perthshire did become an agricultural celebrity for a period in the 1830s. Smith had managed to convert a sodden marsh near his home into productive land, a spectacle that attracted flocks of agriculturalists. His method lay in the systematic excavation of deep, parallel, stone-filled trenches, which he'd covered with soil. Smith was examined by a parliamentary committee and his report, *Remarks on Thorough Draining and Deep Ploughing* (1831), was a bestseller. Not everyone in Britain had access to stones, however, and the real breakthrough came in 1843, when John Reade, a self-taught mechanic, produced a cylindrical clay pipe which could be laid beneath fields. A couple of years later, Thomas Scragg worked out how to mass-produce the pipes using industrial kilns. Mass production and state subsidies for underdrainage changed the economics of farming on clay. Less saturated soil provided longer seasons; it reduced the effort of cultivation; it increased the efficacy of manures; it increased yields and it lowered costs. Clay landscapes could be turned to profit.

Gusts of change transformed agriculture, and the countryside. Some at the time referred to it as 'high farming'. Enterprising farmers experimented with fertilisers, drainage, rotations and crops, pooling their finds in journals like *The Farmer's Magazine*, whose pages covered matters agricultural ranging from new designs of plough to the blight of ridge-and-furrow and the role of farming in feeding Britain's multiplying millions while extending the Empire. While all of these advances were occurring on parallel fronts, coal smoke was beginning to blow in transformative clouds across the countryside as a new generation of machines came to places unfamiliar with any sound louder than thunder.

Silhouettes changed in the Fens as steam-powered beam-engines took over from wind-pumps. Languid sails turning like clock-hands were replaced by rectangular brick engine houses with coal yards and belching chimneys. Steam pumps were huge investments but they could run independently of the wind and so promised controlled drainage. By 1830, low-lying land was being drained with massive 80-horsepower beam-engines which turned wheels fitted with long troughs or scoops. Some of these scoop-wheels could be 15 metres in diameter. From the distance, they looked like Brobdingnagian cartwheels, slowly turning on the spot. When eight wind-powered drainage mills were replaced by a steam engine on the fenland estate of Thorney, the agent declared:

> I am preparing to take down and sell as fast as I can dispose of them, the windmills formerly used for draining on this estate . . . it is impossible that they can ever be required again for the purpose of drainage.

Out in the middle of the Fens between March and Ely, a windmill on the Hundred Foot Drain was replaced by a three-storey engine house, boiler room, workshop, coking shed and wheelhouse concealing an eight-scoop, 41-foot wheel. It was a substantial structure of 300,000 bricks, sitting on a raft of 600 piles and its completion prompted the

Littleport and Downham District Commissioners to commemorate
the event with a triumphant ditty on a plaque:

These *Fens* have oft times been by *Water* drown'd
Science a remedy in Water found
The power of *Steam* she said shall be employ'd
And the *Destroyer* by *Itself* destroy'd.

ERECTED A.D. 1830

Across the Fens, windmills were dismantled or left to collapse. It was
estimated that by the 1850s, seventeen steam engines had replaced
around seven hundred windmills on the fenland between Cambridge
and Lincoln. As Percy Bysshe Shelley exhorted Britain's oppressed
working people to 'Shake your chains to earth like dew', the earth
itself was being bound by iron.

While the last great wetland in Britain was being scooped dry, canal
engineers were burrowing further and deeper underground than ever
before. The highpoint of canal development came with the completion in 1811 of a 3.25-mile tunnel beneath Standedge, a saddle high on
the country's Pennine backbone. The Standedge Tunnel allowed the
Huddersfield Canal to connect with Ashton under Lyne and a web
of canals and navigations reaching across to Preston, Liverpool and
Macclesfield. It was both the longest tunnel and highest section on
Britain's entire canal network. If Standedge was the highpoint, the
Caledonian Canal was the endpoint.

Of all the canals constructed in Britain, this was by far the most
spectacular, although for reasons unrelated to its engineering. The
course of the Caledonian Canal was directed by the great geological
fault which crossed Scotland from north-east to south-west. Running fifty-five miles from sea to sea, it was bounded on both sides by
mountains and puddled with three lochs: Loch Ness, Loch Oich and
Loch Lochy. If the three lochs could be connected to each other, and

to the seas to east and west, merchant ships and fishing vessels could cut across Scotland without having to sail the long and dangerous sea route around the top of Britain. Thomas Telford had been sent to explore the canal's feasibility and had reported in 1802 that it would cost £350,000 and take seven years to construct. He envisioned a cut deep and wide enough to allow passage for ships the size of a 32-gun frigate. Telford's cut was a comprehensive failure: the eventual cost rose to £1 million and it wasn't opened until 1822. Worse still was the fact that the canal had only 12 feet of water in its cuts instead of the intended 20, so it was suitable only for small sailing vessels and fishing boats, while the lack of tow-paths made it impossible to make headway into the winds that frequently funnelled along the Great Glen. And then there was steam. By the time the canal opened, horse-power and wind were being overtaken by engines. Later, Telford would reflect that no one had foreseen 'that steamboats would not only monopolise the trade of the Clyde, but penetrate into every creek where there is water to float them, in the British Isles and the continent of Europe, and be seen in every quarter of the world'.

By 1830, the great age of canal construction had run its course. In the 1730s there had been around 1,200 miles of navigable rivers in Britain, many of them improved. One century later, canals had extended that total to around 4,000 miles of navigable waterways. It had been a remarkable – if partial – transformation. Canals were regional, and they did not constitute an all-embracing system. Most of Scotland, Wales, the West Country and Pennines still lay beyond the reach of navigable waterways.

Britain was accelerating. Speed mattered. Roads were getting straighter, smoother, faster. Turnpikes had proved a transformative model. Pay-as-you-go roads may have been resented for their tolls, but the combined efforts of turnpike trusts had dragged communications into the nineteenth century. This long, piecemeal process of upgrading reached a climax of sorts in the 1820s and 1830s, due in large part to Thomas Telford and John McAdam. Telford dealt with obstacles and McAdam with surfaces. Roads that had once been interrupted by rickety trestles, alarming fords and temperamental ferries were rerouted over bridges. To many road travellers, rivers ceased to exist. Roads that were prone to flooded potholes and mires became so smooth and comfortable that hours were lopped off journeys. The achievements of Telford and McAdam were symbols of the age, evidence that diligence and ingenuity could overcome the impediments of Nature.

As designer, consultant or builder, Thomas Telford worked on thousands of road bridges. In Scotland alone, he contributed to 1,100 bridges and about 1,200 miles of road, 'advancing', he claimed with some truth, 'the country at least a century'. Telford's greatest bridge opened in 1826 and spanned the most formidable void on any British trunk route. The straits between the coast of north Wales and island of Anglesey were familiar to all travellers using the main route linking Britain and Ireland. Over the centuries, thousands had drowned at the crossing. Bridging the void was not easy. The Admiralty insisted that the roadway be at least 100 feet above water level to provide clearance for warships to sail beneath and even at the straits' narrowest point, any bridge would have to be at least 1,700 feet (518 metres) long; nearly twice the length of the new bridge Charles Rennie was then building across the Thames. Adding to Telford's challenges were the ferocious tidal rips of the straits, which would restrict access to the water while building the supporting piers. Telford's solution evolved from a design he'd worked on for a new Mersey crossing at Runcorn Gap. In a bold, futuristic leap, he planned to span Menai's chasm with a flexible, wrought-iron suspension bridge.

Instead of suspending his bridge on cables, he used flat, chain-bar links hung from two immense stone towers and anchored at each end within tunnels bored into the bedrock. So remarkable was the bridge's design, that the complicated process of construction was itself a cause for wonder. Work began in May 1819, with the blasting and removal of rock to provide a level foundation for the west main pier, the rubble being used to build a causeway to the shore, along which sledges could be hauled by horses. To carry the roadway to the bridge, four arches were raised on the Anglesey side, and three on the Carnarvon side. Limestone for the masonry came from quarries at Penmon. Nearly 36,000 iron plates and bars had to be manufactured. Six years after work started, the structure was ready for its suspension chains. It was a Tuesday in April 1825. One of those watching was a Dr Pring of Bangor: 'An immense concourse of persons, of all ranks, began to assemble on the Anglesey and Carnarvonshire shores about twelve o'clock at noon, to witness a scene which our ancestors had never contemplated.' Along with 150 workmen was Telford, of course, and a host of local dignitaries, while the straits were busy with pleasure craft 'arrayed in all their gaudy colours'. Working with clockwork precision (they'd been rehearsing), a 450-foot raft bearing a section of chain was towed out into the straits. Bolts were driven home, ropes lashed and two capstans turned by thirty-two men each slowly raised the first chain from the raft into the air. Carried away by the spectacle, three of the workmen – a carpenter, a labourer and a stonemason – teetered along the dipping curve of chain from tower to tower. The moment wasn't lost on Pring: 'Thus concluded a day, which linked the reciprocal interests of the counties of Anglesey and Carnarvon in a union, which, "it is devoutly to be wished," will never be broken.'

The remaining chains were raised in the same manner over the next couple of months. The bolting of the sixteenth and final chain on Saturday, 9 July 1825 'completed the entire line of suspension . . . of this truly marvellous and sublime work'. The bridge was far from complete, but already it looked like nothing on earth: its two limestone piers rising like the Pillars of Hercules from the turbulent straits, and between those piers, a princess necklace of wrought iron. A band played the national anthem from a scaffold platform bolted to the centre of the suspended chain, the steam packet *St. David* passed beneath them and all the workmen marched to the brass in single file along the chains from the Anglesey to the Carnarvon side. To Dr Pring and the multitude on the seashore below, it had 'a most picturesque effect . . . the altitude diminishing the natural size of the objects, and giving them the imaginary appearance of "aerial beings"'.

The next phase of construction required the hanging of 444 vertical rods from the chains and the attachment to them of 111 transverse bars upon which the deal planks of the suspended roadway rested. By 5 p.m. on 24 September 1825, sufficient planks had been laid for workmen to safely cross the straits, an event marked by a royal salute of twenty-one guns and flag raising.

After seven years of mounting expectation, Dr Pring was running out of superlatives: at the official opening to the public of 'this stupendous, pre-eminent and singularly unique structure' on a blustery Saturday in January 1826, a crammed mail-coach headed a procession of private carriages, stage coaches, 'numerous gentlemen's carriages, landaus, gigs, cars, poney-sociables, &c. &c., upwards of one hundred and thirty in number, and horsemen innumerable.' The strong southerly wind was causing the central section of the bridge to undulate gently but the cavalry of horses seemed unperturbed. Cannons fired all day, the royal standard was raised on each of the main piers, the band played, crossing from one side to the other. Dr Pring must be given the last words: 'Joy, admiration and astonishment seemed depicted in every countenance on beholding the proportion, symmetry and grandeur apparent to the most common observer in every part of this unrivalled structure.' He saw it as a 'grand national work', not least because the three workmen killed during its construction had come to the Menai Straits from Scotland, Wales and England.

Among bridges, the suspended roadway over the Menai Straits was the most spectacular feat of engineering ever accomplished in Britain and the celebrations that punctuated its creation were a measure of the wonder it aroused. The Menai bridge was a monument of the modern age and the spectators and pageantry that accompanied its construction must have had parallels in earlier times. The various phases of construction on the Orcadian isthmus by Loch Stenness, and at Stonehenge and at Silbury must have caused their creators to marvel at their own work. Completing the span of a bridge or the orb of a henge was a moment of finality that could never be recovered. You had to be there. The wonder experienced by subsequent viewers was in part an act of remembrance. Such structures became monuments the day they were finished.

Telford was big box-office: the 'first architect of the age', as Dr Pring put it. John Loudon McAdam was remembered for his road surfaces. As the youngest of ten children, he'd watched the solidities of life washed to the gutter. His father, a lesser laird and banker, lost his ancestral lands, and then their home was accidentally burned down and the family suffered the ignominy of moving to rented – albeit grand – accommodation at Whitefoord Castle. While his father's finances dissolved, young John was said to have amazed the pupils at Maybole parish school with a model of the road to Kirkoswald; whether the model included cross-sections is unrecorded. His father's bank crashed. Then his father died. John McAdam, aged fourteen, was sent across the Atlantic to live with his uncle, a wealthy New York merchant. Over a decade later, John was back in Ayrshire with sufficient funds to buy the estate and house of Sauchrie, between Maybole and Ayr. As a landed proprietor, he was obliged to serve as a Commissioner for turnpike roads and it was this experience that revealed to him the disparity between public expenditure on roads and the subsequent improvement in their condition. More misfortunes followed, involving the British Tar Company and the Muirkirk Iron Works, and then, for the third time in his life, John Loudon McAdam had to begin again, this time in the West Country, where his first passion resurfaced. Underpinned by his role as a trustee of the Bristol Turnpike Trust, he joined the rising clamour being directed at the state of the nation's roads. There was no model of road construction. In McAdam's view, funds raised from tolls were ineffectually spent; the 'defective' state of the roads was 'oppressive on agriculture, commerce, and manufactures, by the increase of the price of transport, by waste of the labour of cattle, and wear of carriages, as well as by causing much delay of time.' Surveyors were 'altogether ignorant of the duties of the office they were called upon to fill'. Far too much legislative attention was being paid to regulating the size and loading of wheels whereas the real cause of road wear was the structure of the road itself. Repeating his (possibly apocryphal) childhood feat of modelling the perfect road, McAdam created in print the highway of his dreams. It was his first book.

Remarks on the Present System of Road Making; with Observations, Deduced from Practice and Experience, With a View to a Revision of the existing Laws, and the Introduction of Improvement in The Method of Making, Repairing, and Preserving Roads, and Defending the Road Funds from Misapplication was a bestseller. Published in Bristol by J. M. Gutch on Small Street, the first print run was lifted off the press in 1816. The passage that restored the integrity of road surfaces could be found near the end of McAdam's masterwork, on page 30. Twenty-six years of road travel had provided the author with the means to make comparisons between the different materials used on road surfaces and the way they'd been applied. In McAdam's view, size mattered. Far too often, stones were too large, forcing the wheels of carriages to slip sideways or bounce upwards. The problem, he pointed out, lay with the inconsistency and wrongheadedness of construction contracts: some stipulated that stones the size of hen's eggs should be used, others that they should be of half-a-pound in weight. McAdam pointed out that hen's eggs come in different sizes and that the weight of a stone depends upon its density:

> The size of stone used on a road must be in due proportion to the space occupied by the wheel of ordinary dimensions on a smooth level surface; this point of contact will be found to be, longitudinally about an inch, and every piece of stone put into a road, which exceeds an inch in any of its dimensions, is mischievous.

One inch: the golden mean for roads. To cut a long and bumpy story short, McAdam's ideas gained traction. Three years after his road-making manual was published, he updated it with a new appendix which described how to standardise road maintenance throughout Britain. He'd relaxed his one-inch rule and returned to the idea that weight – six ounces – was a simpler measure. The layer of stones on a road should be ten inches thick, and to allow for drainage, the surface should be three inches higher in the centre than at the edges (assuming a standard 30-foot road). To avoid frost damage, no material other than broken stone should be used; earth, clay, chalk and the like being liable to 'imbibe water'. The underlying principle was that 'broken stone will combine by its own angles into a smooth solid surface that cannot be affected by vicissitudes of weather, or displaced by the action of wheels, which will pass over it without a jolt, and consequently without injury'. McAdam was suddenly in demand, and by 1819 his methods had been taken up in fifteen counties along 700 miles of road.

McAdam's simple solution to a problem that had endured since the departure of the Romans worked on every level: it boosted the economy; it was cheap; it required elementary oversight by surveyors and the stone-breaking provided unskilled employment for children, women and workhouse paupers. By the time the eighth edition of his book was lifting off London presses in 1827, John Loudon had more than rescued the family name. Practitioners of his road-surfacing system were known as 'Macadamites'. People spoke of 'Macadamization', and referred to those who used Macadamized roads as 'Macadamizers'. A cartoon published in 1827 showed a tartan-clad John – or possibly his road-building son – clutching bags of sovereigns while standing astride two signposts pointing to the Great West Road and Great North Road. Underneath, the caption read: 'MOCK-ADAM-IZING – the Colossus of Roads'.

Macadam spread beyond trunk roads, to parish roads, and into towns and cities. In London, main streets were upgraded from gravel to Macadamized pieces of granite. The general improvement to roads through bridging and resurfacing through the 1820s and '30s lifted the network to a condition not seen since the days of marching legions. By the 1830s, around 1,100 turnpike trusts were responsible for 22,000 miles of road, varying from urban streets and byroads to trunk routes. Controlling access were nearly 8,000 side-bars and toll gates. It had been a remarkable transformation. Countless parish roads were still subject to seasonal sloughs, but in the main, trunk routes no longer presented travellers with the kinds of horrors recorded a century earlier by Celia Fiennes. By 1836, London could be reached from Manchester in only eighteen hours, down from three days in 1750. The Great North Road had been improved so much that travelling times between London and York had dropped from four days in the 1750s to twenty hours by the 1830s, while London–Edinburgh by 'flying coach' had been slashed from ten days to four. By the 1830s, journey times between most large towns in Britain had reduced by three-quarters since 1750. But no matter how smooth the road surface, travel time and capacity was constrained by the speed of a horse and the size of road wagons. To many, the future lay with metal rails.

Observant miners and scientifically minded engineers with a grasp of rolling resistance knew that a horse could pull far greater loads on smooth, hard rails than on roads. But rails were an unfinished project. For three hundred years they'd been used for shifting rocks at quarries and collieries, but their evolution had been limited. The basic 'tram road' consisted of flat plates about 10 centimetres wide with a raised lip on the inner side to direct the revolving wheels. The raised lip was known as a 'flanch' or 'flaunch', an old French word, mangled later by the English tongue to 'flange'. One of the problems with tram roads was the continual spillage of cargoes such as coal and ore, together with the 'creep' of grit and stones from the ground on which they were laid. Without regular cleaning, debris accumulated, which impeded the wheels. A second, thinner type of rail that was gaining adherents by the early 1800s usually measured between three and six centimetres wide and projected above the ground, where it was less exposed to dirt and stones. 'Edge-rails' had no flange. Instead, the wheels were directed by a flange on their inside edge, or by a groove around the rim. Soundly laid edge-rails offered lower rolling resistance than tram roads, which meant heavier loads could be moved by the horse or capstan. There was however little standardisation. The gap between the two rails – the gauge – varied, too, although 4 foot

8 inches was common on the horse-drawn wagonways, because that was the width of a standard equine backside. Lengths of rails varied, too. Some operators used cast-iron rails only one or two metres long; others used wrought iron rails of six metres or so. The rails were joined at timber 'sleepers' or at blocks of squared stone.

Into the imperfect world of rails stepped Mr Henry R. Palmer. Born in Hackney, Palmer had spent his formative years – his apprenticeship – in Bermondsey with Bryan Donkin and Co., whose several achievements included the manufacture of steel writing pens by a new, patented procedure, collaboration in the invention of a rotary printing press and development work on the Henry Fourdrinier paper-making machine. From an early age, Henry Palmer had been exposed to the inner workings of cutting-edge machinery. He understood the relationships between load, resistance and power. After Bermondsey, he worked with Telford and by his late twenties he'd been accepted into the Institution of Civil Engineers. With a devotion peculiar to those of mathematical inclinations who find it difficult to ignore life's inefficiencies, Palmer became obsessed with railway lines. He saw them as the weak link in Britain's transportation systems; less developed than river navigation, canals or roads, yet carrying the potential to be more versatile than all three. He carried out a number of controlled tests on tram roads and railways and was able to demonstrate that dust on a tram road could increase rolling resistance by one-fifth and that bends in the course of lines also increased rolling resistance. Palmer asked the right questions, found the right answers, and then proposed the wrong way forward.

The project was well researched and the subject of a book Palmer published in 1823. *Description of a Railway on a New Principle* argued that any railway should be as straight as possible; it should be 'nicely adjusted to that plane which is most profitable'; its sections should be as few as possible; its 'touching surfaces' should be hard and smooth; it should not be exposed to 'extraneous matter'; it should be firmly fixed to the ground, and should any adjustment be necessary it should be achievable accurately and quickly. In a few lines, Palmer had

Wicken Fen, a surviving fragment of the Great Level, where draining for conversion to farmland intensified from the seventeenth century onwards.

The Trent and Mersey Canal cuts through Staffordshire beside the meandering River Trent.

Cromford, Derbyshire, site of the world's first successful water-powered cotton spinning mill.

Above right Edinburgh. The grid of the New Town spreads before the precipitous burgh.

Right Embankments and a viaduct of 1.5 million bricks allow the edge-rails of the Settle–Carlisle line to maintain a gentle gradient across the valley of the Ribble.

Twentieth-century back-to-back terraced housing in Leeds.

Semis in Oxford. The semi-detached house is the most common type of dwelling in Britain.

An early outing for Modernism in Britain, the Isokon flats of 1934, in Hampstead, north London.

Left Snowdonia National Park was designated in 1951. Its 823 square miles include the highest peak in Wales and England.

Top right Motorway interchange. M25/A2.

Right Farming electricity. Wind and solar plants in Oxfordshire.

Below Torness nuclear power station on the east coast of Scotland, 33 miles from Edinburgh. Construction started in 1980. It began generating electricity in 1988 and is expected to cease operations in 2030.

From Ice Age to Shard in 12,000 years. Behind the Shard, London Bridge leads to the heart of Roman Londinium.

summarised the essential qualities of a railway line. He then stated that the best way of satisfying these objectives was to raise the railway above the ground on posts set at ten-foot intervals, where it would be clear of debris and winter snows which might encumber the rails. By varying the height of the posts according to undulations and obstacles on the ground, the railway could be kept on a steady plane without recourse to embankments and cuttings, bridges, culverts and drains. Because the cost of elevating two rails would be so great, he argued for a single rail from which hung wheeled carriages, with a 'recept-acles for goods' each side, 'the centre of gravity being always *below the surface of the rail*' (his italics). Carriages would be connected to each other in a line, and then connected by a towing rope to a horse. With one eye on future technology, Palmer suggested that the horse could be replaced by 'steam, either as a locomotive or stationary power'. Henry Palmer's 'suspension railway' was the right idea in the wrong century. He'd invented the monorail.

How much George Stephenson was guided by Henry Palmer is unclear, but when Stephenson became involved in a new railway line between Stockton and Darlington, he too settled on the principle that a railway should be as straight as possible, and that where curves were necessary, they should be very gradual. The S & D was formally opened in September 1825. In an event that aroused widespread at-tention, Stephenson drove a steam-powered 'loco-motive' along the 12-mile railway at speeds of up to 15 mph. Three months earlier, in fields near Cheshunt in Hertfordshire, Henry Palmer had demon-strated his 'suspension railway' to guests drawn from 'respectable families in the neighbourhood'. There was a band and flags, refresh-ment booths and a signal gun, and seven carriages linked by chains were towed by a horse for nearly one mile. In the carriages were forty passengers and – to demonstrate the suspension railway's efficacy for freight – 'an immense weight of bricks'. Happily, nobody died and the guests departed 'highly delighted with . . . an invention which prom-ised such advantages to the country'.

In the race for rapid transit, the main movers had barely exceeded the speed of a horse. Steam was generally agreed to be the most advantageous source of power, but there were widely differing views on the means by which vehicles should cross landscapes. Was the ground-level railway best, or the suspension railway? In Brighton, a brewer called John Vallance had found a third way: tunnels bored beneath the land surface, through which passenger vehicles balancing on a single line of rail would be blasted between towns by 'pneumatic transmission'. Once the tunnels were proven, he envisaged a system of 'Insular Defence' in which troop-carrying carriages would hurtle at speeds of 100 mph ('or even more, if necessary') along tunnels up to 300 miles long to any part of the coast requiring military forces. Vallance published a pamphlet on the subject, and took out patents, but he appears to have been more successful in the art of brewing than in the science of pneumatic transmission.

Neither suspension railways nor subterranean tubes turned out to be the solution. The future was eloquently outlined in a work of 1825 titled *A Practical Treatise on Rail-Roads, and Interior Communication in General*. Nicholas Wood used his experience as a colliery manager at Killingworth, and his knowledge of physics, metallurgy and mathematics, to argue the case for the 'steam loco-motive' over the horse, or stationary engines with cables. But Wood knew that the efficiency of motive power was dependent upon the rails along which the wheels were rolling. Having explored all the options, he concluded that the edge-rail was 'decidedly the best'.

By 1830, the world's first twin-tracked, steam-hauled railway had opened between Liverpool and Manchester. Canals could not compete with steam. Where canals had made it possible to shift high volumes of freight at walking pace, railways promised speed and mass mobility. Applications to construct new waterways dropped sharply after 1830, while those for railways rocketed.

London came late to the show, but did so in style with a line that included an elevated section borne upon 878 arches and 60 million bricks. The London & Greenwich Railway was opened in 1836 and was the first in Britain built solely for transporting passengers; a commuter line south of the river from Greenwich through Deptford to London Bridge. Raised upon the longest run of arches in the country, the line was – like Palmer's suspension railway – able to maintain an efficient gradient while minimising disruption to existing streets. It may have been a record-breaking exercise in bricklaying, but the intrusion of such monstrous edifices into street life – and the roaring steam machines they carried – was regarded by many as desecration. The year after the London & Greenwich opened its elevated railway, the parish of St Botolph Without, Aldgate, fought a rearguard action to prevent the Commercial Blackwall Railway Company extending their line into the city of London on the grounds that the

> intersection of the public street by masses of brickwork and contracted archways, of which the viaducts would be composed, would be detrimental to health, and destroy the comforts of the inhabitants by intercepting the free circulation of air, obstructing the King's highways, and by the noise and stench and vibration render most of the houses in the vicinity uninhabitable.

The petitioner, a Mr Christie, felt that the parish would be doing the country a service because 'the railway mania required a wholesome check' and there were 'no less than 77 petitions to Parliament presented for bills for railways this session'. Nevertheless the extended line from Blackwall went ahead and although cable haulage from a stationary engine was used to prevent the risk of sparks causing fires, the gauge was widened in 1850 so that steam locomotives could run on the line.

Edge-rails crept across England, southern Scotland and southern and northern Wales. Having taken to railways later than the north, London had become by 1840 the most defined single hub of the network, with lines radiating out to Brighton, Southampton, Bristol, Birmingham, Liverpool and Manchester. By 1845, Dover and Exeter had been connected by rail to London, and Colchester, Norwich and Yarmouth, too. In the north, new lines linked Birmingham to Sheffield,

Leeds, York and Newcastle, and in Scotland, a self-contained network linked Ayr, Glasgow and Edinburgh. By 1852, most major towns and cities in Britain were connected, with eastern and western extensions to Glasgow and Edinburgh, while a line from Gloucester ran west through the industrial south of Wales. Over 7,000 miles of track had been laid in twenty years, far outstretching the eventual 4,000-mile total for inland navigation and canals.

Steel rails brought new landscapes. The basins and warehouses that characterised the canal network were matched on the railways by stations, engine sheds, coal yards, water towers and sprawling marshalling yards. Like canals, railways introduced entirely new linear habitats to the countryside, while the transport of people and materials fostered dramatic surges of industry and urbanisation at key locations. In Wiltshire, the ancient Domesday village of Suindune (or 'pig hill') had already been transformed by the Wilts & Berks Canal and the North Wilts Canal, but in the 1830s it found itself on the route of the proposed Great Western Railway linking London and Bristol. By 1842, it had become the centre for repairing and maintaining GWR locomotives, with workers housed in a dedicated estate. Between 1801 and 1841, Swindon's population doubled to 2,495.

In Cheshire, the old manor of Crewe – population 295 in 1831 – experienced a similar transformation when the Grand Junction Railway came across the fields. The GJR's primary purpose was to create a link between England's four largest cities: Manchester, Liverpool, Birmingham and London. Crewe started as a cross on a map; the place where the turnpike connecting the Trent and Mersey Canal with the Shropshire Union Canal crossed the GJR's new railway line. Crewe's genesis as a railway town was recorded by the *Chester Courant* in 1843:

> About two years ago only, the site could boast but a few detached farm houses. The company (and a few others) have imported to it a very different aspect. Their own land . . . is about thirty acres, and the whole is laid out in streets, and nearly covered with comfortable

cottages in varied and distinctive styles . . . There are also schools,
an assembly room, committee room for magistrates etc.

Crewe's railway station was the first in Britain to be accompanied by
a railway hotel – the Crewe Arms – and was followed by a 30-acre
locomotive works. Demand was so high that the workforce increased
from 600 to 1,600 in two years and by 1848 they were turning out
a new locomotive and tender every Monday. In the fields, another
820 houses were built in just two years. By 1851, Crewe was a town
of 5,431.

Railway 'mania' shifted more earth and bonded more bricks than
any previous constructional enterprise. Since the aim of any railway
builder was to minimise gradients, the laying of track had to be
preceded by a repertoire of engineering projects – cuttings and embankments, tunnels, viaducts and bridges – devised to overcome the
unhelpful undulations of the natural landscape. Putting their backs
into this gargantuan endeavour were the 'navvies' who had created
the canals, a tribe who did more to change the surface of Britain than
Celts, Romans or Saxons. From the 1840s, there were perhaps 100,000
of them. Most were drawn from the districts through which the
railways passed, but there was also a core of travelling navvies who
moved from site to site selling expertise and buying ale. Mechanisation had created a hungry underclass of agricultural labourers, for
whom an embankment or a cutting might be a godsend for a season.
The sod huts and shanties – or shants – that had accumulated along
the routes of canals now collected along the routes of railway lines.
Shanties varied from flea-infested dens burrowed into hill-slopes to
large, timber huts with felted and tarred roofs. As many as 120 people
could be crammed into huts built for 80. (Long after the age of the
navvies, pubs called The Shant lingered near railway lines.)

Overcrowded shanties accumulated at the mouths of railway tunnels and airshafts. When the social reformers Edwin Chadwick and
John Roberton visited the tunnel workings on the line being pushed
beneath the Pennines to link Sheffield to Manchester, they found five

shafts dropping hundreds of feet into the moor. On the windblown peat and mud were clusters of huts:

> They are mostly of stones without mortar, the roof of thatch or of flags, erected by the men for their own temporary use, one working man building a hut in which he lives with his family, and lodges also a number of his fellow-workmen. In some instances as many as fourteen or fifteen men, we were told, lodged in the same hut, and this at best containing two apartments, an outer and an inner, the former alone having a fire-place. Many of the huts were filthy dens, while some were white-washed and more cleanly; the difference, no doubt, depending on the turn and character of the inmates. In stormy weather, and in winter, this must be a most dreary situation to live in, even were the dwellings well-built and comfortable.

The Woodhead Tunnel took six years to construct, cost the lives of over thirty navvies and seriously injured well over a hundred. In countless other locations, gangs of navvies hacked and loaded and hauled soil and stone.

While tunnels were the least visible manifestation of railway mania, cuttings and embankments were monstrous surface exercises in recontouring. The choice by George Stephenson of a ruling gradient of 1:330 for his line from London to Birmingham provided for fast-running trains but it also led to a stupendous amount of digging. 'There is scarcely a portion', wrote the engineer Francis Whishaw at the time, 'which is not either carried by embankment above the general surface of the country, or sunk below it by means of excavation.' Whishaw rated it as 'one of the most difficult works of the land in the kingdom'. Over the entire length of the line, nearly 10 million cubic metres of soil and stone were shifted. To put that in a landscape perspective, the navvies were constructing the equivalent of a massive henge on every mile of the line, and like the henge-builders of the third millennium BC, they were dependent on hand-tools.

Cuttings opened chasms in the countryside. Near Tring in

Hertfordshire, a slot up to 20 metres deep and over two miles in length was incised into undulating farmland below the Chiltern Hills. A lithograph by John Cook Bourne, who recorded the construction of the railway, shows the earth's surface rent by a long, straight gash filled with ant-like figures toiling with picks while horses high above the cutting's floor haul ropes fed through a pulley system to barrows of spoil being guided up inclined ramps by navvies. For visitors like Bourne, these were transformative spectacles. Bourne also recorded work on the line at Park Village in Camden, where shoring was being installed on cuttings so steep they looked like cliffs in housing estates. Charles Dickens used the spectacle in *Dombey and Son*:

> The first shock of a great earthquake had, just at that period, rent the whole neighbourhood to its centre. Traces of its course were visible on every side. Houses were knocked down; streets broken through and stopped; deep pits and trenches dug in the ground; enormous heaps of earth and clay thrown up; buildings that were undermined and shaking, propped by great beams of wood. Here, a chaos of carts, overthrown and jumbled together, lay topsy-turvy at the bottom of a steep unnatural hill; there, confused treasures of iron soaked and rusted in something that had accidentally become a pond. Everywhere were bridges that led nowhere; thoroughfares that were wholly impassable; Babel towers of chimneys, wanting half their height; temporary wooden houses and enclosures, in the most unlikely situations; carcasses of ragged tenements, and fragments of unfinished walls and arches, and piles of scaffolding, and wildernesses of bricks, and giant forms of cranes, and tripods straddling above nothing. There were a hundred thousand shapes and substances of incompleteness, wildly mingled out of their places, upside down, burrowing in the earth, aspiring in the air, mouldering in the water, and unintelligible as any dream . . . In short, the yet unfinished and unopened Railroad was in progress; and, from the very core of all this dire disorder, trailed smoothly away, upon its mighty course of civilisation and improvement.

The railway builders brought a new level of topographical shock. They were one of Thomas Carlyle's 'Signs of the Times': 'We remove mountains, and make seas our smooth highway; nothing can resist us. We war with rude Nature; and, by our resistless engines, come off always victorious, and loaded with spoils.'

The worst of the wounds were well decorated. Tunnel mouths, viaducts and stations were frequently adorned with architectural references which diverted attention from the brute physicality of the intrusion. A completed line was an exhibition of engineering and a triumph of utilitarianism. Adornment was both an apology and a roar of prowess. However much you shied from the belch and grunt of coal smoke and compressed steam it was difficult not to be awed by a station like Euston's, with its Doric portico and temple steps. Railway lines were laid as processional avenues terminated by monuments. When rails reached Huddersfield, they raised a neo-classical colossus with a gigantic pedimented portico borne upon towering Corinthian columns. To celebrate the laying of the station's six-ton foundation stone, a public holiday was called and church bells pealed across the dale from dawn till dusk. Construction took four years and by the time Huddersfield station was completed in 1850, its stately exterior was decorated with wings and multiple bays, elaborately scrolled consoles, parapets, balustrades and the armorial badges of the two railway companies who had paid the £20,000 bill. Behind the grandiose facade was a single platform and functional spaces designed to service modern transit.

Having killed the canals, flanged rail did the same to turnpikes. Rail fares not only undercut stagecoaches but rail transport was quicker and more comfortable. Stagecoaches went out of business and the toll revenues being collected by turnpike trusts plummeted by one third between 1837 and 1850. With more than one thousand independent turnpike trusts – many of whom were overloaded with debt – charged with the upkeep of 22,000 miles of trunk routes and crossroads, the ascendant railway industry had little competition. For road travellers, it was a return to ruts and potholes as turnpikes were gradually handed over to local authorities who were able to charge a rate on the community through which the road passed. Turnpikes, Telford and McAdam had raised a false dawn over Britain's multifarious roads.

The mid 1800s marked a tipping-point in the 12,000-year story of Britain's continuous occupation. In the fifty years since 1801, the population of England had more than doubled, from 8.3 million to 16.9 million; the population of Wales had nearly doubled, from 590,000 to 1.06 million; the population of Scotland had increased by 56 per cent, to 2.9 million. The rate of increase, and the absolute totals, were on a scale far beyond anything the island had previously experienced. And behind them lay an abrupt tilt of the balance between country and town. Towns had taken a very long time indeed to prosper on this island. A brief flirtation with urbanisation under the Romans had been followed by a long hiatus, and then centuries of gradual restitution. By the late 1600s, around 30 per cent of the population were living in urban centres. By 1801, 42 per cent of Britain's population had taken to towns and cities. By 1841, the balance had tipped and Britain became more than half-urban, with 51 per cent of the population living in towns and cities.

Urbanisation was loaded towards the largest towns and cities, where the number with populations greater than 50,000 had leapt from eight in 1801 to twenty-five by 1841. Manchester, Liverpool and Glasgow had all mushroomed to more than 250,000 people, while the population of London had more than doubled in only fifty years, from 959,000 in 1801 to 2,363,000 in 1851. Industry changed gear, too. Through the 1830s and 1840s, industrialisation matured from manic adolescence into a longer-term exploitation of resources and labour, fuelled almost entirely by coal. Having doubled between 1750 and 1800, coal production had doubled again between 1800 and 1825. Scarcely a town or city in Britain was untouched by industry of some sort. Britain, the island that had always been late to absorb continental innovation, had overtaken not only Europe, but the world, and was embarked upon an uncontrolled experiment in rapid industrialisation.

On the streets, these extreme stats translated into poverty, over-crowding, filth and disease. While Huddersfield's magnificent temple to edge-rails was rising above St George's Square, surrounding communities were crammed into hovels. When the educationalist G. S. Phillips visited the district of Longley in the late 1840s for his *Walks round Huddersfield*, he was appalled by the squalor:

> Of all the sights one meets with in the manufacturing districts, the houses of the mechanics and factory workers are the most distressing. They seem to have been erected after no model; with no design after beauty; but piled together in savage haste, and contempt for the beings destined to dwell in them.

Phillips compared the factory owner to 'a sort of feudal lord' with 'his village struggling at the foot of his factory keep . . .' In many industrial towns, workers were crammed into housing at the factory gates. Most of these unplanned factory 'villages' were poorly built and insanitary. In Leeds, high-density, back-to-back houses drew this vivid description from Mr A. B. Reach, in 1849:

> Conceive acre upon acre of little streets, run up without attention to plan or health – acre upon acres of closely-built and thickly-peopled ground, without a paving stone upon the surface, or an inch of sewer beneath, deep-trodden sloughs of mud forming the only thoroughfares – here and there an open space, used not exactly as a common cesspool, but as the common cess yard of the vicinity – in its centre, ashpits, employed for dirtier purposes than containing ashes – privies often ruinous, almost horribly foul . . .

Manchester too had become a notorious centre of filth. This old Lancashire town was now the world's greatest industrial city; a toiling mass of 300,000, and it was still growing. Visiting in 1835, the French political thinker Alexis de Tocqueville wrote of 'the crunching wheels of machinery, the shriek of steam from boilers, the regular beat of the looms, the heavy rumble of carts ... noises from which you can never escape in the sombre half-light of these streets.' Manufacturing had obliterated earlier landscapes: 'Thirty or forty factories rise on the tops of the hills,' wrote de Tocqueville; 'Their six stories [*sic*] tower up; their huge enclosures give notice from afar of the centralisation of industry.' A reeking vignette of the city was provided by a young German who had arrived in the city in 1842 to work in the family cotton firm of Ermen and Engels. Already a radical, Friedrich, twenty-two, first walked Manchester's dank cobbles in the month of December and it was these wintry first impressions that set the tone for a passage he wrote a couple of years later once he'd returned home to Wuppertal. Near the city centre, Ducie Bridge provided the vista he needed:

> ... a coal-black, foul-smelling stream, full of débris and refuse, which it deposits on the shallower right bank. In dry weather, a long string of the most disgusting, blackish-green slime pools are left standing on this bank, from the depths of which bubbles of miasmatic gas constantly arise and give forth a stench unendurable even on the bridge forty or fifty feet above the surface of the stream. But besides this, the stream is checked every few paces by high weirs, behind which slime and refuse accumulate and rot in thick masses. Above the bridge are tanneries, bonemills, and gasworks, from which all drains and refuse find their way into the Irk, which receives further the contents of all the neighbouring sewers and privies.

The River Irk, added Engels, was 'characteristic for the whole district'. The source of the river's name, the Old Norse word, *yrkja*, had been used in Viking times to mean 'work' or 'cultivate' or 'compose', but by the time Engels peered into this black industrial gutter, Anglo-Saxons had changed irk's use to mean weariness or disgust.

In London, animals were dropping more than 39,000 tons of dung on the streets every year. Surfaces were so sticky with *slop* – dung, dirt, mud and blood – that cartwheels could lift a paving slab. In wet weather, parts of the world's greatest city were beset with slurry. Underfoot, many of the capital's thoroughfares were no more than farm lanes, trampled and splattered by hundreds, thousands, of users a day. In the ten years to 1841, the death rate in London doubled. The human geographies of poverty, cholera and crime seeped through the streets, while the physical city swelled on lungfuls of finance and migrants. If the creation of parks and squares can be seen as an early effort to carve sanitary – even aesthetic – reserves from the cramped metropolis, the gradual upgrading of the city's roads, river and sewers can be regarded as a second phase during which the urban form was prised further away from its rustic antecedents. New streets were seen as a means of improving traffic flow and of infiltrating the stench and gloom of close-packed buildings with air and space. While Regent's Street had created a new north–south axis, the main east–west axis of Oxford Street was extended by hacking a passage through the slums of St Giles. New Oxford Street now connected Holborn with the West End. New kinds of road surface were sought. Cobbled streets had been common since pre-Roman times, but granite setts were noisy, bumpy, slippery and difficult to clean thoroughly. Wood paving was quieter and smoother, but also slippery. A road surface of tarred gravel had appeared outside Nottingham in 1840, and five years later the same surface had been spread on Huntingdon's High Street. In Paris, a mix of bitumen and inert mineral matter – asphalt – had been used successfully since around 1850, but it wasn't until 1869 that the Val de Travers Asphalte Co. Ltd resurfaced Threadneedle Street in London and opened a new era of urban road surfaces. Asphalt was smooth, easy to clean and retained some grip in the wet. Records kept by London's paving authority, the Commissioners of Sewers, showed that 60,802 yards (35 miles) of asphalt had been laid in the City of London by 1873.

The sporadic conversion of London's streets to a new, metropolitan model was accompanied by the taming of the Thames. London's great river had been its creator, its defence, its waterway, its sewer, its open space and its commercial artery. But it was also an impediment to north–south movement. A spate of bridge building raised the total number of crossings to eleven by 1850, together with the world's first underwater tunnel, a pedestrian walkway built by a French engineer, Marc Isambard Brunel, and his son, Isambard Kingdom Brunel. Symbolically, the greatest modification to the Thames at this time was the canalisation of the river's banks. The old landing beach that had given Saxons a toehold on the Thames, and led to the emergence of thriving Lundenwic, was buried beneath a road-bearing artificial embankment that extended far into the river. Upstream, the Chelsea Embankment tamed another stretch of the river, and on the south bank the Albert Embankment walled the section between Westminster Bridge and Vauxhall Bridge. No longer was the Thames a broad, natural, double-meander delimited with tidal beaches. Instead, the river ran deep and narrow behind new stone walls which had reclaimed 21 hectares of riverside land for roads and building plots. The practician behind the canalisation was the descendant of another French immigrant. Joseph Bazalgette was the engineer of the age, a lean man who rode for two or three hours a day ('splendid exercise' he recalled in old age 'for counteracting the effects of a sedentary life') during a career devoted to towns and cities. Bazalgette was the practical Utopian, the engineer who understood that real urban progress would not be made with fanciful *grands projets* but with drains and walls. The project that led him to knighthood at Windsor Castle in 1874 was the world's greatest sewer system: 1,300 miles of tunnels connected to west–east intercepting sewers that led to remote outfalls downstream of London. In an era that saw the raising in Hyde Park of the world's first great international exhibition in a glittering complex lit by 300,000 panes of glass, the enduring architectural legacies were Bazalgette's four monumental sewage pumping stations, his embankments and drains.

The Thames had become a dead river. A symbol of the life that had once thrived in England's greatest waterway was wrapped around lamp standards that still decorate Bazalgette's embankments. They were designed by George Vulliamy, the ancestor of an immigrant clockmaker. The sturgeon was a magnificent fighter that could grow to more than 3 metres in length; a 'royal fish' that had to be presented to the monarch when caught. Decimated by pollution and over-fishing, it survived in the Thames until Victoria's reign. A 66-pound specimen was caught at Putney in 1867 by an angler called Lewis Gibson. After that, the Thames sturgeon lived only as a lamp standard, cast in iron.

What of the space that was once wildwood? Viewed through the prism of invention, the countryside had become an exhibition for new technologies. Places that had lived for centuries as cottage backwaters found their views dominated by monumental intrusions. Silvery railways sliced through fields with unwavering linearity, bringing with them cuttings and embankments which realigned landscapes as emphatically as the ditches and banks of Britain's first Iron Age. In the Fens, the low rumble of windmills had been drowned by the belching chunter of coal-fired steam pumps. Systematic, piped underdrainage and artificial fertilisers were changing Britain's soils. Steam was responsible for the final act of transformation, the culmination of a process that had begun nearly 6,000 years earlier when fields were pioneering, hand-dug plots for imported seed.

Since 1830 or so there had been a rising surge of patents relating to steam machinery which could be used for cultivation and in 1858 a competition run by the Royal Agricultural Society of England resulted in a £500 prize for John Fowler whose machine – a portable steam engine that dragged a wheeled plough to and fro across the field on cables – was judged the best for turning soil 'at a saving compared to horse labour'. Cable-steam ploughing was fast and it avoided the compression of soil by horses' hooves. It also cut to a greater depth. Agriculture had been wrenched from its organic roots. The perfect, steam-ploughed, deep-turned field was large, rectangular, underdrained, smoothed of ridge-and-furrow and enhanced with chemical fertiliser.

Lust for land also drained the largest remaining natural lake in southern England. Whittlesey was a natural wonder, an inland sea with a will of its own. Drainage of the surrounding fens had reduced the mere in the couple of centuries since Celia Fiennes had gazed across it from Stilton on the Great North Road. On Robert Morden's county map of 1722, Whittlesey appeared as an irregular lake no more than four miles long. By the time it was more accurately drawn in 1786 by John Bodger, a resident of Stilton, the mere measured some 12,000 feet in length and 6,000 broad; just over two miles by one mile, so roughly the same surface area as Derwent Water in the Lake District. By this time, Whittlesey was less than a couple of metres deep, and in places far shallower, but it was still a renowned landmark on the Great North Road, and one that travellers using Lieutenant-Colonel Paterson's road book of 1803 were advised to view from the highpoint near Norman's Cross. In summer, it was popular for sailing regattas and in winter when the mere froze, this icy mirror became the scene of skating competitions. When the Cambridge geologist Professor Adam Sedgwick came by in January 1841, he 'saw thousands, and, I think, tens of thousands whirling on the ice'. So the local landowners drained it.

A new steam machine, the Appold Centrifugal Rotary Pump, helped to kill Whittlesey. It had enjoyed a good showing in the Machinery Department of the Great Exhibition in Hyde Park and by the 1860s it was out on the Fens. John Appold's invention used curved vanes revolving at high speed to lift prodigious volumes of water; one minute and 788 revolutions would raise 1,236 gallons nearly 20 feet. If the lift required was only 8 feet, Appold's pump could discharge 2,100 gallons per minute. Since it had no valves to be blocked, it was ideal for draining marshland, where muddy water and suspended debris were commonplace. The Appold Pump began lifting Whittlesey's water in June 1850 and by autumn the following year, the job was done. Revealed on the bed of the dead mere were seven blocks of stone bearing masons' marks; dressed oolitic limestone lost from a barge that may have been destined for the abbey at Ramsey. The new land was divided by dykes into farms and fields, the peat was carted away and burned, the soil was clayed, and pioneer crops of clover and grasses were planted. Wheat and oats followed, and coleseed and mangel-wurzels for animal feed. Within eight years of the draining, the value of the drained land rose more than tenfold. In hungry Britain, the Fens were agricultural gold mines, producing yields that were 30 per cent higher than the national average. Back in the 1830s around half of the peat fens were under cultivation; by the 1870s, it had risen to 75 per cent. There was no place for wetlands in a steam-powered, profit-driven Britain.

The draining of Whittlesey was part of a great surge of agricultural expansion which ran through the first seventy-five years or so of the nineteenth century. Population growth and rising incomes brought marginal land to the plough. Chalk uplands, moors and hills were turned over to cultivation. Coastal marshes were banked, drained and converted into fields. Three, possibly four million acres of arable land were created between 1800 and the 1860s. The use of field drains, fertilisers, new crop rotations and new machinery such as threshers and reapers all increased productivity and reduced the need for manual labour. Between 1851 and 1871, the agricultural workforce of 2 million withered to 1.5 million. Despite the doubling of Britain's population in the fifty years from 1811, some 80 per cent of the food being consumed in the 1860s was still grown within the United Kingdom.

Rural Britain was a field of inequality. Encrusted with marble and alabaster, the country palaces of Houghton and Holkham shared a county with countless wattle-and-daub cottages whose reed roofs were no more watertight than those being laid five thousand years earlier. A staff member of the *Norfolk News* who toured villages in the winter of 1863–4 with a sanitary inspector from Norwich provided this description:

> A stranger cannot enter the village without being struck with surprise at its wretched and desolate condition. Look where he may, he sees little else but thatched roofs – old, rotten and shapeless – full of

holes and overgrown with weeds; windows sometimes patched with rags, and sometimes plastered over with clay; the walls are nearly all of clay, full of cracks and crannies; and sheds and outhouses – where there are any – looking as if they had been overthrown very early in the present century, and left in hopeless confusion where they fell.

They commonly found seven or eight people sleeping in a single room, which might have to be reached by a ladder and trapdoor. Throughout the country, government inspectors had little trouble finding cottages in similar states.

Things got worse in the 1870s when a run of bad harvests coincided with the opening of the North American prairies for food production and a more rapid conversion from sail to steam power. The tonnage of sailing ships arriving at British ports was overtaken by that of steam ships, which could carry larger cargoes faster than sailing ships. By 1881, the tonnage of steam arrivals was double that of sail. By 1913, over 90 per cent of British merchant tonnage was being powered by steam and the total tonnage had risen from 2.5 million tons in 1815 to over 11 million. Ports boomed while farms went bust. Arable farmers were hit hardest. If the mid-century shift in balance to a mainly urban population could be seen as a first tipping-point, the agricultural depression of the century's closing quarter was the second. Between 1861 and 1901, the total number of male labourers in the countryside of England and Wales fell by over 40 per cent. By 1901, agriculture accounted for only 6 per cent of the national income, down from 20 per cent fifty years earlier.

Britain was the oak-keeled three-master that sailed on a following wind into the eye of the storm to emerge as a careering steamship whose starving, ragged crew shovelled coal while the officers sipped burgundy in the saloon. Not everyone in late-nineteenth-century Britain saw thunderous industrialisation and runaway urbanisation as their kind of Utopia. In *After London*, published in 1885, the nature writer Richard Jefferies unleashed on England an unspecified disaster which wiped out most of the human population and allowed Nature to retake its territory. In the first year, couch grass overran fields of stubble and roads. During the second year, fields once rich with wheat and barley were overtaken by docks and thistles, oxeye daisies and charlock. 'By the thirtieth year there was not one single open space, the hills only excepted, where a man could walk, unless he followed the tracks of wild creatures or cut himself a path.' This was the revenge of the wildwood writ across England. It took just three decades for the island at the heart the world's largest empire to revert to its natural state. 'Where London formerly stood, there is a vast, stagnant swamp.' Jefferies was an early practitioner of post apocalyptic fiction, tapping into a collective social fear that humanity had gone too far, that humans had destroyed their one and only habitat through a combination of carelessness and over-exploitation of natural resources. His unspecified catastrophe was not one that affected the capacity of flora to colonise and reproduce; Jefferies just removed humans from the biosphere and let Nature get on with it.

By 1900, the countryside had become a minority interest. In fifty years, the rural–urban balance had tilted emphatically towards high-density living. Urbanisation had triumphed. A roughly equal distribution of population between town and country in the 1850s had become by 1900 an extreme imbalance, with only a fifth of people living in what could be called 'rural landscapes'. The shift in distribution was accompanied by spectacular population growth. The doubling of Britain's population between 1801 and 1851 was followed by a near-doubling over the next fifty years. In one century, Britain's population had more than tripled from 10.5 million to 37 million.

No other country in Europe had moved so far towards urbanisation. Big was better. In the machine world, more capacity delivered more might. Since 1851, the number of small towns (those with populations of 10,000 or less) had actually decreased, from 958 to 851, while the numbers of large towns and cities had increased. There were now 74 towns with populations over 50,000, up from 25 in 1841. Liverpool,

Glasgow, Manchester and Birmingham each had more than half a million inhabitants. Over a fifth of English and Welsh people – and 15 per cent of Scots – were living in towns and cities of between of 100,000 and 500,000. But among all of these teeming centres, one was in a class of its own.

London was now home to more than 6 million, up from 2.4 million in 1851. Indeed, the capital had become so big that it defied definition. The 'Heart of the Empire' had expanded to become a place of multiple meanings. Right at the centre, and approximating to the original Roman town, the 'City of London' was set within a far vaster 'Greater London' that had been defined in the 1870s as the Metropolitan Police District, which extended fifteen miles outward from Charing Cross. Midway in size between the City and Greater London was the new County of London, an administrative enclave of 117 square miles which had been carved from the counties of Middlesex, Surrey, Kent and Essex following the Local Government Act of 1888. London County Council – an elected body – was the largest municipal authority in Britain and it had the biggest headaches: how to clear its industrial-era slums, and how to house its rapidly lengthening ranks of workers.

For every Jefferies fearing the present, there were planners and architects drafting the future. Britain had been built with imagination, ingenuity and doggedness. The people who constructed canals, bridges, monorails and Palladian palaces began with an idea and pursued it with inventive application until it was complete. By the closing decade of the nineteenth century, the monuments of the deep past had been joined by – and, in many cases, overlaid by – the monumental structures of the industrial age. The built landscape was spangled with wonders. And yet for each Doric portico there were thousands of sub-standard back-to-backs and tenements. The challenge was humanitarian: building a habitat fit for the twentieth century.

A preview of future landscapes could be seen outside Birmingham and Liverpool. In 1879, the Cadbury brothers Richard and George moved their cocoa and chocolate business out of town to a greenfield site four miles south of central Birmingham beside a brook called the Bourn. In these healthy, rustic surroundings, they built a more efficient factory and laid the foundations for a model housing estate; a small cluster of semi-detached houses set in large gardens upwind of the new works. Over twenty years, the village developed according to the Cadburys' guiding principle that each house should occupy no more than a quarter of its plot and should have a garden of no less than one-sixth of an acre planted with a minimum of six fruit trees. By 1898, 'Bournville' had a triangular village green framed by Hazel Road, Sycamore Road and Linden Road. The Bourn trickled through a public park and an area of woodland – Stock Wood – had been retained between Acacia and Laburnum Roads, the road names a daily reminder to Bournville's residents that the countryside ran through the community.

Meanwhile, another industrialist, William Hesketh Lever, was building a model village beside the estuary of the Mersey on the Wirral peninsula. Port Sunlight was named after one of Lever's soap brands (he also came up with Lux soap powder and Vim) and it was a place utterly unlike the packed terraces of his boyhood home in Wood Street, Bolton. The success of Port Sunlight was founded in part upon the gradual development of the site, from 28 cottages by 1889 to 800 dwellings by 1914. Lever wanted to build a spacious, attractive, healthy community: houses were clustered in groups, with gardens to front and back. A community assembly hall large enough to seat 800 was opened in 1891 and other public buildings followed: schools, a hospital, a museum and a library. In the architecture at Port Sunlight and Bournville could be seen the accommodating, idiosyncratic ideas of the 'Arts and Crafts' movement, a late-nineteenth-century reaction to the repetitive machine-made landscapes of the Industrial Revolution which sought to recover a sense of place by incorporating local materials and 'lost' vernacular styles. Tall chimneys, gables, spreading roofs and a generally picturesque composition gave new model villages an appearance far removed from the serried factory terrace.

The likes of Port Sunlight and Bournville were rare oases. As a proportion of Britain's housing stock, they were immeasurably small.

But they embodied an idea that had been explored intermittently for centuries: that cities had got too big and that access to the country-side was essential for healthy living. The leading proponent of 'garden cities' by the late 1800s was Ebenezer Howard, the London-born son of a confectioner who had emigrated in his early twenties to America, where he'd tried homestead farming in Nebraska then moved on to Chicago, which was being rebuilt after a fire. It was here that the young Londoner watched new public parks being incorporated into an expanding urban grid. Through places, people and books, Howard constructed his own vision for the future. *The Age of Reason* by Tom Paine played its part, and so did *Hygeia: a city of health*, by Sir Benjamin Ward Richardson. In *Looking Backward* by Edward Bellamy, Howard read of a Bostonian who woke from a trance in the year 2000 to discover that the United States of America had been transformed through a combination of technology and state capitalism into the perfect community. 'I determined', wrote Howard, 'to take such part as I could . . . in helping to bring a new civilisation into being.' Howard's book *To-morrow! A Peaceful Path to Real Reform* was published in 1898 and the following year he founded the Garden City Association to promote his ideas. International conferences were held at Bournville and Port Sunlight and Howard's book was republished in 1902 as *Garden Cities of Tomorrow*. The most astonishing aspect of the Howardian dream was that it led to the building of a prototype garden city. With a blissful disregard for existing market towns, the Garden City Association bought 25,000 acres in Hertfordshire, close to Baldock and Hitchin. Letchworth was already a village of 500 or so, and plans called for it to be swamped with housing, industry and designated open spaces. The first of the new houses was completed in 1904 and although a school was finished the following year, the building of a garden city was considerably more ambitious than a model village. By 1930, the population of Letchworth had yet to reach 15,000.

The real success of the garden-city movement was its effect on city housing estates. In London, the young and ambitious London County Council (LCC) led the way with modern developments like the Bourne Estate in Clerkenwell, where 2,642 people were rehoused in five-storey blocks of flats separated by garden strips and decorated with Arts and Crafts flourishes. Similar blocks sprang up in the twelve inner-city districts identified by the LCC for slum clearance. In 1903, the year the inner-city Bourne Estate was completed, the LCC began work on its first 'cottage estate' on a spacious, 15-hectare site out at the end of the tramline in Tooting. At Totterdown Fields, the break from rank, cramped slums began with a street grid that curved slightly on its long axis. Pre-existing trees were kept where possible and young planes planted. More pragmatic than Port Sunlight with its low density of eight cottages per acre, the new world of Tooting was planned at 31.8 per acre, which was achieved by narrow frontages and short setbacks from the street. But each two-storey cottage had a small front and rear garden, and indoors there was a kitchen with a food cupboard, dresser and plate rack. Some had a bath. The LCC's principal architect, Owen Fleming, gave his cottages a sense of identity by decorating them with Arts and Crafts features like tall chimneys and a variety of porch styles. By the time the estate was completed in 1911, 1,229 cottages had been built for 8,788 people. With only four shops and an absence of community facilities or employment within the estate, this was a 'garden suburb' rather than a Howardian 'garden city' but for London it was a landmark development. Totterdown Fields Estate took the garden city ideal and adapted it to the more restrictive city fringe. Further LCC cottage estates took advantage of fresh legislation that allowed councils to build beyond their boundaries. For the LCC, this opened the door to far grander schemes than were possible within the more confined spaces of the County of London. Beyond the county boundary to the south, work started in 1901 on a new cottage estate at Norbury and, in the same year, builders moved onto a site in Tottenham, also outside the LCC area.

Arts and Crafts cottages set back from tranquil, tree-lined streets were the foreground of a dream. A bigger reality could be observed from the basket of one of the manned gas balloons beginning to appear above British landscapes. Housing estates and factories oozed like molten tar, infilling districts of least resistance and merging with other urban engulfments to form seamless, grey overlays. Seven of these vast, built-up continua now existed in Britain. Greater London was the most enormous, sprawling by 1900 from Acton to Stratford, Finsbury Park to Clapham Common. Over 2 million now lived in a continuous urban zone in south-east Lancashire; 1.6 million in the West Midlands; 1.5 million in West Yorkshire and on central Clydeside; 1.2 million in Merseyside and 0.8 million on Tyneside.

No other country had succeeded in producing so many of these aggregated urban areas. Their unprecedented scale excited revulsion, wonder, confusion, resignation. It took an eccentric Professor of Botany at Dundee University to finally find a word that could be used to describe this urban phenomenon peculiar to Britain:

> Some name, then, for these city-regions, these town aggregates, is wanted. Constellations we cannot call them; conglomerations is, alas! nearer the mark, at present, but it may sound unappreciative; what of 'Conurbations'?

Patrick Geddes had grown up in Aberdeenshire amid heather and gardens, and after a false start at Edinburgh University had travelled to London where he found a mentor in Thomas Huxley, the most vociferous proponent of Darwinian evolution. Geddes adopted the theory of evolution as the basis for his own thoughts on history, ethics, sociology and town planning. Green space was vital to human existence. Urban parks had been 'the best monuments and legacies of our later nineteenth-century municipalities', but they were modelled by 'prosperous city fathers' on the landscaped grounds of stately homes rather than the countryside. Where, wondered Geddes, was the scope for the 'natural activities of wigwam-building, cave-digging, stream-damming'? Children in town parks were watched over as 'potential savages', to be 'chevied away, and are lucky if not handed over to the police'. In his book *Cities in Evolution* (1915), Geddes urged 'our friends the town planners and burgh engineers' to reorientate their thinking. Instead of the city taking to the country, the country should be brought to the city. Towns, wrote Geddes, 'must now cease to spread like expanding ink-stains and grease-spots'. A 'tree development' would see towns 'repeat the star-like opening of the flower, with green leaves set in alternation with its golden rays'. Behind this verdant allusion lay the principle of preserving within built-up areas wedges of countryside or green corridors that would allow cities and towns to breathe.

Geddes' ideas were formed as the road system struggled to recover from what Sir James MacAdam had called the 'calamity of railways'. While edge-rails had tautened their hold on towns and cities, piecemeal turnpikes had been replaced by a complicated and largely ineffective jigsaw of local authorities whose responsibility it was to handle highway administration. In England alone, road administration was split between nearly 1,900 local authorities. And there had been virtually no investment in new roads for half a century: In England and Wales, the total extent of public roads, at 120,000 miles, was little changed from its pre-railway total. Pedalling furiously to the rescue of Britain's roads came a new breed of road user.

For decades, there had been occasional sightings of *velocipedists* careering along on their eccentric devices, but in 1863 a Parisian coach repairer called Pierre Michaux built a machine powered by pedals attached to the end of cranks. By 1869, *The Times* was able to report that two young men had undertaken a 'velocipeding' journey on 'bicycles' from Liverpool to London in four days. It was, reported the paper, 'the longest bicycle tour yet made in this country, and the riders are of opinion that, had they been disposed, they could have accomplished the distance in much less time'. Bicycles took off so quickly that by the mid-1870s there were probably around 50,000 of them in Britain. The use of roads for pleasure introduced a new user-group to Britain's rutted highways, and by 1886 the National Cyclists' Union and Cyclists' Touring Club had combined forces to form the Roads Improvement Association. Armed with their manifesto, *Roads, Their Construction and Maintenance*, the RIA began to make a difference.

Warning signs were erected on dangerous hills and parish surveyors were taken to court by velocipedists brandishing the 1835 Highways Act, under which 'the Oath of One credible Witness' was sufficient to instigate proceedings for road repairs. Writing in 1913, the celebrated social reformers of the age, Beatrice and Sidney Webb, described how cyclists had established 'a new ideal as to what a road should be'. In the world of the Webbs, cyclists represented the poorer, younger end of the social spectrum, the end 'despised' by those mounted upon horses who looked down upon people who 'rode ironmongery'. But these were the people who had reinvigorated Britain's road system:

> These new users of the roads required no rugosities for foothold: they were shod, not with iron but with air-inflated india-rubber. In the story of the King's Highway their advent has opened up an entirely new era . . . What the bicyclist did for the roads, between 1888 and 1900, was to rehabilitate through traffic, and accustom us all to the idea of our highways being used by other than local residents. It was the bicyclist who brought the road once more into popular use for pleasure riding; who made people aware both of the charm of the English highway and of the extraordinary local differences in the standards of road maintenance; and who caused us all to realise that the administration, even of local byways, was not a matter that concerned each locality only, but one in which the whole nation had an abiding interest.

Cyclists provided the long-awaited voice for road improvements. Between 1890 and 1902, road expenditure in England and Wales beyond London and the County Boroughs increased by no less than 86 per cent. During those twelve years, virtually all of the wheeled traffic was either horse-drawn or pedal-powered. For a few short years, an egalitarian, low-carbon machine sped silently along Utopian roads.

Meanwhile, intimations of a very different future were appearing in newspapers. Powered by the light internal combustion engine, the 'horseless carriage' would – according to *The Times* of 17 February

1896 – 'rival in importance the railway and the electric telegraph'. A share issue that spring for The Great Horseless Carriage Company (Limited) promised 'no limits to the bounds of the new industry' and that the time would come when 'every man has his motor car in his backyard'. Few inventions have been recognised so early for their potential to alter the human habitat.

The humble velocipede had been celebrated as the 'King of the Road', but it was a short reign. The motor car was bigger, faster, noisier, more dangerous and it promised a future of high-speed, effortless, independent mobility for those unable or unwilling to pedal, ride or take a train. Britain's roads, of course, were completely unsuited to high-velocity, heavily loaded wheels, and the most common complaint during these early years of motorisation was pollution: choking vortices of fine, pulverised dust billowing and swirling in the wake of every bouncing car. Hedges and trees were coated in pale drifts that lifted with every breeze. The Webbs wrote eloquently of the horrors inflicted by motor traffic on shires that had been used to walking pace, of roadside gardens depreciated in value, of cottages 'rendered almost uninhabitable', of frightened horses and terrified pedestrians, of the King's Highway that had 'ceased to be a place in which people could saunter, or children play, with any degree of safety'. By 1910, there were 144,000 motor vehicles on Britain's roads, and over the next four years that figure had more than doubled to 389,000. This explosion of internally combusting traffic accelerated the roads-improvement programme started by cyclists and changed the colour and texture of every major through-route. The pale road of the past became the blacktop of tar and asphalt. By 1913, the Webbs reckoned that England had 'a larger mileage of dustless roads than any other country'.

This Britain of blacktop and conurbations was stilled by the outbreak of war in Europe. Compared to the utter devastation along the entrenched front lines on the continent, the effect on British landscapes was negligible. In common with most overseas wars, the drain in manpower and resources caused a hiatus in civilian development at home; expenditure on roads dropped from £18 million in 1914 to

£12.5 million in 1918. Meanwhile, a rash of war manufactories and facilities appeared on sites more familiar with slumber. At Gretna on the Solway Firth, a vast explosives factory was operational by 1916. Covering a total area some 9 miles long and 1.5 miles wide, it employed 20,000 (mainly women) workers. On uncultivated land at Holton Heath on Poole Harbour, the Royal Navy built a cordite factory which was so large that it included 14 miles of narrow-gauge railway line. A military port was constructed at that age-old Channel bridgehead, Richborough. Where legions, Saxons and Christians had landed, the War Office acquired 1,600 hectares beside the lofty walls of the shore fort built nearly 2,000 years earlier. The huge Richborough site included a power station, slipways, railway sidings and, by 1918, accommodation for nearly 16,000. Under the Munitions of War Act 1915, around 240 National Factories were constructed while a great many existing factories were expanded. The workforce at the Royal Ordnance Factory at Woolwich Arsenal grew from 10,000 in 1914 to 75,000 by 1917. But very few of the factories, storage depots and vast military facilities were of any value beyond the Armistice in 1918. Most were converted to civilian use. The legacy of the war was not brick and military railways but shock: between 1914 and 1918, 750,000 Britons were killed and 2.5 million wounded. They would be remembered by name in countless villages, parish churches, market towns and cities. The 'war memorial' became a new civic place.

TWENTY-TWO

Interland

1920–2016

On 8 March 1920, Pathé News released a 55-second silent cinema newsreel. It opened with a wide shot of a long, straight road. There were no markings on the road, which had a flawless surface and impeccable camber. This picture of Roman linearity was framed by wide, cropped verges. A large, open-topped touring car flickered soundlessly away from the camera. A smaller motor vehicle approached. The camera panned right to follow the approaching car as it turned into a slightly elevated slip-road and halted inside an immaculate timber structure resembling an open-sided barn with an attached summerhouse. A uniformed man sprang from the summerhouse and thrust a tube through the car's door. Twenty seconds after halting, the car pulled away and rejoined the long, straight road. The title of the newsreel was: <u>*Roadside Petrol Supply Station*</u> *The first of its kind in Great Britain proves a great boon to motorists.*

The petrol station was located halfway along the main road between London and Bristol and it was constructed as the number of licensed motor vehicles in Britain surged above half-a-million. Peace had unleashed a passion for the private car, whose sales soon overtook the total for goods vehicles, buses and coaches. At the top end, the big cars cost a small fortune: in 1920, you needed £700 (over

£30,000 today) to buy a flash Austin '20' or a six-cylinder, five-seater Buick, secondhand. For double that price, you could drive away from the British Mercedes Motor Co. in Long Acre, Covent Garden, in an 'overhauled by the makers' six-seater with a Silent Knight engine and Zeiss acetylene headlights. But manufacturers already knew that the future lay in the 'Lure of the Small Car'. In January 1920, when the public flocked to the Scottish Motor Show in Glasgow's Kelvin Hall, they were able to inspect the latest Morris-Oxford, and two new cars called the Hammond and the Bean, small yet perfectly formed motor vehicles manufactured for the middle market. There was – as one motoring correspondent put it – something alluring about 'big cars in miniature'. Two years later, Herbert Austin Morris began producing his Austin Seven, an economy-car that was robust, reliable and so tiny that it wobbled in gusts of wind. The 'Baby Austin' swelled the careering, vehicular armada on British roads.

From the off, the internal combustion engine was destined to inflict a far greater impact on the landscape than steam. Motor vehicles were lighter, faster, more agile and more versatile. They could reach places steam could not. The road was reinvented as an axis of development. At first, the modifications were relatively unobtrusive. Workshops specialising in maintenance, repairs and sales began to appear. Reflective marker posts were planted along the edges of some arterial roads to prevent vehicles running into ditches in poor visibility. In bad weather, the Automobile Association – which had accumulated 100,000 members by 1920 – deployed warning signs. Resurfacing spread beyond major arterial routes, where Macadam's crushed stones had become a dangerous liability. The bouncing wheels of speeding cars, lorries and coaches not only raised clouds of white dust but also punched holes in surfaces laid for horse-drawn vehicles. Spraying Macadamized surfaces with tar helped, but entirely new 'flexible roads' were the long-term solution. By 1920, annual road expenditure had recovered from its wartime dip, £27 million being spent that year, compared to £12 million in 1917. By 1921, expenditure had catapulted to £43 million and by 1922, to £49 million. Road-rollers were in constant use. 'State of the Roads' reports in *The Times* repeatedly warned of remetalling and of greasy surfaces, dangerously bumpy sections and works to overhead wires, drains and tramlines (the newspaper also printed warnings about 'police activity'). New kinds of materials were trialled. In Southwark, the Borough Engineer experimentally laid a section of rubber road and although the results were applauded (he reckoned the noise reduction doubled the rateable value of adjacent properties), the cost was very high – £4 per yard laid – because of the difficulties encountered while attaching the rubber slabs. Oxford Street was rebuilt on an 18-inch foundation of concrete laid with wooden blocks topped with layers of creosote and tar. The most promising material was asphalt.

'The road' became a different kind of place. Junctions that for millennia had performed a role as social intersections where horse- or foot-travellers paused to check routes, or to interact with passers-by, or to rest, became accident blackspots where motor vehicles collided. The first manually operated signal-lights were installed in 1926 in London's Piccadilly, and by the following year, a junction in Wolverhampton was protected with automated traffic lights. Motoring growth was dramatic. Membership of the AA more than doubled in five years, to a quarter of a million by 1925. By 1929, the total number of motor vehicles had passed the two million mark, nearly half of them private cars. Annual expenditure on roads had more than doubled in ten years. In London, the number of car licences being held rose from around 100,000 in the early 1920s to 260,000 a decade later. As roads became more dangerous, special crossing-places for pedestrians were designated, with lights that could be activated to halt oncoming traffic. The first were installed in 1932 in Croydon on the road between London and Brighton. With the vehicular furniture came schemes to widen busier sections of road. Roads became a battleground between potentially lethal motor vehicles and vulnerable pedestrians. Road-related fatalities and injuries broke new records in 1934 when there were 7,343 deaths and 231,603 injuries, half of them to pedestrians and three-quarters of them in built-up areas. More Britons died on roads in 1934 than at the Battle of Jutland eighteen years earlier. As pedestrians were mown down and marginalised, the pavement assumed a heightened role as a peripheral safe haven. Segregation between the mechanised and muscular became the new urban order.

With cars came aeroplanes and their grass airfields and barn-like clusters of buildings. From 1919, Britain's first aerodrome to be designated with Customs approval for international flights occupied a grass strip beside the Great West Road. A few miles beyond London's suburbs, this stretch of Hounslow Heath was well trodden; Romans had laid a javelin-straight road across the Heath, linking their new town on the Thames with Calleva Atrebatum. And the Heath had been the seventeenth-century lair of highwaymen preying on travellers using the main road between the capital and the West Country. In 1858, Charles Dickens summarised the demise of this wild remnant:

> If any highwayman who galloped to the gallows a century ago, could see Hounslow Heath now, he would wonder where the four thousand acres that covered fourteen parishes had shrunk to ... only a few dozen acres of grass field enclosed for the cavalry reviews on one side of the road, and a few dozen acres of rough furze and bramble on the other for cavalry drill.

The drill ground was used during the 1914–18 war for training pilots and in 1919 it had been the place of departure for several aircraft attempting to make the first flight from Britain to Australia (the winner made it in twenty-seven days). But within months of gaining its Customs shed, Hounslow Heath was overtaken by a site on the Brighton road south of London, a grass strip so well placed for aviation that it was quickly designated 'The Air Port of London' (although frequent flyers knew it as Croydon or Waddon Aerodrome).

Ten miles south of London Bridge, Croydon was closer to France, it was on one of the main roads out of London, it had a better record than Hounslow for fog-free skies and it was already an established centre for the air industry: the Royal Flying Corps Station Beddington had developed on the west side of the Brighton road, and Waddon Aerodrome on the east side. In an unusual example of shared space, civilian aircraft taxied across the main road while traffic was halted by an official holding a red flag. Aviation – like shipping in earlier eras – was well placed to take advantage of Britain's insular circumstances. As early as 1920, the Minister of Roads, Sir Henry Maybury, had talked of 'the great aerodromes which will be planted on the outskirts of all populous centres'. By the late 1920s, Croydon was the busiest international airport in Europe, with a new terminal building, a control tower and an Aerodrome Hotel. From its viewing deck spectators could watch twenty aeroplanes a day lifting off for foreign fields.

The number of people in Britain had leapt by 16 per cent in twenty years; up from 37 million in 1901 to 43 million by 1921. Three-quarters of the population now lived in towns and cities and the gigantic urban conglomerations of the industrial revolution had grown even larger. Just over 40 per cent (a figure unchanged since the 1890s) of the population were concentrated in the conurbations of the south-east, central Clydeside and in the industrial belts of Tyneside, Merseyside, West Yorkshire, the West Midlands and south-east Lancashire. The woeful living conditions in many parts of these conurbations had been revealed during a 1917 inquiry into the health of the nation and into the condition of its housing stock. In every conurbation there were districts that had changed little since the mid-nineteenth century. This testament from a 1920s Liverpool mother would have been familiar in thousands of neighbourhoods:

It was a little two-bedroomed house and there was my mother, aunt and cousin in one bedroom and me, my husband and the baby in the other. It was an old house, very cold and damp and it didn't have any conveniences at all. No water at all in the house . . . we had to do all the washing outside.

The situation was far worse in Scotland, where more than half the population lived in one- or two-roomed homes, compared to just

7 per cent in England. Scotland's cities had been overcrowded since the mass rural–urban migrations of the early nineteenth century and tenants were still trapped in a rental economy that had no exit door. The old practice of *feuing*, whereby the buyer of land for development had to pay an annual fee to the original landowner, made development more expensive in Scotland than in England. The stricter building standards also pushed up costs; construction in Glasgow cost 45 per cent more than it did in London. To cover their bills, Scottish developers opted for density over comfort, and then charged rents as much as 30 per cent above those of, say, Sheffield or Leeds. This knot of causal factors was tightened by low wages in Scotland, which forced tenants to share their already cramped rooms.

The desperate lack of adequate working-class housing led in 1919 to the Housing, Town Planning, &c. Act – known as the 'Addison Act' – that sought to build a 'land fit for heroes' by using government subsidies to contribute to the construction of new homes. The intention was to build 500,000 houses in just three years, but the wobbling economy of the early 1920s reduced the completed total to 213,000. The Act marked a tectonic shift in accommodating Britain's population. Homes for working people were now a national responsibility, to be pursued through local authorities, whose job it would be to develop clean, modern dwellings. Subsequent Acts revisited the subsidies and duties of local authorities and by 1938, the amount of housing in the public rented sector had risen to 10 per cent, up from only 1 per cent in 1914.

Meanwhile, there had been a shift from the privately rented sector towards owner-occupation, which was up from 10 per cent in 1914 to 25 per cent by 1938. Rising incomes were giving households the chance to get a mortgage, while interest rates and house prices were low and building societies were more alert to the market. On the supply side, many decades of slum clearance had reduced the stock of rented houses and landlords were feeling the pinch, squeezed by competition from local authorities and by rent controls. Most private new-build was intended for owner-occupancy. In Greater London, local authorities provided 153,000 new dwellings between 1919 and 1938, but a far greater number – 600,000 – were built by private developers, mainly for purchase by owner-occupiers.

In the south-east, the Addison Act of 1919 triggered another suburban rampage. The single largest council estate was appended to London's eastern suburbs straddling the borough boundaries of Barking and Dagenham. On a 1,200-hectare site by Beacon Tree Heath, London County Council planned to build 24,000 houses in only five years, but the economic slump delayed progress and it wasn't until 1930 that the first 18,000 houses were completed. By 1938, the Becontree Estate had 25,736 dwellings and 115,652 residents. In terms of population, it was nearly twice the size of Oxford and ten times larger than most small towns in Britain. At the time, it was the largest municipal housing estate in the world. But to the author of a 1934 report, this unprecedented suburb was scarcely noted outside the East End:

> If the Becontree Estate were situated in the United States, articles and news reels would have been circulated containing references to the speed at which a new town of 120,000 had been built. The work of the firm of contractors would have been shown as an excellent example of the American business ideal of Service to the Community. If it had happened in Vienna, the Labour and left Liberal Press would have boosted it as an example of what municipal socialism could accomplish . . . If it had been built in Russia, Soviet propaganda would have emphasised the planning aspect . . . But Becontree was planned and built in England where the most revolutionary social changes can take place, and people in general do not realise that they have occurred.

Estates were infills on a vast scale. Planners scoured maps for gaps. A space in the countryside furnished with a sub-hour railway line to the city was a development-in-waiting. To north, west and south, smaller Becontrees were grafted onto London's fringe. South of the

Thames, the Castelnau Estate was built on a market garden in Barnes. Leafy Roehampton was colonised by the estates of Dover House and Putney Park. Lewisham sprouted the Bellingham Estate. Down south at Croydon, where the capital's tendrils were beginning to tickle the North Downs, the pre-war Norbury Estate was extended. A few miles away, between the towns of Mitcham and Sutton, beckoned another gap on the map.

Back in the sixteenth century, the countryside between Mitcham and Sutton had been a spread of meadows, woods and pastures watered by the trout-filled Wandle. It was a favoured bolt-hole for Queen Elizabeth I, who used to drop by Sir Francis Carew's refurbished manor at Beddington, where he'd cultivated the first orange trees in Britain from seeds provided by Sir Walter Raleigh. A couple of miles away, Sutton – *Sudtone*, 'south farmstead or village' – was more down-to-earth, its Domesday assessment noting 8.5 hides, with land for fifteen ploughs, a couple of churches, two acres of meadow and enough woodland for ten pigs. In the 1750s, the village was put on Londoners' mental maps when it became a stagecoach halt on the main road between the capital and Brighton, by then emerging as one of Britain's first seaside resorts. There were two coaching inns in Sutton, the Greyhound and the Cock, but for a century or so the place was little more than a rural road halt where horses were changed and passengers grabbed an ale; in 1801, there were 571 residents. But then the railway came and by 1901, Sutton was a town of 17,223. Three miles up the London road, Mitcham had taken the industrial route (snuff, copper, iron, dye) and grown even bigger. By the Twenties it was home to 35,000 people. Then, in 1926, the City and South London Railway completed its extension of the underground line from Clapham Common to a greenfield terminus outside a little village between the two towns.

The village was called Morden (*Mordone*, 'hill in marshland'). Today it has the largest mosque in Western Europe, but in 1926 it was the sleepy backwater that woke up one September morning and found that it had become the most southerly terminus on the underground railway system. Morden's fields were harvested for the last time. With the powers to compulsorily purchase land suitable for garden-city-style development, LCC bought 825 acres (334 hectares) of farmland adjacent to the new underground station. A light railway was laid to carry building materials and the St Helier Estate began to take shape. In the spirit of Ebenezer Howard, this was a cottage estate, the terraced houses built in red brick with low-pitched roofs and a variety of details that included Georgian-style windows, bays and porches with door-canopies. Exceptionally for estates of the age, over 50 hectares were set aside for green space. Many of the houses faced onto small greens and roads had grass verges. In places, the great elms that had lined the original country lanes were left standing. St Helier estate was the largest local authority development in south London.

Away to London's north-west, the extension of the Metropolitan line opened the development door to rural Middlesex. Marketed as 'Metro-land' in the railway's one-penny guide of the same name, this fresh suburban province promised 'a new residential country, and one especially desirable to those who wish to enjoy a rural or semi-rural life amid beautiful natural surroundings, combined with the regular pursuit of their avocations in the City'. Architecturally, Metro-land varied from bungalow to detached villa, but by far the most popular dwelling was the semi-detached house. With its two living rooms, kitchen, three bedrooms and indoor bathroom, neatly framed by a front garden, side access and a back garden with space for a vegetable plot, the semi was all an inner-Londoner could ask for. It was almost a country house. Readers of the 1920 edition of *Metro-land* were tempted with a new batch of semis going up in Wembley Park, at £1,200 each. The semi became England's most popular type of dwelling and eventually 4.9 million of them would comprise 31 per cent of the country's total housing stock.

By the 1930s, places depicted as secluded rustic nooks on Ordnance Survey maps of the early 1800s had become a conjoined suburban continuum; the fields and hedgerows around hamlets and villages like

'Neesdon', 'Wembly Green', Harrow on the Hill, Pinner and Ruislip were overwhelmed by new-build. Kinking country lanes sprouted new species of side-turn: the 'avenue', the 'close' and the 'cul-de-sac' providing access to ordered clusters of neat, new semis. In the twenty years from 1901, Pinner's population tripled in size from 3,366 to 9,462.

The countryside shrank under the laws of supply and demand. On the supply side, agricultural land was cheap, transport was improving and planning regulations were relaxed. On the demand side, towns and cities were crowded, slums needed clearing, wages were rising and so were living standards. And the number of people in England had jumped by two million in one decade. By 1931, England's population had reached 37.4 million (Scotland's had declined slightly to 4.8 million and Wales had dipped to 2.6 million).

Watching the approaching tsunami of semis in 1928 was an architect called Clough Williams-Ellis:

> We plant trees in the town and bungalows in the country, thus averaging England out into a dull uneventfulness whereby one place becomes much the same as any other – all incentive to exploration being thus removed at the same time as the great network of smoothed-out concrete roads is completed.

One of those parping like Toad of Toad Hall along the open road was the journalist H. V. Morton. The landscapes of Thirties Britain had many biographers; writers who sought, thought and jotted; artists who lingered by hedgerow or factory to make a picture. Some were romantics; others radicals. But none could deny the extraordinary diversity of landscapes which was compressed into a single archipelago. Insularity, a benevolent climate, geological riches, plentiful timber and water had all contributed to the makings of some of the world's most spectacular cities, most beautiful countryside and worst slums. Among the romantics, it was Morton who best captured the extremes of Thirties Britain. Puttering the shires in his Bullnose Morris, he had a quick eye for quaint villages, creaking inns and ruined abbeys, but he was also attracted to the industrial cauldrons:

> Here was New England: an England of crowded towns, of tall chimneys, of great mill walls, of canals of slow, black water; an England of grey, hard-looking little houses in interminable rows; the England of coal and chemicals; of cotton, glass and iron.

In Search of England was published in June 1927 and sold in such huge numbers that there had been another twelve editions by 1931. The frontispiece of the battered hardback I'm looking at now has a sepia photograph of a farmworker riding a shire horse past a thatched and beamed country cottage. Morton kept on driving. *In Search of Scotland* (twenty-five editions by 1939) produced his perfect British city, a city where the new had not been superimposed on the old, but built beside it, a modern city of rectangular streets and tramcars on level land gazing up at the intact 'spectre' of its own past:

> Is there another city in the world which marches hand in hand with its past as does Edinburgh; which can look up from its modernity and see itself as it always was, upon a hill intact, impregnable, and still in arms? Salisbury could have done so had the hill of Old Sarum not been ruined; but I know of no other.

Onto this 12,000-year accumulation of human landscapes fell 74,000 tons of bombs.

The landscapes of my childhood were pocked with wartime relics: parts of a Messerschmitt cockpit in a local quarry; bits of rusting 'Doodlebug' in a grassy scar near my grandparents' house in Sussex; dank pillboxes and disused airfields. A twenty-minute bike ride from home in Norfolk, there was an abandoned 8th Air Force bomber field with an eerie control tower and concrete runways that were the biggest flat spaces I'd ever pedalled. Strangled by briars in the surrounding woods were ghostly huts and bunkers stacked with empty ammunition boxes. The fighting had come no closer to Britain than its skies and yet the land bore the weight of memory.

Rearmament began to exert a physical presence from 1936 as Baldwin prepared the country for the 'ghastly possibility' of war. A sudden, devastating air attack was expected to create carnage and panic. As early as 1925, the military writer Basil Liddell Hart had asked the readers of *Paris; or the Future of War* to imagine 'London, Manchester, Birmingham, and half a dozen other great centres simultaneously attacked, the business localities and Fleet Street wrecked, Whitehall a heap of ruins, the slum districts maddened into the impulse to break loose and maraud, the railways cut, factories destroyed.' The view of Liddell Hart – and many other air-power theorists – was that Britain's will to resist such an attack would be vaporised by 1,000-kilogram bombs.

In February 1936, Cabinet approved a military budget that devoted the lion's share to the Navy and RAF: four new aircraft carriers and seven battleships were to be built, and the number of RAF aircraft in the UK was to be increased to 1,500. The plan was deterrence, not a repeat of the Somme. An unready Britain was to be converted with the greatest urgency into the semblance of an impregnable island. Building, equipping, maintaining and operating new fleets of ships and planes demanded new factories, storage facilities, improved ports and new airfields. In terms of acreage, it was the biggest concerted construction exercise ever undertaken.

At sites thought to be beyond the reach of air attack, work began on munitions works, factories and armament facilities. Most had rail connections to the national network. Places that had barely changed for generations were subjected to tectonic upheavals: villages evacuated, roads closed, fields ripped by an onslaught of trucks and diggers. Seven miles from Coalbrookdale in Shropshire, the village of Donnington had a long history of coal-mining and iron-making and so it was familiar with industrial backdrops. An area south of the village was scarred with old shafts, disused collieries and brick works and Donnington itself was home to the Midland Iron Works, famous for producing gas holders. But the new Army Ordnance Depot was on another level of intrusion, eventually covering over 120 hectares. In the suburbs of Nottingham, the old National Shell Filling Factory at Chilwell became a Central Ordnance Depot for the Royal Army Ordnance Corps. Three new underground Central Ammunition Depots were designated, one in the north, one in the Midlands and one in the south. To protect them from accidental explosion and from enemy attack, these ammunition depots were spread across extended areas, their clusters of buildings separated by grass laid with concrete access roads. Deep in the Wiltshire countryside at Monkton Farleigh, quarries that had provided so many buildings with Bath stone were converted into subterranean bunkers capable of storing 350,000 tons of ammunition. Networks of conveyors, rail lines, tunnels and even an underground station facilitated movement of the ammunition. The other two CADs – at Nesscliffe (Shropshire) and Longtown (Cumberland) – were not completed until war had started. Another huge rearmament site was hidden ten miles up the railway from Southampton Water in the chalk folds of the downs. On the secluded upper reaches of the River Dun, the government bought 500 acres (200 hectares) of farmland. For three years a small army of civilian labourers – including miners from Ireland and fitters from Scotland – burrowed and hewed the gently sloping site between the village of West Dean and the abrupt scarp of Dean Hill to create a network of tunnels, bunkers, roadways and narrow-gauge railways, together with a range of workshops and offices. An aerial photograph taken in the summer of 1944 revealed a vast tract of grassland crossed by strange, geometric lines and specked with clusters of buildings. The pilot would have been able to make out the discreet incisions of bunker-entrances cut into the scarp. The incongruity of these establishments was illustrated by this aerial image, which made the place look like a settlement for aliens set in a medieval patchwork of fields patterned with corn stooks, hedges and woodland. The Dean Hill Royal Naval Armaments Depot was so large that the city of Salisbury (less its outer suburbs) would have fitted onto the site. An even larger armaments site was built in South Wales, just off the A48 between Chepstow and Newport. Cordite manufacture required huge volumes of fresh water (150 tons of cordite a week needed 14,000 cubic metres of water every day), and this could be provided by drainage from the 'Great Spring' which had been encountered during the boring of the Severn Tunnel. The gradient of the site was ideal for explosives manufacture; it was adjacent to a main road and railway for transportation and it was a long way from enemy airfields. At 1,580 acres, the Royal Navy Propellant Factory, Caerwent, was three times larger than the Dean Hill facility.

The prospect of war reinvigorated slumbering industrial districts as the government sought the means to mass-produce weapons. Existing factories were expanded and new ones built. New machine tools accelerated production and enabled less-experienced workers to turn out complex components. Under a government plan that came to be known as the 'Shadow Factory' scheme, existing civilian companies deployed their experience and skills to set up duplicate factories for weapons and machines. Britain's car industry was an enormously valuable war asset. Companies like the Nuffield Group, Rootes Group, Austin, Ford, Morris, Standard, Rover, Vauxhall, Daimler and Leyland switched their energies from the manufacture of cars and commercial vehicles, to mass-production in shadow factories of aircraft, aircraft engines, tanks, military vehicles and weapons. Initially, shadow factories were built beside their parent companies, but from 1938, anxieties about their exposure to bombing – and logistical requirements ranging from the need for space to the availability of airfields – meant that new factories began appearing on the edges of manufacturing towns and cities, and in more remote rural locations.

One of the shadow factories that quickly became central to Britain's war preparations sprang from a 140-hectare sliver of land sandwiched between the Birmingham and Fazeley Canal and the Birmingham–Derby railway line just beyond the eastern suburbs of Birmingham. It was an ideal site: level, with excellent communications, access to a skilled engineering workforce and it lay just a few hundred metres from the Castle Bromwich Aerodrome. Here, Lord Nuffield, designer of the Bullnose Morris and by now Britain's most celebrated industrialist, began work on a factory for the mass-production of aircraft. The first sod was cut in July 1938 by the Air Minister, Sir Kingsley Wood, who announced an initial order of a thousand Vickers-Supermarine Spitfires. The factory would be, reported *The Times*, 'the biggest unit of its kind in the country'.

On this threatened island, naval bases and airfields were the points of delivery for the mass-produced weaponry clattering out of factories. Having been closed down in 1925, the 600-hectare Rosyth Dockyard on the Firth of Forth was fitted out in 1938 with new workshops and reopened. Its defences were upgraded, as were those around the naval bases at Scapa Flow in the Orkney Islands and Invergordon on the Cromarty Firth. But it was the RAF's extraordinary programme of airfield building that converted the most hectares to military use. On the same day that the Air Minister had flown to Castle Bromwich to open the Spitfire factory, he'd also opened a new aerodrome at Luton in Bedfordshire, a location he welcomed for being 'little more than a mile from the centre of the town, and easily reached from a railway station less than three-quarters of an hour from the Metropolis'. Luton was to be a centre for the rapidly expanding Royal Air Force Volunteer Reserve, whose young, spare-time trainees would be the fighter pilots of tomorrow. A far larger training base run by the RAF was already in use on the coast of Wales west of Cardiff; RAF St Athan sprawled over 400 hectares and, by the summer of 1939, its three grass runways were accompanied by hangars, a headquarters block, workshops, stores, accommodation quarters, a church, gymnasium, swimming pool and 1,200-seater cinema. Airfields were built the length and breadth of Britain, tripling the total available from 52 in 1934 to 158 by 1939.

The bombing onslaught was expected to destroy huge swathes of towns and cities. The smallest and most common construction project on the eve of war was a 'building' that measured just over 2 metres long and 1.4 metres wide. Inside, it was just high enough for a six-footer to stand. This prefabricated bomb-shelter was supplied as a pack of 14 corrugated, galvanised steel panels, 6 steel channels and sections, 4 steel rivets, 26 nuts, 26 bolts, 52 washers, 2 clips and a 'spanner-tommy bar'. Printed instructions (*Directions for the Erection and Sinking of the Galvanised Corrugated Steel Shelter, February 1939*) described how each shelter could be self-assembled using the tool provided. Anyone failing to first read the instructions would find that assembly had to be preceded by the excavation of a hole in the garden, within which the shelter would be sunk. Nowhere did the instructions state how much time was required to attach the 130 separate steel parts to each other in the correct manner, although there was a warning that one particular phase of erection was 'a two-man job'. In Birmingham, Theresa Bothwell described how the family shelter got built:

> Mother was the handywoman. Dad was useless with a hammer and nail so she put up the Anderson shelter and erected two bunks on the left-hand side of the shelter. On the other side she put a bench for us to sit on, and a shelf for putting tins and packets of food. We'd got a paraffin heater and a bucket for toilet use.

If the shelter had been correctly sited and recessed four feet into the ground and then covered by a minimum of 15 inches of soil, its occupants would – in theory – be able to survive the blast of a 200-pound bomb exploding 20 feet away, or a 100-pound bomb exploding at 10 feet. In the instruction booklet, a section headed 'Finishing Off' advised that clinker or duckboards would make the shelter more habitable and a strong box placed as a step just inside the entrance would make it easier to enter and exit. These do-it-yourself, semi-subterranean, family-sized, bomb-proof, garden sheds came to be known as 'Anderson shelters' after the Lord Privy Seal who had been placed in charge of air-raid precautions. Between February 1939 and the outbreak of war in September, around 1.5 million Anderson shelters were issued, thus making them the most ubiquitous type of standardised 'dwelling' ever to have risen from British soil.

Following a seven-month hiatus – the 'phoney war' – during which there were no major military clashes on the Western Front, the next phase of construction kicked off following the German invasion of France and the Low Countries in May 1940 and the subsequent evacuation of the British Army through Dunkirk at the end of that month. Invasion of Britain seemed imminent. Destruction of the RAF in the air would be followed – it was believed – by airborne landings then a beach invasion which the Commander-in-Chief, Home Forces, General Sir Edmund Ironside, believed would be 'pushed forward with the utmost brutality'. Among many weaknesses facing Britain at that moment was its geography: 'Our country is a small one', wrote Ironside, 'and armoured troops can penetrate at a prodigious speed.' On 16 June, Marshal Pétain had to announce over the radio to the French nation that fighting should cease. Britain stood alone in Western Europe as the last bastion of resistance.

To delay the onrush of German armour across the shires, Ironside devised a pattern of interlocking 'stop-lines' that utilised obstacles like rivers, canals and steep hills. Gaps between these existing defensive features would be plugged by V-shaped anti-tank ditches 1.7 metres deep and 3.6 metres wide, with a rampart on the attackers' side. It was back to the Bronze Age, with an explosive twist: Ironside's 'stops' would be strengthened by 'necessary defensive weapons such as anti-tank obstacles, pillboxes, wire, static anti-tank guns and, where suitable mines'. Waterways were particularly crucial, not only those facing attacking German forces but those 'at right angles to the general line of the front . . . as they hinder lateral methods of attack'. Bridges would be prepared for demolition. Behind the 'defensive crust' along the coast, Ironside hoped his 'stops' would interrupt the flood of German armour for long enough to allow mobile defensive forces to plug the gaps. Along the south and east coast, it was reckoned that nearly 500 miles of beach were suitable for landing armoured vehicles, and around a third of those landing beaches were within reach of German air support.

The main line of defence – the GHQ (General Headquarters) Line – stretched from Somerset to Essex and was intended to protect the industrial heart of England and the capital. An evolving concept, the GHQ Line was then extended from Essex to the Firth of Forth. Between May and late July 1940, some 8,000 pillboxes were built and another 17,000 were either under construction or planned. Seafronts were blocked with barbed wire and minefields. Bus shelters were turned into fortified anti-tank positions and roads blocked with concrete anti-tank bollards. In the event, it wasn't blocks of concrete that were most important, but a chain of relatively flimsy Radio Direction Finding masts erected along Britain's east coast from the Orkney Islands to Kent and then west along the Channel coast to Portsmouth and beyond. The Chain Home radar was the most sophisticated air defence system the world had ever seen, linked through control rooms to fighter airfields equipped with Hurricanes and Spitfires.

The year-long *Luftschlacht um England* – Air-battle for England – was the most destructive military assault Britain had ever endured. As a prelude to the despatch of a cross-Channel invasion fleet, the Luftwaffe had been directed by its Supreme Commander to 'dislocate English imports, the armaments industry, and the transport of troops to France'. From July 1940, bombs and incendiaries fell on radar stations, aircraft factories, ports and RAF airfields. Bases vital to Britain's defence reeled as runways were cratered, hangars blown up and support buildings destroyed. The largest conflagration since London's fire of 1666 was ignited by hits on oil-storage tanks outside Pembroke Dock. There was collateral damage. The difficulties of hitting relatively small targets with free-falling explosive from fast-moving planes meant that houses were frequently destroyed. Selected targets in London were also hit in August, and on 7 September the docklands downstream of Tower Bridge were attacked by 350 planes. For a week, by day and night, London shuddered under relentless raids. In Cambridgeshire, an Aircraftwoman at RAF Duxford heading for her night shift as a plotter and tracer was startled by a strange glow on the horizon, 'like the biggest sunset you ever saw'. Duxford was forty miles from London's docklands. Parts of London's cityscape had been rendered formless: in less than three weeks, more than 3 million tons of debris accumulated and some 1,800 roads were blocked.

By the end of October 1940, the London region had been pounded by around 20,000 bombs. The wharves, warehouses and factories of the docklands suffered most, but bombs and incendiaries also fell among the tight-packed terraces of the East End and on seemingly random parts of the capital all the way from Holborn to Fulham and Chelsea. In two months, the air assault of September and October killed 13,281 civilians and injured 17,437. Most were Londoners. At this dark hour of need, the old capital revealed how much it had expanded downward into the earth: in Stepney, an underground goods yard became a nocturnal sanctuary for as many as 14,000 people on some nights. The basements of department stores were opened as shelters, and so were church crypts and cellars beneath factories and pubs. The stations of London's Underground filled with up to 150,000 people every night. Conditions were frequently disgusting, with excrement scattered in tube tunnels and beneath railway arches where families tried to sleep. By the beginning of December 1940, over 32,000 houses had either been destroyed completely or were too badly damaged to repair. Railway bridges had been smashed and over 4,000 water mains ruptured. Severed gas and electricity mains were constantly being mended and thousands of unexploded bombs had to be defused.

In the midst of these horrors, the Ministry of Information released a film for American audiences. *London Can Take It!* was voiced with soft authority by war correspondent Quentin Reynolds, who portrayed the capital as 'a great fighter in the ring' who would get up from the floor after being knocked down: 'London does this, every morning.' There was a strong sense in the narration that the adversary 'Jerry' faced was more than flesh and blood. It was a place:

> London doesn't look down upon the ruins of its houses, upon those made homeless during the night, upon the remains of churches,

hospitals, workers' flats. London looks upwards toward the new dawn and faces the new day with calmness and confidence . . . It is hard to see five centuries of labour destroyed in five seconds. But London is fighting back.

Reynolds could be forgiven for compressing in the heat of the moment nearly two thousand years of urban history into a mere five centuries, and of course he was speaking for 'every other British city and town, where resistance to the intense aerial attack and powers of endurance are every bit as heroic'.

From mid-November, the bombing spread to the Midlands, where the factories of Coventry were blitzed in a raid that killed 568 and left the city centre shattered. Through that long, cacophonous winter, Birmingham was attacked, as were the inland cities of Manchester, Sheffield, Nottingham and Norwich. Merseyside was hit for eight successive nights in May 1941, killing over 1,700. On the coast, Southampton and the ports of Cardiff, Bristol, Swansea, Portsmouth and Hull were bombed. A raid in March 1941 on Clydebank killed 528 and another the following month on Belfast killed 900. In mid-April, Plymouth was bombed and then Merseyside again, in a week-long onslaught that killed nearly 2,000. Key centres were struck repeatedly.

By summer 1941, the war had moved on. The first two months of bombing had punished London in particular, but during the prolonged raids that followed, deaths and injuries suffered in other parts of Britain were roughly equivalent to those of the capital. Between September 1939 and the end of 1941, a total 43,685 civilians were killed and 51,694 seriously injured. A further 186,611 were slightly injured or treated at First Aid Posts.

Destruction affected centres from the Clyde to the Thames and from East Anglia to the West Country. After London, Liverpool was Britain's most bombed city. The blitz on Merseyside had destroyed 66,000 houses. Birmingham suffered the third-worst intensity of bombing. Clydebank and Coventry lost around a third of their houses. In Sheffield, nearly 3,000 houses and shops were destroyed.

The sudden, cataclysmic destruction of familiar landmarks inflicted various levels of topographic trauma. Many of Goering's targets were embedded within neighbourhoods of tightly packed streets where a stray stick of bombs or drifting mine could destroy a place known to generations. Close-knit communities were linked from street to street by cousins or aunts or uncles. The neighbourhood was more than a home; it was a life. The violent, random demolition of houses, shops and churches was profoundly shocking. 'To see the result of years of work swept away in a second,' recalled the driver of a mortuary van, 'leaves one with an awful feeling of instability.' In Coventry, a visitor to the city centre the morning after the big raid spoke of his shock at being able to see right through the devastated cathedral, from one side to the other. The ubiquitous bomb site, livid with purple Buddleia and butterflies, became a new kind of urban space.

The Blitz was followed by a surge in construction as the RAF – and from 1942, the USAAF – expanded their bombing offensives on the continent. Many more airfields were needed, with stronger runways in concrete which could bear the weight of heavy, four-engined bombers. The standard Class A base covered around 400 hectares and was serviced by 10 miles of roadway, 20 miles of drains, 10 miles of electrical conduits, 6 miles of water mains and 4 miles of sewers. On a typical base, over four million bricks were needed for the various buildings, and around 500,000 cubic yards of concrete. Around 1,000 labourers would take 18 months to complete all the works. At the peak of construction in 1942, a new airfield was being completed in Britain every three days.

The Second World War concluded for London and the south-east with a return to sudden destruction from the air. The first cruise missile was launched towards London on 13 June 1944. Guided by an auto-pilot and armed with a 850-kilogram explosive warhead, the V1 – known to its recipients on the ground as the 'doodlebug' or 'buzz bomb' – had a range of 150 miles but this was increased to 250 miles as launching ramps were built in the Netherlands. Around 10,000 V-1s were launched at Britain, causing 6,184 deaths and 17,981 serious injuries. Nine-tenths of those deaths and injuries occurred in the London Civil Defence Region, which was struck by around 2,500 V-1s. Within London, the southern boroughs of Croydon, Wandsworth and Lewisham were hit the hardest, each receiving over a hundred V-1s.

From September 1944, the 'doodlebug' was accompanied by the V-2, a guided ballistic missile carrying 910 kilograms of explosive. With a range of 120 miles, this weapon was also limited to south-eastern targets. Fewer of them crossed the Channel, but they killed another 2,754 people and seriously injured 6,523. Travelling faster than the speed of sound, the rockets were inaudible until they exploded. The single worst V2 missile attack killed 168 people when it came down on a Woolworth's in New Cross. Some 517 fell in the London Civil Defence Region, but the county of Essex was hit by 378 and Kent by 64. Norfolk received 29.

The human cost of aerial bombardment and its physical effects on centres of population overshadowed a more peaceable aspect of war-time Britain. There was another landscape in this war, one that was not a maelstrom of fear, loss, rubble and dust. It could be seen in the languid meanders of Cuckmere Haven by Eric Ravilious, whose river relaxed between undefiled downs; it could be seen in any stretch of countryside that lay beyond earshot of a city or port or airfield. Away from this localised war slumbered a Britain that was in many ways more tranquil than it had been for decades. By 1943, fuel rationing and the diversions of conflict had cut the volume of cars on British roads to levels not seen since the mid-Twenties. There was so little traffic that my father and uncle spent the school summer holiday of 1943 riding their bicycles on a 2,000-mile round-trip of England and Wales on arterial roads that were not much busier than country lanes. Horses tugged ploughs; in the green folds of Sussex my mother and her sisters foraged for blackberries beneath the vapour trails of aerial combat.

The audit for those in the blast-zone was rather different. By virtue of being an island, Britain had been beyond – just – the reach of land war, but nevertheless it had been a 'war-zone'. By the end of the conflict in 1945, the number of civilians killed in Britain by enemy action had risen to 59,628, while 85,504 had been admitted to hospital, mostly with serious injuries. Another 315,000 had been slightly injured or required treatment of some sort at First Aid Posts. Nearly half a million civilians had been killed or injured, but the total number of people affected by the bombs and rockets was several multiples of that figure. For many of those who sailed and died in the Navy and Merchant Navy, British ports had been their last terrestrial 'home', just as army camps and military airfields had been 'home' to soldiers and airmen who died in battle. Losses in war leave a trail of physical associations which track back from place of death through ports and airfields and camps to schools, universities and places of birth. For forces based in Britain, airfields were closest to combat and in many respects were weighed with the same kinds of tensions and tragedies that were experienced on the front line of the land and sea war. The total number of RAF personnel who died in the war was 101,223, and virtually all of them had been based at some point at airfields in Britain. The total number of American airforce personnel lost flying against Germany alone was 30,099. For many, Britain's airfields would always be places of particular melancholy.

The effect of so many place-related deaths, injuries and wartime experiences was to memorialise thousands of sites, both through collective association and through physical structures. The Second World War is recalled from southern England to northern Scotland, from Wales to East Anglia, in countless memorial plaques, listed wartime buildings and protected ruins. They range from the inconspicuous to the monumental. All are moving. The plaque to the bomber crew on the cliff path west of Clovelly is not visited by many, but the 'Airman's Grave' in Ashdown Forest is set within its own stone enclosure and every Remembrance Sunday is the focus of a service. The windowless walls of Coventry Cathedral stand as a memorial to events on the night of 14 November 1940. Close to St Paul's Cathedral, Christ Church Greyfriars was one of eight Wren churches destroyed on the night of 29 December 1940. After the war, its ruins gathered wildflowers and eventually they became a memorial garden. In scores of villages and churches, names were added to war memorials. One of Britain's more remote war memorials stands on the shore of Loch Ewe, which was used as a collecting haven for ships about to undertake perilous 'Arctic Convoys'. A few years ago, I was there one wet and windy autumn evening when a piper in tartan walked slowly up the rise playing a lament, accompanied by an elderly gentleman dressed in blazer and tie. Like so many monuments, the chiselled stone of Loch Ewe had become the place to recall events that strain comprehension.

Half of those killed on British soil – some 29,890 – lost their lives in London. Damage in the capital exceeded by far that inflicted on other cities. Over 73,000 buildings were totally demolished in the London regional area and another 43,000 damaged beyond repair. Some 14 million tons of debris had to be cleared in the city. As much was recycled as possible for rebuilding the capital: around 340,000 tons of metals were collected and 140 million bricks were cleaned and stacked. The useless debris was taken to dumps whenever they became available. Blackheath and Mitcham were recipients, and Hackney Marshes and Crystal Palace. The northern part of Regent's Park was spread with debris (which still reappears in dry summers) and Cumberland Basin on Regent's Canal was filled in (and is now the car park for London Zoo). More debris was taken down the Thames estuary in barges.

Post-war Britain was a tired land wreathed in loss, patched with hastily laid concrete and dotted with hundreds of thousands of military buildings and air raid shelters. A total of 444 airfields were constructed in the UK between 1939 and 1945. Around 28,000 pill-boxes had been built. Before the war, the three military services had held 102,000 hectares, but by 1945 the total was nearly fifty times greater, at 4.6 million hectares, or roughly a fifth of Britain's land area.

The dank, mass-produced Anderson shelter was superseded by the pristine, mass-produced prefab. Back in July 1944, the Deputy Prime Minister, Mr Attlee, had informed the House that it wouldn't be possible for some years to build enough permanent houses and that the interim measure would be 'emergency factory-made houses'. Built by government-approved manufacturers to Ministry of Works specifications, each prefabricated dwelling would provide 59 square metres of floor space, an indoor lavatory, piped hot water and a fitted stove. Walls were pre-painted in magnolia and skirting in green gloss. Surrounded by a small garden, the basic two-bedroom layout was intended to suit young couples eager to start families. A Labour politician, Neil Kinnock, remembered the 'remarkable dwelling' that had been his boyhood home: 'With our inside bathroom and our inside toilet, and our fitted kitchen with our refrigerator . . . a fitted electric stove, fold-down table, it was a place of wonder.'

The prefab was the Carley float of the cul-de-sac; a basic, life-changing, factory-line solution to a crisis. Designed for speedy deployment, all components had to measure no more than 7ft 6in wide (2.3m) so that they could be moved on the back of a lorry. At the heart of every prefab was the 'service unit', an integrated plumbing system that fitted into a void between the kitchen and bathroom. Springing like mushrooms from the edges of parks and green belts, prefabs symbolised renewal, efficiency and industry. They provided the first rung on the ladder to post-war family life. The intention had been that prefabs would play a bigger role in the post-war housing crisis. Under the Housing (Temporary Accommodation) Act of 1944, at least 300,000 were meant to be built in two years. Relatively few were constructed – just 156,623 between 1945 and 1949 – but they could be seen snug behind their flowered gardens all the way from Clapham to the Isle of Lewis, where they were fitted with insulated walls.

Peacetime Britain had a housing crisis. Some three-quarters of a million houses had been destroyed or badly damaged during the war. Added to that was the pre-war deficit: almost five million dwellings had no bath installed, six million lacked an inside WC and seven million had no plumbed hot water. By 1945, the total shortfall of dwellings may have been as much as 1.4 million. The problem was not evenly spread. London may have suffered the greatest war damage, but Glasgow had endured longer-term deprivation. In Scotland, well over a fifth of the population was trying to live in sub-standard housing. In Glasgow, the situation was notoriously appalling, with half of the city's dwellings having no more than two rooms (the figure for Greater London was just over 5 per cent).

Prefabrication was the new vernacular. War had prepared the ground for uniformity. Britain had become an island of repeated forms: of similar berets and sandbags, of identikit trucks and planes, of pillboxes cast in concrete from standard moulds. Pop-up housing was an aftermath of war-effort. Five years of experimentation in materials and structures had produced a reservoir of engineering knowhow, and the cessation of hostilities had left factories with surplus capacity. There was a wide variety of models. The Airey house used a pre-war design of concrete posts clad in reinforced concrete blocks that were manufactured in Royal Ordnance Factories. Five lorries could carry the parts for two houses, which could be erected in a couple of weeks. Some 20,000 Airey houses were assembled after the war. The British Iron & Steel Federation produced a 'metal house', with a frame of steel and much of its cladding in steel, too. The BISF house was planned to last sixty years and a pair of them were included in a set of thirteen experimental semis erected in 1944 by the Ministry of Works at Northolt in the suburbs of west London. Another pair of semis on the Northolt site were constructed from 'no-fines concrete': concrete made without aggregate. Eventually, over 30,000 BISF houses were built, while the no-fines concept was reproduced in even greater numbers. In Cornwall, the English China Clay Company developed the Cornish Unit house, whose concrete unit walls used waste sand from the china clay pits. In 1945, the first Cornish Unit bungalow was erected in a village five miles out of St Austell and shortly afterwards, a two-storey house was put up in the town itself. Over the next ten years, another 40,000 were raised by local authorities across Britain. The John Laing 'Easiform' house was another pre-war concept that found its market after 1945. The Easiform used standardised shuttering to form cavity walls of poured concrete. Because the walls were cast on site, the basic configuration could be adapted with bay windows, extra storeys, and with hipped or gabled roofs. In the decade from 1945, local authorities built 47,000 of them. The versatility of the Easiform process led to it being used for a variety of structures, from bungalows through to parades of shops with overhead flats. The 'pier and panel' system used by Orlit Houses was also suitable for flats. The Orlit was assembled by attaching pre-cast concrete slabs to reinforced concrete frames and could be fitted with a pitched or flat roof. In the London borough of Poplar, the Orlit system was used to build the first block of pre-cast concrete flats in Britain. All the flats had three bedrooms, and those on the second and third floors also had recessed sun-balconies. The three-storey prefabs in Mellish Street were ready for letting by March 1948.

Away from the scramble for pop-up homes, dreams were being committed to print. At the general election of July 1945, the UK had voted for the road signposted to the sunlit uplands of industrial regeneration, full employment, free healthcare and public housing. Railways, roads and civil aviation were taken into public ownership, along with coal and gas, electricity, cable and wireless. Reorganisation on such a scale offered a window – a gigantic, never-to-be-reopened picture window – of opportunity for town planners and architects. Patrick Abercrombie had been poised for action long before the war ended. Since 1935, he'd been professor of town planning at University College, London. With his pre-war experience in regional planning, his skills as a consultant architect and the vision that founded the Council for the Preservation of Rural England, Abercrombie was – according to one of his contemporaries – the 'one man' who had 'a truly synoptic view of the physical planning problems of the British Isles'. In 1943, Abercrombie was let loose as a planning consultant for London County Council. Working with the LCC's chief architect, John Forshaw, Abercrombie came up with the County of London Plan.

A month after the Allies under General Montgomery invaded

mainland Italy, Abercrombie took to the stage at the Royal Geographical Society to outline his vision for the run-down, battered capital of the British Empire. The professor began by reminding his RGS audience of London's unusual geography: a city dominated by 'the capricious but masterful' Thames but a topography that was otherwise 'not particularly exciting'. The same could not be said for the city's urban geography. London did not 'emanate from a single focal point'. It wasn't 'a single radiation from one centre'. Instead, London had three centres: a commercial hub in the City; a political and administrative centre at Westminster, and a third centre in the West End: 'the centre of the amusement or the meeting place of the whole of the Empire'. Abercrombie suggested that his audience consider the West End as 'the Agora' (a classical comparison that must surely have raised discreet smiles from RGS Fellows familiar with Soho). This triple-ventricle conurbation had four characteristics: it was overgrown, congested, squalid and beautiful.

The professor was a realist. He wasn't planning Utopia, but 'an adequate centre of the Empire'. The principle guiding Abercrombie's plan was the creation of 'a balanced form of environment'. That wartime autumn evening at the RGS, he repeatedly used the word 'balance', a return to his understanding of Feng Shui, that harmony should be sought between humanity and the environment. There were, he explained, six 'determinants' that had to be addressed if balance were to be imposed on a city as disparate as London. The first was the control and planning of road and rail traffic. To this had to be added preparations for post-war congestion of air traffic. Road traffic in particular was a 'menace' that 'slaughters so many men, women, and, particularly, children'. The second determinant was housing. Too many were living in 'overcrowded and bad houses'. London was sprawling and there was a case for arguing that 'some sort of decentralization' was necessary. Abercrombie's third determinant was industry, which he suggested had congregated excessively within the London area. Fourthly, he saw the relic villages of London as the basis for future urban communities. He wanted 'to return to a smaller unit than the whole city'. His fifth determinant was open space. London's admirable parks were badly distributed, leaving outlying areas like Shoreditch, Stepney and Bethnal Green with hardly any open space. Like Ebenezer Howard, Abercrombie saw green space as essential to the urban balance: 'One longs to see a break, with green trees, where nature, even if only Nature under some confinement of man's devising, asserts herself. Open spaces are essential to break up that continuity of the urban mass.' The sixth and final determinant of city planning was the largely subterranean mass of conduits carrying water, gas, electricity and so on. Abercrombie spoke of the 'dignity of sewers' and of the 'somewhat negative character' of this 'perfect maze of underground services'. He was thinking in particular of the difficulties likely to be encountered by pushing through new underground railway lines.

Lord Latham, leader of London County Council, was at the RGS that night. As the man who would have to find a way of implementing a plan so ambitious that it made the urban visions of Wren and Hooke look like doodles, Latham had to be circumspect. As he pointed out, there were about eighty statutory authorities that were entitled to be consulted. The President of the RGS was rather more effusive, praising the planners' 'audacity', 'scope' and 'imagination':

> Their dreams may not take practical shape as they have dreamed them, but in what they have done they have started the younger generation on the road that they have to travel if their London is to be a London that is more worth living in than the old slummy, ramshackle London that older men like myself have known and loved.

The County of London Plan was followed in 1944 by the Greater London Plan. Where the former had been directed at those parts of London under LCC authority, the latter covered the outer areas into the Home Counties, and had been commissioned by the standing conference on regional planning. The two plans were complementary, with Abercrombie guiding both. Looking at the London region, Abercrombie visualised a greater metropolitan area that would consist of 'rings': outside the tightly packed administrative core would be an inner urban ring, and then a suburban ring, and then a green belt. Beyond the green belt would be an outer country ring. Within the green belt, agriculture would be prioritised and other forms of development strictly controlled. The suburban spread of London would be stemmed at its 1939 boundaries, and new development would focus on orbital new towns set within the 'country ring'. Traffic would be channelled along nine arterial roads that entered the metropolitan area to join a ring road. Railways – loathed in London for their smoke, noise, space-wasting viaducts and ugly bridges – would be banished underground.

In 1946, a government-sponsored film was released in which LCC – aided by Abercrombie, Forshaw and Lord Latham – made the case for reinventing 'the greatest city the world has ever known'. It was now or never: 'Here is our chance to do something . . . we saw beyond all the suffering and destruction a great opportunity to build a new London, a better, finer, more spacious city than the old.' It would, explained Latham, 'cost a great deal of money, but not more than unplanned building, and a lot less than war'. *The Proud City* was a Churchillian rallying cry: 'In a way, you know, this is London at war, against decay and dirt and inefficiency. And in the long run, a plan such as this is the cheapest way to fight those enemies.' All too well, Latham knew London's historic resistance to major urban planning:

> What a grand opportunity it is. If we miss this chance to rebuild London, we shall have missed one of the great moments of history. We shall have shown ourselves unworthy of our victory. So let's start now to rebuild London where everyone can live a full, happy, healthy life. A London her citizens can be proud of.

The actual business of rehousing Londoners in permanent homes within the metropolitan area could not be achieved through houses alone. Abercrombie and Forshaw worked out that the densities required to allow 60 per cent of residents to remain in their own neighbourhoods would mean accommodating two-thirds of them in blocks of flats up to ten storeys high. The architects were ready.

Churchill Gardens estate was an early outing for public authority Modernism, the architectural movement that had rooted on the continent during the 1920s. Severing all ties with tradition and ornament, Modernism sought to bring functional purity to the built landscape with cubic forms and modern materials, particularly concrete. The stage chosen was the north bank of the Thames between Vauxhall Bridge and Chelsea Bridge. In the early Thirties, this part of Pimlico had been a tight jam of terraced houses, factories, warehouses and wharves. Adjacent to the seat of government in Westminster, it was a prime district for redevelopment. In 1935, a significant start had been made with the demolition of an Army clothing factory and construction of the largest self-contained block of flats in Europe. Ten storeys high and over 600 feet long, the 1,250 flats of Dolphin Square were arranged around an internal garden designed by Richard Suddell, president of the recently created Institute of Landscape Architects. While bombs, aerial mines and V-1s had fallen on all sides of Dolphin Square, the block had emerged from the war virtually unscathed. Next door were the bomb sites that became Churchill Gardens, one of the first post-war estates to adopt a Modernist solution to inner-city housing. A competition run by Westminster City Council in 1946 was won by a pair of young architects, Philip Powell and Hidalgo Moya, who brought the clear, austere, interwar housing of Germany, France and the Netherlands to the Thames, with a combination of medium-rise blocks of flats and low-rise terraces set spaciously on uncluttered panels of grass. Britain's first district heating system was provided by waste water pumped from Battersea Power Station. Work began in 1947 and by the time it was complete, the estate's 1,800 dwellings were home to 6,500 Londoners.

Abercrombie's capricious Thames was also the setting for the first great peacetime showcase of British strength and imagination. A derelict, bomb-wasted site on the south bank was cleared and then covered with futuristic structures displaying – as a contemporary film put it – 'the essence of Britain'. The South Bank Exhibition of 1951 was the centrepiece of a nationwide Festival of Britain on the centenary of the Great Exhibition that had filled Hyde Park in the heady summer of Empire and Industrial Revolution. The show on the south bank was architecturally choreographed by Hugh Casson, who had to work with an asymmetrical site split by the elevated approach to the Hungerford rail bridge. Casson's team resurrected pre-war Modernism with an informally arranged set of buildings that included the temporary Dome of Discovery and permanent Royal Festival Hall with an open, liberated interior and ascending staircases. At the time, nothing in Britain could compare. Pedestrian spaces on various levels shared the riverbank with glass pavilions, fountains and concourse cafes. Not everyone loved it: the conductor Sir Thomas Beecham (who once described the sound of a harpsichord as 'two skeletons copulating on a tin roof') regarded the South Bank Exhibition as 'a monumental piece of imbecility', but in five months the site attracted 8.5 million visitors. Rising above the site like an exclamation mark was the Skylon, a 250-foot-high aluminium 'vertical feature' whose name had been provided by a Mrs Fidler (her husband was the chief architect to Crawley Development Corporation). The brain behind this extraordinary, suspended apparition was Hidalgo Moya, architect of Churchill Gardens. Rising from the rebuilt bank of the Thames, the slender, reflective sheathing of the Skylon reclaimed the capital's airspace.

Outside London, Abercrombie's vision was explored through the creation of 'New Towns' that would – it was hoped – help to rehouse the inhabitants of cleared slums as well as providing new homes for a booming population. In a remarkable surge, fourteen New Towns were designated between 1946 and 1950, eight in southern England and six in northern England, Scotland and Wales. It was the single greatest exercise in systematic 'town planting' since Alfred founded his *burhs* one thousand years earlier. Whether Alfred was as hands-on as Attlee's Labour government is hard to say; contrary to Ebenezer Howard's ideal of self-governance, the New Towns were to be run by government-appointed and funded development corporations. They would also be engineered to produce a strong sense of community. Greyhound racing would be discouraged and it was hoped that 'civic cinemas' would provide the social benefits not apparent in commercial cinemas, with their 'limited cultural range and American productions'. 'Our aim', explained the new Minister of Town and Country Planning, Lewis Silkin, 'must be to combine in the new town the friendly spirit of the former slum with the vastly improved health conditions of the new estate, but it must be a broadened spirit, embracing all classes of society.' Populations would be allowed to rise to between 50,000 and 80,000 and each New Town would be girdled by a green belt of protected countryside.

The eight southern New Towns encircled London (more or less) at a radius of between 21 and 35 miles. Here at last was a simulacrum of More's *Utopia*, with pre-eminent Amaurot at the centre, surrounded by sub-cities no more than a day's walking apart. Here too, was Ebenezer Howard's 'Group of Slumless Smokeless Cities' that he'd depicted in a diagram that looked a bit like a Ferris wheel, each set within a green belt of contributory land-uses: a 'Home for Inebriates', an 'Insane Asylum', 'Convalescent Homes', 'Industrial Homes', a 'Home for Waifs' and a district of 'Epileptic Farms'. In fact, the similarity to Howard's concentric 'ring' was limited: of the eight New Towns, Crawley lay to the south of London, Basildon to the east and Bracknell to the south-west, and the other five – Hemel Hempstead, Hatfield, Welwyn Garden City, Stevenage and Harlow – all lay to the north. It could be argued, however (and has been, enthusiastically, by Peter Hall and Colin Ward in their book *Sociable Cities*), that Letchworth, Welwyn Garden City and Stevenage formed an equidistant trio of Howard's Slumless Smokeless Cities, connected by his 'rapid rail transit' and surrounded by green belt furnished with the requisite cemeteries, colleges, convalescent homes and so on. The first of the eight new designations was Stevenage. Like the other seven, this was already a 'place' with inhabitants, and they were unimpressed by their planned promotion to New Town status. When Silkin turned up to address them, he found the station signboard altered from 'Stevenage' to 'Silkingrad'. But by 1952, the first residents were moving into the Monkswood and Broom Barns estates.

Not for the first time in Britain's landscape story, the planner's realm was removed from public reality. The push for homes failed to meet demand. From 1949 until 1951, the number of permanent homes built each year in the UK was running at just over 200,000, jumping to 248,000 in 1952. Over 80 per cent of the annual total was social housing, but council waiting lists were still far too long. Abercrombie joined Wren and Hooke in watching his ambitious plans for London falter at the first hurdles (although his green belt did become formalised in 1955 and an element of his road plan surfaced eventually as the M25). The reality for London, as always, was not planning but 'semi-planning': piecemeal regeneration paid for on the whole by public authorities.

Beyond the urban fringes, a landscape more amenable to manipulation was recast for peacetime. Among those with a claim to the National Parks idea is an American painter called George Catlin, who wrote to New York's *Daily Commercial Advertiser* in 1833 suggesting that a 'nation's park' be designated at Yellowstone. But twenty-three years earlier, the Revd Joseph Wilkinson had published in London his *Select Views in Cumberland, Westmoreland and Lancashire*, which opened with an eloquent introduction by an unnamed poet. William Wordsworth infiltrated his evocation of lake and fell with critical asides about the 'red tile', the 'gentleman's flaring house' and the 'mania of ornamental gardening'. And he appealed to his readers that the Lake District and its ancient ways of life be protected from discordant, modern encroachments and be regarded as 'a sort of national property, in which every man has a right and interest who has an eye to perceive and a heart to enjoy'. Wordsworth's 'national property' took a century to materialise. By the end of the Second World War, increasing leisure time, a widening popular appreciation of 'natural' landscapes and the surge in motoring were putting the squeeze on Britain's most treasured tracts of countryside. It took another visionary to seize the moment.

Ilkley-born John Dower had contracted TB while surveying installations at Dover Harbour early in the war. A man of great energy and
practicality, his pre-war work as an architect had varied from youth
hostel design to aerodromes and by 1942 – after being invalided out
of the Royal Engineers – he had been tasked by the Ministry of Works
and Buildings with helping to plan post-war reconstruction. His 'one-
man White Paper' of 1945 defined a national park as 'an extensive area
of beautiful and relatively wild country in which, for the nation's benefit and by appropriate national decision and action', four objectives
would be met: the 'characteristic landscape beauty' would be strictly
preserved; access and facilities for 'public open-air enjoyment' would
be amply provided; 'wildlife and buildings and places of architectural
and historic interest' would be protected; and 'established farming
use' would be effectively maintained. By the end of the Fifties, maps
of England and Wales were marking the boundaries of national parks
in the Peak District, the Lake District, Snowdonia, Dartmoor, the
Pembrokeshire Coast, the North York Moors, the Yorkshire Dales,
Exmoor, Northumberland and the Brecon Beacons.

Rolled-out at astonishing scale and speed, the national parks
were a last-ditch measure to halt the advancing diggers. The spokes-
person for Fifties England was an economic historian from Exeter
who 'detested' London and who refrained from learning to drive a
car. In his own words, *The Making of the English Landscape* (1955) by
Professor W. G. Hoskins was 'an attempt to study the development
of the English landscape much as though it were a piece of music, or
a series of compositions of varying magnitude, in order that we may
understand the logic that lies behind the beautiful whole'. Hoskins
opened his story with Anglo-Saxon villages in the middle decades
of the fifth century and concluded in the mid-1950s with an eru-
dite howl of anguish. He identified the later years of the nineteenth
century as the beginning of the end: 'Since that time, and especially
since 1914, every single change in the English landscape has either
uglified it or destroyed its meaning, or both.' Abuses included the

demolition of country houses, the loss of ancient landscapes to overspill ('a word as beastly as the thing it describes'), the spread of 'ranch-farming' and open-cast mining, the bulldozing of hedgerows and stone walls and the eruptions of prefabs, barbed wire and battle-training areas. It was this 'Barbaric England of the scientists, the military men, and the politicians' that the Devonian professor urged his readers to turn from and 'contemplate the past before all is lost to the vandals'.

Hoskins was writing at one of those moments of accelerated change that characterise the story of Britain's landscapes. England's population had grown by 3.8 million in the twenty years since 1931 (Scotland had seen a slight increase to 5.1 million, but Wales had remained static at 2.6 million). In the decade from 1945 until 1955, the number of vehicles on the road had jumped from 2.6 million to 6.5 million. Air traffic was soaring: in 1955, the Queen opened a new terminal building at Heathrow, and Gatwick opened for full-scale passenger traffic. The numbers of passengers carried on UK airlines increased from 1.2 million in 1950 to 4 million in 1958. Hoskins wasn't the first to see modernity as the enemy of history and his views on aviation were delivered as if the good professor was swatting at passing planes with a cricket bat:

What clsc has happened in the immemorial landscape of the English countryside? Airfields have flayed it bare wherever there are level, well-drained stretches of land, above all in eastern England. Poor devastated Lincolnshire and Suffolk! And those long gentle lines of the dip-slope of the Cotswolds, those misty uplands of the sheep-grey oolite, how they have lent themselves to the villainous requirements of the new age! Over them drones, day after day, the obscene shape of the atom-bomber, laying a trail like a filthy slug upon Constable's and Gainsborough's sky.

In the seven years from 1958 through until 1965, the number of passengers carried on UK airlines went up from 4 million to 11 million.

*

Cities ascended. From the mid-Fifties, buildings began reaching for the sky. The most celebrated – for a while, anyway – was founded on the Festival of Britain site. By the time the Shell Centre topped out at 107 metres in 1961, it had become the first building in London to clear the Victoria Tower of the Palace of Westminster, also out-reaching Liverpool's Royal Liver Building as the tallest lay building in Britain. (Born of parents who had known London during the Blitz, I was taken as a small boy to the viewing terrace at the top of the Shell Centre to gaze across a sooty city patched with pale new concrete.)

Sixties London saw a high-rise boom. One of the first American-style 'skyscrapers' to climb above the capital's roofs was built speculatively by Harry Hyams, a property developer who'd made millions on Lon-don's rising rents. Hyams used the architectural practice of R. Seifert & Partners to implant a spectacular landmark on London's busiest shopping street. The site at the eastern end of Oxford Street lay at the meeting point of Bloomsbury, Soho and Covent Garden. Triangulated by three of London's most distinctive cultural districts, this was already a conspicuous plot by the time Richard Seifert's well-drilled team ne-gotiated their way through height restrictions and got permission to build a 33-storey tower. Seifert had just completed a 22-storey tower on Kingston bypass and wanted to push twentieth-century Modern-ism further. Following Le Corbusier, Centre Point was a 'point-block' raised on free-standing supports so that traffic could pass through its ground level. But Seifert – or rather his partner George Marsh – broke from the clean regularity of International Modernism and assembled a tower from pre-cast, concrete mullions whose facets and recessed glazing created angular patterns of light and shade which tapered skyward. By decreasing the depth of the mullions from the bottom to the top of the building, an impression of even greater height was achieved. Centre Point was original, ingenious, confident and widely despised. To the architecture critic Nikolaus Pevsner, it was 'coarse in the extreme'; to architect Ernő Goldfinger, it was 'London's first Pop-art skyscraper'. By the end of the Sixties the City and West End were being overlooked by one hundred or more office blocks standing taller than 100 feet and another sixteen higher than 300 feet. St Paul's Cathedral, the tallest structure in London for 300 years, ceased to be the silhouette that drew the eye.

When high-rise was adopted for housing, the results were less enduring. The tower-block of stacked flats was seen by local authorities as a neat solution to the housing crisis. Using prefabricated concrete panels, 'system-built' blocks could be bolted together Meccano-style. Slums could be cleared and their inhabitants resettled vertically in point-blocks occupying a fraction of their original footprint. It was a solution encouraged by the Housing Subsidy Act of 1956, which increased the amount paid to councils per flat in relation to height from the ground. On the Brandon Estate in Southwark, the LCC built six 18-storey point-blocks and forty low-rise blocks for residents cleared from Bermondsey slums. In Paddington a tract of land north of the railway line had been conveniently cleared by a V-1 flying bomb, and this became the site for the Warwick Estate, which included six 21-storey tower-blocks. In Deptford, the Pepys Estate pushed higher still, to 24 storeys. In 1965, when the Greater London Council inherited the LCC's housing stock, the push for height accelerated and in just one decade the GLC put up 384 tower-blocks.

Glasgow went further still. In what turned into the most intense multi-storey campaign seen in any British city, nearly three-quarters of Glasgow's housing completions between 1961 and 1968 were tower-blocks. At the time, many British cities had acute housing problems, but in terms of numbers and deprivation, Glasgow's was still the most extreme. The 1957 *Report on the Clearance of Slum Houses, Redevelopment and Overspill* had recommended the building of 60,000 overspill houses and within Glasgow itself 40,000 new dwellings. This represented a mammoth building programme that pitched advocates of Abercrombie's dispersed New Towns against those who favoured the rehousing of people from slums within Glasgow. David Gibson, Convenor of the Housing Committee from 1961 until 1964, saw the report's overspill plan as an attempt by the Westminster/Edinburgh elite to hollow out Britain's second city. In his Chairman's address at the Annual Housing Inspection of 1962, he explained why inner-city high-rise was the answer:

> In the next three years the skyline of Glasgow will become a more attractive one to me because of the likely vision of multi-storey houses rising by the thousand ... It may appear on occasion that I would offend against all good planning principles, against open space and Green Belt principles – if I offend against these it is only in seeking to avoid the continuing and unpardonable offence that bad housing commits against human dignity. A decent home is the cradle of the infant, the seminar of the young and the refuge of the aged!

When he was sixteen, Gibson had moved from Ayrshire to Glasgow's East End. He knew how bad the tenements could be and was impatient to improve living condition for his fellow Glaswegians; he was, acknowledged one of the planners, 'a man in a hurry'. And tower-blocks could be built far quicker than low-rise overspill developments, which meant communities could be rehoused with minimal disruption. In Royston Area 'A', Wimpey showed what was possible by putting up three 20-storey blocks in eight months. From 1961, immense tower-blocks erupted above Glasgow's skyline: 26-storey blocks at Wellfield Street; a pair of 31-storey blocks at Bluevale Street that became the tallest buildings in Scotland; ten 20-storey slab blocks on a reclaimed chemical wasteland. On green-belt land at Barlornock, a scheme to build four-storey maisonettes was superseded by a cluster of eight 28- and 31-storey tower-blocks with dwellings for 4,700 people. Blocks sprouted from off-cuts of golf courses, from reclaimed prefab sites, from patches of wasteland. 'Gap-site' opportunism on this scale was unprecedented and created an apparently random distribution of blocks. In their book *Tower Block* (1993), Miles Glendinning and Stefan Muthesius described Glasgow as 'the new "shock city" of the Modern housing revolution'. By 1972, a staggering 48,000 new dwellings had been built within the city boundaries since the publication of the 1957 slum clearance report.

The residential tower-block surge came to an end in the late Sixties. In 1968 the partial collapse of a 22-storey tower block in Newham, east London, killed four people. Ronan Point had been occupied for only two months. A gas explosion in an eighteenth-floor kitchen took out the four floors above, and caused the flats below to fall. Although the block was rebuilt, the architectural dream of 'streets in the sky' had become a nightmare for many. Glasgow's high-rises came down, too. Costs of maintenance were underestimated, crime and anti-social behaviour undermined the strength of their community; some, like the Red Road blocks, were clad in asbestos; lifts in others proved too small for coffins. But Gibson's high-rise revolution gave thousands the new life they craved, in clean, well-lit homes with bathrooms, their own WCs and double-sinks in the kitchen. They may have been improperly planned and spatially opportunistic, but as provisional solutions to a housing crisis, they did the job.

Residential tower-blocks have risen again. The lesson learned from the post-war blocks was that high-rise was high cost: lift systems, complex wiring looms, spaghettis of plumbing, fire protection and security have to be maintained if a tower-block community is to live in comfort and safety. It is not a surprise that the most resilient residential high-rises are privately owned. In London, the resurgence of high rise came after the year 2000; by 2014, there were forty-one buildings in the city taller than 100 metres. As this book leaves my desk for publishing, a report by New London Architecture reveals that a total of 436 'tall buildings' (defined as blocks of over twenty storeys) are due to be built across London. In the NLA's survey, the mean number of storeys is thirty and 73 per cent of them are destined for residential use, while office use has dropped to a mere 4 per cent.

Traditional industry has fled the landscape. I remember its fading embers: the ruddy glow of Sheffield steel mills through the windscreens of articulated lorries while hitch-hiking in my teens; Manchester and

Leeds in their soot-black make-up; the rust-puddled setts of Clydeside shipyards; the swinging cranes of Surrey Docks; the winding gear of collieries through the back windows of my parents' Made-in-Britain Morris Oxford. Norwich – my home town from the age of eight – had factories making mustard, shoes and chocolate. There were several huge breweries and the Boulton and Paul steel factory that had produced air raid shelters and the frames for tank transporters. On Fridays the whole city reeked sweetly of Caramac being cooked in the Mackintosh confectionery factory. Somewhere in a box I have the school project I wrote on the port of Norwich. Canoeing on the River Yare, you could meet a North Sea freighter twenty miles from the sea. Much later, while walking the length of England following a line of longitude, I found myself beside a hole in a Black Country brick wall by one of Telford's canals. Crawling through, I emerged in the Chance glassworks, a vast, derelict complex of sheds and kilns rugged in grit and splinters. Chance had supplied all the glass for the 'Crystal Palace' built in Hyde Park for the Great Exhibition of 1851. A day or two later I was escorted (incongruously wearing a rucksack and wielding an umbrella) through the Longbridge car factory during a night shift. Orange robots jerked and sparked as cars rolled off the line towards the open roads of Britain. I was twelve when the millionth Mini was steered off the Longbridge production line. So many landscapes have been scrubbed, recycled or demolished.

During these post-war decades the British economy mutated from manufacturing to services; from mine and factory to desk and telephone. It was as if the bass player in the band turned down the amp, little by little, while a young dude with a Moog took over the tune. It was a long-drawn-out – and frequently bitter – process of modernisation that cost millions of jobs and flooded the land market with redundant industrial sites. The coal pits that had mesmerised Celia Fiennes in the seventeenth century and then founded the Industrial Revolution fell silent. Between 1954 and 1993, the number of collieries plummeted from 850 to 50. Steel production cooled from 28 million tonnes in 1970 to 14.9 million tonnes in 1992. Upper Clyde

Shipbuilders, created by the government in 1968, with five yards and 8,500 workers, was in receivership by 1971. The numbers in manufacturing employment fell from 7.1 million in 1979 to 4.5 million in 1992.

And so 'industrial heritage' is the most recent addition to Britain's back-catalogue of architectural treasures. They can be found in all cities and towns: the piano factory converted into flats; the warehouse turned into an office; the textile mill that became a superstore. Many of the larger industrial sites were cleared down to ground level: when the steel works between Sheffield and Rotherham finally closed in the early Eighties, the derelict site was replaced by the 230 shops and 11-screen cinema of the Meadowhall retail complex. On the north bank of the Thames estuary, the Shell Haven refinery was closed in 1999, but has now become London Gateway, a deep-water container port. The closure in 1992 of Ravenscraig Steel Works in North Lanarkshire created one of the largest derelict sites in Europe, a tract of mangled land twice the size of Monaco. Bit by bit, Ravenscraig is being recolonised, by a college, a housing estate, a sports complex. At Longbridge, much of the plant I walked through in 1997 has been demolished and the old car-works have sprouted the multi-storey wedge of Bournville College, shops and homes. Where riveters once hammered ships into shape, call centres now line the Clyde. Chance glassworks is still derelict.

Extraction sites like quarries and open-cast mines can take longer to restore. It took several hundred years for the flooded medieval peat diggings in Norfolk to find a new vocation as recreational lakes that eventually became the Broads National Park. More recently, the abandoned gravel extraction pits of the Thames valley have become species-rich wetlands that provide local communities with bird reserves, fishing, walking and dinghy sailing. The Eden Project began as a redundant china clay pit in Cornwall and in its first ten years of operation, 13 million visitors explored its geodetic biome domes. And although they are not everybody's idea of beauty, the jagged craters in Welsh mountains where slate was hacked free of bedrock have become 'heritage features' in the Snowdonia National Park. More problematic are the contaminated sites and mines that have yet to be restored to further use, and which have unfunded liabilities. The reclaiming of industrial landscapes for new uses has already been underway for decades. Britain was first into the Industrial Revolution and first to embark upon the protracted recycling of redundant sites. There was no template to follow. The real and troubling industrial legacy has been the human cost of closures and the contemporary stranding of entire communities far from sources of employment. Much of Britain's housing is now in the wrong place.

The canals, edge-rails and roads that kept the Industrial Revolution moving have endured, although for different reasons. It was the work of the Inland Waterways Association from the 1940s that revived many canals. Silt and earth were cleared from blocked cuts, new locks constructed and water returned to systems that had not seen narrowboats for decades. Britain now has around 2,200 miles of navigable waterway. The total length of the rail network has been lopped from 19,471 miles of track in 1950 to 9,792 miles in 2015, but journeys have doubled in twenty years. One-third of the network is electrified and in recent decades it has been integrated with new light rail and tram systems: two in London (the Docklands Light Railway and London Tramlink), Manchester Metrolink, Midland Metro (between Birmingham and Wolverhampton), Sheffield Supertram, Tyne and Wear Metro, Nottingham Express Transit, Blackpool Tramway and Edinburgh Trams.

The most enduring network is the oldest: the web of overland routes that has evolved from the foraging and hunting paths of Britain's early migrants. By 2015, there were 246,000 miles of roads in Britain, more than half of the network (54 per cent) being a dense capillary system of rural minor roads totalling 132,498 miles. Urban minor roads make up a further 33 per cent (81,994 miles) of the network, leaving 9 per cent for rural 'A'-roads and 3 per cent for urban 'A'-roads. Just 1 per cent (2,265 miles) is provided by the most recent type of road to be added to the network, motorways. Back in 1955, one of the evils Hoskins had hammered with his typewriter keys was 'the arterial by-pass, treeless and stinking of diesel oil, murderous with lorries'. Three years after his book was published, Britain's first motorway was opened to traffic by the Prime Minister, Harold Macmillan. Now a section of the M6, it ran for just over eight miles past Preston in Lancashire and had nineteen bridges. The following year, a second motorway – the M1 – began to carve a passage through ancient field borders from just outside Watford in Hertfordshire to Crick in Northamptonshire where it intercepted Roman Watling Street – the A5. In common with arterial bypasses (and Roman roads), the new motorways were entirely new creations that took little account of existing roads or valued landscapes. By 2015, there were 2,300 miles of motorway, a tiny yet intrusive proportion of the quarter-million miles of British roads. With a road surface eight times wider than a country lane, most motorways are also flanked by extended verges, cuttings or embankments, while junctions consume enough land to build a hamlet. Noise and light pollution extend the carriageway's reach even further.

The road network is busier than ever. In the sixty years since 1955, when Professor Hoskins published *The Making of the English Landscape*, the number of motor vehicles on British roads increased from 6.5 million to 36.5 million. The latest figure for vehicle miles travelled on British roads is 317.8 billion, which is the highest 'rolling annual total' ever recorded and a 19 per cent increase in two decades. The most common vehicle in this pungent, racketing metallic armada is the car. There are 30.2 million of them. Road culture has created linear motoring landscapes. Road surfaces and verges are daubed with vehicular instructions and advertising hoardings sized to catch the eyes of passers-by in top gear; intersections are punctuated with traffic lights and roundabouts; slip-roads and clover-leaves isolate islands of intoxicated grass; service stations have grown into drop-in centres that offer a car wash, coffee and fast-food. Every town and city is blighted by unsightly car parks. Out-of-town shopping centres have replaced the bustling quest of High Street browsing with low-exertion, sanitised uniformity. Public space has been overrun with cars. And apart from the light rail and tram systems, there has been very little effort to curb further increases. The 'right to drive' is exercised regardless of landscape cost. Historic buildings are listed and then despoiled with vehicles; ring-roads and one-way systems carve through historic districts, severing communities, interrupting pedestrian flows and spewing noise and exhaust pollution onto pavements and buildings. Oxford Street is a fetid gulch. In London, cars are so absurdly inefficient that off-peak average speeds dropped from 12 mph in 1904 to 10 mph by 1996 – slightly slower than a bicycle.

A century after the bike was nudged into the gutter by the motor vehicle, bicycle lanes are beginning to reclaim marginal road space. Since 2008, cycle traffic in Britain has increased every year; but is still less than one quarter of the 1949 figure of 14.7 billion bicycle miles; bike traffic is, however, higher than both motorcycles and buses. In a woefully late response to the rediscovered convenience, healthiness and low-carbon credentials of the bike, cycle lanes are slowly appearing, but most are inadequate and don't begin to compare to Dutch or Danish cycle paths of forty years ago. Far more impressive is the work of the charity Sustrans, which began in Bristol in 1977 with the conversion of a disused railway track between Bath and Bristol. By the year 2000, Sustrans had created a 5,000-mile network of cycle routes and by 2005, the network had passed 10,000 miles.

As this book approaches its end-papers I'm going to turn my handlebars back towards the beginning; towards the unconstrained green spaces of the early chapters. I write in a small room near the centre of a large city and in order to think and to restore energy I leave this desk at least twice a day and join the slow, green web: the leafy pavements and public parks, the cycle routes, canals and towpaths, the national parks and nature reserves, the 140,000 miles of public rights of way in England and Wales, the permissive paths, the 'open access' land and the glens and straths and reptilian arêtes of Scotland where there's a 'right of responsible non-motorised access, for recreational and other purposes, to land and inland water'. The slow green web is the filament reminder of the wildwood and its glades and the grassy chalk plains.

And there is much more of it than you might think. In England, 89 per cent of the land area is not 'urban'. In Wales, the figure is 96 per cent and in Scotland, 98 per cent. Neither are towns and cities 'concrete jungles'. More than three-quarters of the urban space in England – 79 per cent – is taken up by non-urban land-use like gardens, parks, allotments, canals, lakes, rivers and so on. If the urban 'green space' is added to the countryside, the total 'natural' cover in England rises to 98 per cent. In Scotland and Wales, the figure is over 99 per cent. The challenge is to improve its condition and accessibility.

The rate of landscape change during Britain's 11,600 years of continuous human occupation has been uneven. Overall, it has been a process of acceleration, but there have been peaks and troughs. For half of those twelve millennia, the only significant recorded human impact on the environment was the felling of wildwood. The first episode of accelerated change came with the boatloads of immigrant farmers who opened the glades of Britain to agriculture. After that, 4,000 years passed before landscapes were recording settlements that showed sufficient scale and nucleation to be regarded as 'town-like'. By then, Britain had been continuously occupied for nearly 10,000 years. Much of the island was still 'wild', with extensive tracts of woodland prowled by bears and wolves. The next human shock-wave was Roman. Over a period of 150 years, Britain was propelled at gladius-point into Europe. Tribal landscapes of hillfort and farm were superimposed with rectilinear roads and walled towns and an entire colonial suite of lesser features ranging from villa to public bath. Roman colonisation proved unsustainable and Britain relapsed. Several centuries passed before towns reappeared in a form that would prove enduring. It was from the proto-urban 800s that the upward curve of development took off, interrupted only by plague. As recently as the 1600s – 96 per cent of the way through the story of human occupation in Britain – our habitat could be described as 'balanced', with a reasonably intact breadth of biodiversity and a range of landscapes from city and town, through to village, farmland, uncultivated wilderness, extensive forest and wetland. The final 4 per cent – the last 400 years – tipped the balance. Britain led the world into industrialisation and for a while, no other landmass of its size was so intensively exploited for its natural resources nor so thoroughly converted to production. London was the biggest city on the planet. Industrial Britain was a gigantic, greasy orrery whose cogs of population growth, urbanisation and intensifying farming span faster by the decade. Growth on a finite landmass with limited resources could never be sustainable and although Britain's industrial machine began to decelerate before the cogs flew off, the transition to a sustainable economy is in its infancy.

For the landscapes of Britain, this will be an era of adjustment. It is difficult to imagine transformation on past scales. The great ages of henges, farms, castles, canals, railways, factories and suburbs, when a single form was replicated from coast to coast, altering landscapes so thoroughly that they would become transformative chapters in our island story, were facilitated in part by space. In this era of contested landscapes, dramatic new structures are local redevelopments of recycled places. The pleas to build new housing on 'brownfield' sites arise because of the rising value – whether financial or aesthetic – now placed on green fields. Recycling is the modern virtue. The Angel of the North stands upon one of the North East's redundant coal fields. The Turner Contemporary art gallery in Margate occupies an old seafront car park. The Shard was built on a site occupied by a 1970s, 25-storey block called Southwark Towers. There are too many people and not enough spaces for big, physical ideas to root and multiply. The prolonged attempts to provide a single new inter-city railway (HS2), a single new airport runway for London, a single nuclear power station, are indicative of an island with diminished spending power and competing landscape interests.

The manic hacking, burning and making of the Industrial Revolution has been succeeded by a more reasoned attitude towards our habitat. The core components of a rich landscape are known and defended (not always successfully). The value of biodiversity is acknowledged, and of sustainable agriculture. The negative effects of pollution are known. After a century of experimentation with urban housing, the relationships between structure, space and social cohesion are better understood. The idea that designated National Parks and reserves work well to protect and enhance local cultures and species is proven. In short, we have learned a lot about looking after our landscape. Despoilers of our habitat do so in the knowledge that they're sabotaging Britain's future. Unfortunately, 12,000 years of increasing carelessness have denuded our ecosystems of their original riches. Soils are degraded and run-off has savaged aquatic ecosystems with nitrogen and phosphorus from fertiliser. The Farmland Bird Index – a measure of biodiversity of farmland – plummeted by 43 per cent between 1970 and 1998 and has not recovered.

Right from the start of this story, accelerated episodes of landscape change have been driven by two external factors: the south and the sun. Change brought from the south (or rather, south and east) began with bands of immigrant forager-hunters and then European farmers and Roman armies. It continued with Saxons, Christians, Danes, and Normans. They all brought relatively abrupt transformation on a territorial scale. Subsequently, plague, Daimler Benz and the Luftwaffe were significant agents of landscape change. These days, invasion seems improbable and the 'south' has extended its compass to become a global wellspring of people and ideas. Meanwhile, the solar energy that reopened the door to human habitation back in 9,600 BC must now be understood in conjunction with an atmosphere and oceans whose characteristics have been altered by humanity in ways that are notoriously difficult to model. With climate change, it's always easier to record what has already happened than to predict the future. We do know that global sea level rose between 17 and 21 centimetres since 1900 and that the rate of rise over the last twenty or thirty years was greater than the twentieth-century average. Around the south and east of England (where land is sinking slightly), sea level is rising by around 20 millimetres a decade. We also know that global surface temperatures have risen by 0.85 degrees Centigrade since 1880 and that Earth is warmer than at any time in the past thousand years.

While I was writing this book, the amount of CO_2 in the atmosphere passed through 400 parts per million for the first time in 4 million years. Atmospheric concentrations of methane and carbon dioxide – the two most significant greenhouse gases – are higher than they have been at any time in the last million years. As we've already found, changes in sea level and temperature can have profound effects on the landscape, from the crops we grow to the types and locations of buildings we inhabit. To the external drivers of landscape change must be added the main internal driver: population. Back in 1900, there were 37 million people in Britain. As this book goes to press, the most recently available population figures for England are 54.8 million. Scotland has reached 5.4 million and Wales 3.1 million. At 8.6 million, London, the town with a habit of disproportion, now exceeds the combined populations of Scotland and Wales.

Landscape matters because it is our habitat. It is the only place we have. I wrote this book because I wanted to tell its story. I wanted to explore how we modified this island, from the hearth of the first reindeer hunter to the glass spire of the Shard. I wanted to know where the ideas came from; who built them; what was treasured and kept; what was lost and regretted. I wanted to find the turning points. It is a story of ups and downs. And this paragraph has to end part-way through a chapter; beyond the end of the industrial era yet not far enough into the 'sustainable era' to know whether we have the time and initiative to confront the accelerated pace of environmental change and population growth. We occupy a land that is between; an 'interland'. Standing back from this manuscript, I see an island richer than I ever imagined. It's been a long book with a short message: that to care about a place, you must first know its story.

ACKNOWLEDGEMENTS

This book has been a lifetime in the making and many years in the writing. My engagement with British landscapes began with family outings and holidays, and so the first people I would like to thank are my parents, Naomi and Hol, for early exposure to field-paths, woods and winterised peaks in the western Highlands. Successive geography teachers and then geography lecturers had little difficulty convincing me that study of places and people would create a resource for life. It was a module on historical geography at what is now Anglia Ruskin University that introduced me to W. G. Hoskins and *The Making of the English Landscape*. My thanks to all those on the frontline of state education.

The growing hoard of books that eventually turned my study into a bibliographic cave has been augmented by two libraries, without which I could not have undertaken the necessary research. As ever, I am indebted to the staff of the British Library and of the London Library. I'm very grateful to my daughter Imogen for several field trips to the British Library and for her expertise in sorting out the book's bibliography. The Royal Geographical Society has been a continual source of inspiration, both for the wealth of its collections and for the community of geographers who regard it as 'home'. I've been a Fellow for more than thirty years, and am currently President. This book has been vastly enriched by conversations, lectures, books and papers originating at the RGS-IBG.

More years ago than I care to admit, this book was commissioned by Alan Samson at Weidenfeld and Nicolson. I could not have asked for a more loyal, patient ally. The processing of a large, late manuscript has been handled with attentive care by Paul Murphy, and the book will be presented for publication by Group Publicity and Communications Director, Helen Richardson.

This geographical journey into Britain's deep past has been supported by many friends. Very often you helped without realising, or knew but were kind enough to put up with the months (years) of silence. Derek Johns, my ever-supportive literary agent, was there on Day 1 of this project and on retirement handed on the baton to Jim Gill at United Agents. Books of this kind are family projects. It's impossible for those closest to the author not to become drawn into the force-field of endeavour. Connie, Kit, Imo and Annabel, you have been very patient. Annabel, without *you*, this book would have remained a dream in a Mesolithic glade. Thank you.

BIBLIOGRAPHY

My aim has been to provide a clean story rather than a reference book, so I have not inflicted you with the 2,721 footnotes I used as scaffolding during my research. The sources listed below are selective but include the most significant books and journals that I turned to. Apart from the general references, I have tried to place each source in its chapter of 'first use'.

GENERAL SOURCES

Aalen, F. & O'Brien, C. (eds), *England's Landscape, The North East*, 2006
Ackroyd, P., *London: The Biography*, 2000
Atkins, P., Simmons, I. & Roberts, B., *People, Land and Time: An Historical Introduction to the Relations between Landscape, Culture and Environment*, 1998
Bagwell, P. & Lyth, P., *Transport in Britain 1750–2000: From Canal Lock to Gridlock*, 2002
Barber, P. (ed.), *The Map Book*, 2005
Barker, G., *The Agricultural Revolution in Prehistory: Why did Foragers become Farmers?*, 2006
Barratt, N., *Greater London: The Story of the Suburbs*, 2014

Behringer, W., *A Cultural History of Climate*, 2007

Bird, E., *Coastal Geomorphology: An Introduction* (2nd edn), 2008

Black, J., *London: A History*, 2009

Black, J., *Metropolis: Mapping the City*, 2015

Blair, J. (ed.), *Waterways and Canal-building in Medieval England*, 2007

Bradley, R., *The Social Foundations of Prehistoric Britain: Themes and Variations in the Archaeology of Power*, 1984

Bradley, R., 'Time Regained: The Creation of Continuity', *Journal of the British Archaeological Association*, 1987

Bradley, R., *Altering the Earth: The Origins of Monuments in Britain and Continental Europe*, 1993

Bradley, R., *An Archaeology of Natural Places*, 2000

Bradley, R., *The Prehistory of Britain and Ireland*, 2007

Bradley, R. & Edmonds, M., *Interpreting the Axe Trade*, 1993

Brooke, J. L., *Climate Change and the Course of Global History*, 2014

Cameron, K., *English Place-Names*, 1961

Carey, J., *The Faber Book of Utopias*, 2000

Christie, P. M. L., *Chysauster and Carn Euny*, 1993

Clark, P. (ed.), *The Cambridge Urban History of Britain, Vol II: 1540–1840*, 2000

Clayre, A. (ed.), *Nature and Industrialization*, 1977

Clifton-Taylor, A., *The Pattern of English Building*, 1972

Coad, J., *Dover Castle*, 2007

Crawford, H. (ed.), *The Sumerian World*, 2013

Cresswell, T., *Place, a short introduction*, 2004

Cummings, V., *A View from the West: The Neolithic of the Irish Sea Zone*, 2009

Cummings, V., Jordan, P. & Zvelebil, M., *The Oxford Handbook of the Archaeology and Anthropology of Hunter-gatherers*, 2014

Cunliffe, B., *Iron Age Communities in Britain: An account of England, Scotland and Wales from the seventh century BC until the Roman Conquest* (4th edn), 2005

Cunliffe, B., *Britain Begins*, 2013

Cunliffe, B. (ed.), *The Oxford Illustrated Prehistory of Europe*, 1994

Cunliffe, B. (ed.), *England's Landscape: The West*, 2006

Darby, H. (ed.), *A New Historical Geography of England after 1600*, 1973

Darvill, T., *Prehistoric Britain* (2nd edn), 2010

Daunton, M., *The Cambridge Urban History of Britain, Volume III: 1840–1950*, 2001

Davies, J., *A History of Wales*, 1994

Davies, J., *The Making of Wales*, 1996

Davis, B., Brewer, S., Stevenson, A. & Guiot, J., 'The temperature of Europe during the Holocene reconstructed from pollen data', *Quaternary Science Reviews* 22, 2003

Day, L. & McNeil, I. (eds), *Biographical Dictionary of the History of Technology*, 1996

Delano-Smith, C. & Kain, R., *English Maps: A History*, 1999

Diamond, J., *Guns, Germs and Steel: The Fates of Human Societies*, 1997

Diamond, J., *The World Until Yesterday: What Can We Learn From Traditional Societies?*, 2012

Dorward, D., *Scotland's Place-names*, 1995

Dow, K. & Downing, T., *The Atlas of Climate Change, Mapping the World's Greatest Challenge* (3rd edn), 2011

Dyer, C. & Jones, R. (eds), *Deserted Villages Revisited*, 2010

Dyos, H. J. & Aldcroft, D. H., *British Transport: An Economic Survey from the Seventeenth Century to the Twentieth*, 1969

Earle, J., *Black Top: A History of the British Flexible Roads Industry*, 1974

Edwards, K. J. & Ralston, I. B. M. (eds), *Scotland After the Ice Age: Environment, Archaeology and History, 8000 BC–AD 1000*, 2003

Fagan, B., *Beyond the Blue Horizon: How the Earliest Mariners Unlocked the Secrets of the Oceans*, 2012

Field, D., *Use of Land in Central Southern England during the Neolithic and Early Bronze Age*, 2008

Finlayson, B. & Warren, G. (eds), *Landscapes in Transition*, Levant Supplementary Series, Volume 8, 2010

Fleming, J., Honour, H. & Pevsner, N., *The Penguin Dictionary of Architecture and Landscape Architecture* (5th edn), 1998

Fortey, R., *The Hidden Landscape: A Journey into the Geological Past*, 1993

Friel, I., *Maritime History of Britain and Ireland*, 2003

Frost, H., *Sailing the Rails: A New History of Spurn and its Military Railway*, 2001

Gaffney, V., Fitch, S. & Smith, D., *Europe's Lost World: The rediscovery of Doggerland*, Research Report No. 160, Council for British Archaeology, 2009

Gardiner, M. & Rippon, S. (eds), *Medieval Landscapes: Landscape History after Hoskins*, Volume 2, 2007

Gibson, A. & Sheridan, A. (eds), *From Sickles to Circles: Britain and Ireland at the Time of Stonehenge*, 2004

Goodall, B., *Dictionary of Human Geography*, 1987

Goring, R. (ed.), *Scotland, the Autobiography: 2,000 Years of Scottish History By Those Who Saw It Happen*, 2007

Hadfield, C., *British Canals: An Illustrated History*, 1952

Haigh, H. E. A. (ed.), *Huddersfield: A Most Handsome Town, Aspects of the history and culture of a West Yorkshire Town*, 1992

Hall, P., *Cities in Civilization, Culture, Innovation, and Urban Order*, 1998

Harris, R., *Discovering Timber-framed Buildings*, 2009

Harris, S. & Yalden, D. W., *Mammals of the British Isles* (4th edn), 2008

Haslam, S., *The Historic River: Rivers and Culture down the Ages*, 1991

Hegarty, C. & Wilson-North, R., *The Archaeology of Hill Farming on Exmoor*, 2014

Hobsbawm, E., 'Inventing traditions', in Hobsbawm, E. & Ranger, T. (eds), *The Invention of Tradition*, 1983

Hooke, D., *England's Landscape: The West Midlands*, 2006

Hoskins, W. G., *Devon*, 1954

Hoskins, W. G., *The Making of the English Landscape*, 1955

Hoskins, W. G., *Local History in England* (2nd edn), 1972

Hunter, J. & Ralston, I. (eds), *The Archaeology of Britain: An Introduction from Earliest Times to the Twenty-First Century*, 2009

Hurcombe, L., *Archaeological Artefacts as Material Culture*, 2007

Hutton, R., *Pagan Britain*, 2013

Jackman, W., *The Development of Transportation in Modern England*, 1966

Johnson, M., *Ideas of Landscape*, 2007

Jones, A., *A Thousand Years of the English Parish*, 2000

Kain, R. (ed.), *England's Landscape: The South-West*, 2006

Kidson, P., Murray, P. & Thompson, P., *A History of English Architecture*, 1965

Kirby, D., *The Earliest English Kings* (2nd edn), 2000

Kunzig, R. & Broecker, W., *Fixing Climate: The Story of Climate Science – and How to Stop Global Warming*, 2008

Lehmberg, S., *English Cathedrals: A History*, 2005

Lloyd, D., *The Making of English Towns*, 1998

Loveday, R., "The Greater Stonehenge Cursus – The Long View', *Proceedings of the Prehistoric Society 78*, 2011

Loveluck, C., *Northwest Europe in the early Middle Ages, c. AD 600–1150: a comparative archaeology*, 2013

Matthew, H. C. G. & Harrison, B. (eds), *Oxford Dictionary of National Biography*, 2004

Mayhew, S., *A Dictionary of Geography*, 2004

McCormick, M., Büntgen, U., Cane, M. et al., 'Climate Change during and after the Roman Empire: Reconstructing the Past from Scientific and Historical Evidence', *Journal of Interdisciplinary History* XLIII:2, Autumn, 2012

McGrail, S., *Boats of the World*, 2001

McOmish, D., Field, D. & Brown, G., *The Field Archaeology of the Salisbury Plain Training Area*, 2002

Miles, D., *The Tribes of Britain: Who are we? And where do we come from?*, 2005

Millman, R., *The Making of the Scottish Landscape*, 1975

Mills, A. D., *Dictionary of British Place Names*, 1991

Mitchell, A., *A Field Guide to the Trees of Britain and Northern Europe*, 1986

Mithen, S., *After the Ice: A Global Human History 20,000–5000 BC*, 2003

Morgan, K. (ed.), *The Oxford Illustrated History of Britain*, 1984

Muir, R., *The English Village*, 1980

Muir, R., *The Stones of Britain*, 1986

Muir, R., *Portraits of the Past*, 1989

Muir, R., *Fields*, 1989

Mumford, L., *The Culture of Cities*, 1940

Murphy, P., *The English Coast: A History and a Prospect*, 2009

Newman, P., *The Field Archaeology of Dartmoor*, 2011

Van de Noort, R., *North Sea Archaeologies: A Maritime Biography, 10,000 BC–AD 1500*, 2011

Oosthuizen, S., *Landscapes Decoded: The Origins and Development of Cambridgeshire's Medieval Fields*, 2006

Palliser, D. (ed.), *The Cambridge Urban History of Britain, Volume I, 600–1540*, 2000

Parker Pearson, M., *Stonehenge: Exploring the Greatest Stone Age Mystery*, 2012

Piggott, S. & Thirsk, J., *The Agrarian History of England and Wales, Volume 1: Prehistory to AD 1042*, 1972

Platt, C., *The English Medieval Town*, 1976

Pollard, J. (ed.), *Prehistoric Britain*, 2008

Pryor, F., *Britain BC: Life in Britain and Ireland before the Romans*, 2003

Pryor, F., *Home: A Time Traveller's Tales from Britain's Prehistory*, 2014

Quartermaine, J. & Leech, R., *Cairns, Fields, and Cultivation: Archaeological Landscapes of the Lake District Uplands*, 2012

Rackham, O., *The History of the Countryside*, 1986

Rackham, O., *The Last Forest: The Fascinating Account of Britain's Most Ancient Forest*, 1989

Rackwitz, M., *Travels to terra incognita: The Scottish Highlands and Hebrides in Early Modern Travellers' Accounts, c. 1600 to 1800*, 2007

Roberts, J. M., *The Penguin History of the World*, 1992

Roberts, J. M., *History of Europe*, 1997

Rose, E. & Nathanail, C. (eds), *Geology and Warfare: Examples of the Influence of Terrain and Geologists on Military Operations*, 2000

Ross, C. & Clark, J., *London: The Illustrated History*, 2008

Rowley, T., *Villages in the Landscape*, 1994

Rumble, A. & Mills, A. (eds), *Names, Places and People: An Onomastic Miscellany in Memory of John McNeal Dodgson*, 1997

Russell, P., *The Good Town of Totnes*, 1964

Shennan, I. & Andrews, J. (eds), *Holocene Land–Ocean Interaction and Environmental Change around the North Sea*, 2000

Shennan, I., Lambeck, K., Flather, R. et al., 'Modelling western North Sea palaeogeographies and tidal change during the Holocene', in Shennan, I. & Andrews, J. (eds), *Holocene Land–Ocean Interaction and Environmental Change around the North Sea*, 2000

Sheppard, F., *London: A History*, 1998

Sheppard, M., *Primrose Hill: A History*, 2013

Short, B., *England's Landscape, The South East*, 2006

Shuman, B., 'Patterns, processes, and impacts of abrupt climate change in a warm world: the past 11,700 years', *WIREs Climate Change*, Vol. 3, Issue 1, 2012

Simmons, I., *An Environmental History of Great Britain*, 2001

Skempton, A. et al. (eds), *A Biographical Dictionary of Civil Engineers in Great Britain and Ireland, Volume 1: 1500–1830*, 2002

Smout, T. C., *A History of the Scottish People 1560–1830*, 1969

Stocker, D., *England's Landscape: The East Midlands*, 2006

Stringer, C., *Homo Britannicus: The Incredible Story of Human Life in Britain*, 2006

Thomas, J., *The Birth of Neolithic Britain: An Interpretive Account*, 2013

Tolan-Smith, C., 'Mesolithic Britain', in Bailey, G. & Spikins, P. (eds), *Mesolithic Europe*, 2008

Tuchman, B., *The March of Folly: From Troy to Vietnam*, 1984

Vansittart, P., *London: A Literary Companion*, 1992

Walsham, A., *The Reformation of the Landscape: Religion, Identity, and Memory in Early Modern Britain and Ireland*, 2011

Watkin, D., *English Architecture: A Concise History*, 1979

Whitten, D. & Brooks, J., *The Penguin Dictionary of Geology*, 1972

Whittow, J. B., *The Penguin Dictionary of Physical Geography*, 2000

Wickham-Jones, C., *The Landscape of Scotland: A Hidden History*, 2001

Willan, T., *River Navigation in England 1600–1750*, 1936

Williams, A., *The Geography of Iron and Steel*, 2015

Williams, A. & Martin, G., *Domesday, A Complete Translation*, 2003

Williamson, T., *England's Landscape, East Anglia*, 2006

Williamson, T., *The Origins of Hertfordshire*, 2010

Winchester, A. (ed.), *England's Landscape: The North West*, 2006

Woodell, S. (ed.), *The English Landscape, Past, Present, and Future (Wolfson College Lectures 1983)*, 1985

Yalden, D. W. & Albarella, U., *The History of British Birds*, 2009

Yates, M. & Longley, D., *Anglesey: A Guide to Ancient Monuments on the Isle of Anglesey*, 2001

Zipf, G., *Human Behaviour and the Principle of Least Effort: An Introduction to Human Ecology*, 1949

CHAPTER ONE

Encyclopaedia Britannica, 11th edn, 1910–11

Anderson, D., Maasch, K. & Sandweiss, D., *Climate Change and Cultural Dynamics: A Global Perspective on Mid-Holocene Transitions*, 2007

Barton, N., Roberts, A. J. & Roe, D. A. (eds), *The Late Glacial in North-west Europe: Human adaptation and environmental change at the end of the Pleistocene*, 1991

Bos, J., van Geel, B., van der Plicht, J. & Bohncke, S., 'Preboreal climate oscillations in Europe: Wiggle-match dating and synthesis of Dutch high-resolution multi-proxy records', *Quaternary Science Reviews* 26, 2007

Dietrich, O., Heun, M., Notroff, J., Schmidt, K. & Zarnkow, M., 'The role of cult and feasting in the emergence of Neolithic communities: New evidence from Göbekli Tepe, south-eastern Turkey', *Antiquity*, Vol. 86, No. 333, Sept. 2012

Eren, M. I. (ed.), *Hunter-gatherer Behaviour: Human Response During the Younger Dryas*, 2012

Evans, A., Wolframm, Y., Donahue, R. & Lovis, W., 'A pilot study of "black chert" sourcing and implications for assessing hunter-gatherer mobility strategies in Northern England', *Journal of Archaeological Science* 34, 2007

Fleitmann, D., Mudelsee, M., Burns, S., Bradley, R., Kramers, J. & Matter, A., 'Evidence for a widespread climatic anomaly at around 9.2 ka before present', *Paleoceanography* 23, 2008

Gibbard, P. L. & Lewin, J., 'Climate and related controls on interglacial fluvial sedimentation in lowland Britain', *Sedimentary Geology* 151, 2002

Hall, S., 'A comparative analysis of the habitat of the extinct aurochs and other prehistoric mammals in Britain', *Ecography* 31 (2), 2008

Hey, G. (ed.), *Later Upper Palaeolithic and Mesolithic Resource Assessment, Solent Thames Research Framework: Regional*, 2010

Hind, D., *Chert use in the Mesolithic of Northern England*, 1998

Johansen, L. & Stapert, D., 'Two "Epi-Ahrensburgian" Sites in the Northern Netherlands: Oudehaske (Friesland) and Gramsbergen (Overijssel)', *Palaeohistoria* 39/40, 1997 & 1998

Leroy, S., Zolitschka, B., Negendank, J. & Seret, G., 'Palynological analyses in the laminated sediment of Lake Holzmaar (Eifel, Germany): duration of Lateglacial and Preboreal biozones', *Boreas* 29, 2000

Lewis, J. S. C., & Rackham, J., *Three Ways Wharf, Uxbridge: A Lateglacial and Early Holocene hunter-gatherer site in the Colne Valley*, 2011

Liran, R. & Barkai, R., 'Casting a shadow on Neolithic Jericho', *Antiquity*, Vol. 85, Issue 327, March 2011: http://antiquity.ac.uk/projgall/barkai327/

Lowe, J. J. & Walker, M. J. C., *Reconstructing Quaternary Environments*, The Loch Lomond Stadial, 1984

Lynch, A. H., Hamilton, J. & Hedges, R. E. M., 'Where the Wild Things Are: Aurochs and Cattle in England', *Antiquity* 82, 2008

Marshall, C., 'Fast-Warming Arctic Proves Deadly to Animals and People', *Scientific American*, 24 November 2014

Morigi, A., Schreve, D. & White, M., *The Thames through Time, The Archaeology of the Gravel Terraces of the Upper and Middle Thames – Part 1, The Ice Ages: Palaeogeography, Palaeolithic Archaeology and Pleistocene Environments*, 2011

Niekus, M., 'A Geographically Referenced [14]C Database for the Mesolithic and the Early Phase of the Swifterbant Culture in the Northern Netherlands', *Palaeohistoria* 47/48, 2005–06

Özkaya, V. & Coskun, A., 'Körtik Tepe, a new Pre-Pottery Neolithic A site in south-eastern Anatolia', *Antiquity* Vol. 83, Issue 320, June 2009

Parker, A. & Goudie, A., 'Late Quaternary environmental change in the limestone regions of Britain' in Goudie, A. S. & Kalvoda, J. (eds), *Geomorphological Variations*, 2007

Pettitt, P. & White, M., *The British Palaeolithic: Hominin societies at the edge of the Pleistocene world*, 2012

Sauer, C., 'Seashore – Primitive Home of Man?', *Proceedings of the American Philosophical Society* 106, 1962

Saville, A., *Mesolithic Scotland and its Neighbours: The Early Holocene Prehistory of Scotland, its British and Irish Context, and some Northern European Perspectives*, 2004

Stapert, D. & Johansen, L., 'Flint and pyrite: making fire in the Stone Age', *Antiquity* Vol. 73, Issue 282, 1999

Watkins, T., 'New light on Neolithic revolution in south-west Asia', *Antiquity* 84, 2010

CHAPTER TWO

Bayliss, A. & Waddington, C., 'Re-calibration of other 8th Millennium cal BC sites in the British Isles', in Waddington, C. (ed.), *Mesolithic Settlement in the North Sea Basin: A Case Study from Howick, North-east England*, 2007

Chisham, C., *The Upper Palaeolithic and Mesolithic of Berkshire, Thames and Solent Research Framework*, 2006

Edwards, K. & Mithen, S., 'The colonization of the Hebridean Islands of Western Scotland: evidence from the palynological and archaeological records', *World Archaeology* 26, 1995

Edwards, K. J., 'Palaeoenvironments of Late Upper Palaeolithic and Mesolithic Scotland and the North Sea area: new work, new thoughts', in Saville, A. (ed.), *Mesolithic Scotland and its Neighbours: the Early Holocene Prehistory of Scotland and its European Context, and some Northern European Perspectives*, 2004

Edwards, P. & Sayej, G., 'Resolving contradictions: the PPNA-PPNB transition in the Southern Levant', 2007: http://www.academia.edu/632205/Resolving_contradictions_the_PPNA-PPNB_transition_in_the_southern_Levant

Harding, P., 'A Mesolithic site at Rock Common, Washington, West Sussex', *Sussex Archeological Collections* 138, 2000

Hardy B. L. & Svoboda, J. A., 'Mesolithic stone tool function and site types in Northern Bohemia, Czech Republic', in Haslam, M., Robertson, G. et al., *Archaeological Science Under a Microscope (Terra Australis 30)*, 2009

Healy, F., Lobb, S. J. & Heaton, M., 'Excavations of a Mesolithic site at That-cham, Berkshire, Radio carbon date 9,100 + 80 BP', *Proceedings of the Prehistoric Society*, Vol. 58, January 1992

Meiklejohn, C., Chamberlain, A. T. & Schulting, R. J., 'Radiocarbon dating of Mesolithic human remains in Great Britain', *Mesolithic Miscellany*, Vol. 21, No. 2, May 2011

Milner, N., Taylor, B., Conneller, C. & Schadla-Hall, T., *Star Carr: Life in Britain after the Ice Age*, 2013

Milner, N. et al., 'A Unique Engraved Shale Pendant from the Site of Star Carr: the oldest Mesolithic art in Britain', *Internet Archaeology* 40, 2016

Scarre, C., *The Megalithic Monuments of Britain and Ireland*, 2007

Watkins, T., 'New light on Neolithic revolution in south-west Asia', *Antiquity* 84, 2010: http://antiquity.ac.uk/Ant/084/0621/ant0840621.pdf

Weninger, B. et al., 'The Impact of Rapid Climate Change on prehistoric societies during the Holocene in the Eastern Mediterranean', *Documenta Praehistorica* XXXVI, 2009

CHAPTER THREE

Bonsall, C., Macklin, M., Anderson, D. & Payton, R., 'Climate change and the adoption of agriculture in north-west Europe', *European Journal of Archaeology*, Vol. 5(1), 2002

Bondevik, S., Mangerud, J., Dawson, S., Dawson, A. & Lohne, Ø., 'Record-breaking Height for 8000-Year-Old Tsunami in the North Atlantic', *Eos, Transactions, American Geophysical Union*, Vol. 84, No. 31, 5 August 2003

Bondevik, S., Mangerud, J., Dawson, S., Dawson, A. & Lohne, Ø., 'Evidence for three North Sea tsunamis at the Shetland Islands between 8000 and

1500 years ago', *Quaternary Science Reviews* 24, 2005

Brown, T., 'Clearances and Clearings: Deforestation in Mesolithic/Neolithic Britain', *Oxford Journal of Archaeology*, Vol. 16, Issue 2, 2002

Bryn, P., Berg, K., Forsberg, C., Solheim, A. & Kvalstad, T., 'Explaining the Storegga Slide', *Marine and Petroleum Geology* 22, 2005

Clare, L., Rohling, E. J., Weninger, B. & Hilpert, J., 'Warfare in Late Neolithic/Early Chalcolithic Pisidia, southwestern Turkey. Climate induced social unrest in the late 7th millennium cal BC', *Documenta Praehistorica* XXXV, 2008

Davies, P., Robb, J. & Ladbrook, D., 'Woodland clearance in the Mesolithic: the social aspects', *Antiquity* 79, 2005

Fraser, S., Murray, H. & Murray, J. C., *A Tale of the Unknown Unknowns: A Mesolithic Pit Alignment and a Neolithic Timber Hall at Warren Field, Crathes, Aberdeenshire*, 2009

Gooder, J., 'Excavations of a Mesolithic House at East Barns, East Lothian, Scotland: an interim view', in Waddington, C. & Pedersen, K. (eds), *Mesolithic studies in the North Sea Basin and beyond: proceedings of a conference held at Newcastle in 2003*, 2007

Innes, J. B. & Blackford, J. J., 'The Ecology of Late Mesolithic Woodland Disturbances: Model Testing with Fungal Spore Assemblage Data', *Journal of Archaeological Science*, Vol. 30, Issue 2, Feb. 2003

Jacques, D. & Phillips, T. et al., 'Mesolithic settlement near Stonehenge: excavations at Blick Mead, Vespasian's Camp, Amesbury', *Wiltshire Archaeological & Natural History Magazine* 107, 2014

James, N., 'Stonehenge: New contexts ancient and modern', *Antiquity* 86, 2012

Kitchener, A. & Doune, J., 'A record of the aurochs, *Bos primigenius*, from Morayshire', *The Glasgow Naturalist*, Vol. 25, Part 4, 2012

Li, Y. L., Tornqvist, T., Nevitt, J. & Kohl, B., 'Synchronising a sea-level jump, final Lake Agassiz drainage, and abrupt cooling 8200 years ago', *Earth and Planetary Science Letters 315–316*, 2012

Renfrew, C., Boyd, M. & Ramsey, C. B., 'The oldest maritime sanctuary? Dating the sanctuary at Keros and the Cycladic Early Bronze Age', *Antiquity* 86, 2012

Rohling, E. J. & Palike, H., 'Centennial-scale climate cooling with a sudden cold event around 8,200 years ago', *Nature* 434, 2005

Tilley, C., 'The Power of Rocks: Topography and Monument Construction on Bodmin Moor', in *World Archaeology*, Vol. 28, No. 2, October 1996

Tooley, M. J. & Smith, D. E., 'Relative sea-level change and evidence for the Holocene Storegga Slide tsunami from a high-energy coastal environment: Cocklemill Burn, Fife, Scotland, UK', *Quaternary International* 133–134, 2005

Waddington, C. (ed.), *Mesolithic Settlement in the North Sea Basin: A Case Study from Howick, North-east England*, 2007

Waddington, C. & Pedersen, K., *Mesolithic studies in the North Sea Basin and beyond: proceedings of a conference held at Newcastle in 2003*, 2007

Weninger, B., Schulting, R., Bradtmoller, M., Clare, L., Collard, M., Edinborough, K., Hilpert, J., Joris, O., Niekus, M., Rohling, E. J. & Wagner, B., 'The catastrophic final flooding of Doggerland by the Storegga Slide tsunami', *Documenta Praehistorica* XXXV, 2008

Whitehouse, N. J. & Smith, D. N., '"Islands" in Holocene forests: Implications for Forest Openness, Landscape Clearance and "Culture-Steppe" Species', *Environmental Archaeology* 9, 2004

Wickham-Jones, C. R. and Pollock, D., 'Excavations at Farm Fields, Kinloch, Rhum 1984–85: A Preliminary Report', *Glasgow Archaeological Journal*, Vol. 12, Issue 12, 1985

CHAPTER FOUR

Anthony, D. W., *The Horse, the Wheel and Language: How Bronze-Age Riders from the Eurasian Steppes Shaped the Modern World*, 2007

Bradley, R., *The significance of monuments: on the shaping of human experience in Neolithic and Bronze Age Europe*, 1998

Collard, M., Edinborough, K., Shennan, S. & Thomas, M, 'Radiocarbon evidence indicated that migrants introduced farming to Britain', *Journal of Archaeological Science* 37, 2010

Collett, H., Hauzeur, A. & Lech, J., 'The prehistoric flint mining complex at Spiennes (Belgium) on the occasion of its discovery 140 years ago', in Allard, P., Bostyn, F., Giligny, F. & Lech, J. (eds), *Flint mining in prehistoric Europe. Interpreting the archaeological records*, 2008

Crane, N., *Coast: Our Island Story*, 2012

Garrow, D. & Sturt, F., 'Grey waters bright with Neolithic argonauts? Maritime connections and the Mesolithic-Neolithic transition within the "western seaways" of Britain, c. 5000-3500 BC', *Antiquity* 85, 2011: http://www.academia.edu/1480948/Garrow_D._and_Sturt_F._2011._Grey_waters_bright_with_Neolithic_argonauts_Maritime_connections_and_the_Mesolithic-Neolithic_transition_within_the_western_seaways_of_Britain_c._5000-3500_bc._Antiquity_85_59-72

Gronenborn, D., 'Climate Change and Socio-political crises: some cases from Neolithic Central Europe', in Pollard, T. & Banks, I. (eds), *War and Sacrifice: Studies in the Archaeology of Conflict*, 2007

Lönze, H., *pers. comm.*, 2010

Maroo, S. & Yalden, D. W., 'The Mesolithic mammal fauna of Great Britain', *Mammal Review* 30, 2000

Oross, K. & Banffy, E., 'Three successive waves of Neolithisation: LBK development in Transdanubia', *Documenta Praehistorica* XXXVI, 2009

Perlès, C., 'From the Near East to Greece: let's reverse the focus – cultural elements that did not transfer', in Lichter C. (ed.), *BYZAS 2. How did farming reach Europe? Anatolian-European relations from the second half of the 7th through the first half of the 6th Millennium cal BC*, 2005

Thomas, J., 'Current debates on the Mesolithic-Neolithic transition in Britain and Ireland', *Documenta Praehistorica* XXXI, 2004

Tipping, R., 'The Case for Climatic Stress Forcing Choice in the Adoption of Agriculture in the British Isles', in Finlayson, B. & Warren, G. (eds), *Landscapes in Transition*, Levant Supplementary Series, Vol. 8, 2010

Todd, M., *The South-West to AD 1000*, 1987

Vanmontfort, B., 'Bridging the gap. The Mesolithic-Neolithic transition in a frontier zone', *Documenta Praehistorica* XXXIV, 2007

Whittle, A., 'Europe in the Neolithic: the creation of new worlds', *Antiquity* 70, 1996

CHAPTER FIVE

Ashraf, Q. & Michalopoulos, S., 'Climatic Fluctuations and the Diffusion of Agriculture', 2013: http://www.brown.edu/Departments/Economics/Papers/2013/2013-3_paper.pdf

Barber, M., Field, D. & Topping, P., *Neolithic Flint Mines in England*, 1999

Bayliss, A., 'Neolithic narratives: British and Irish enclosures in their time-scapes', in Whittle, A., Healy, F. & Bayliss, A. (eds), *Gathering Time: Dating the Early Neolithic Enclosures of Southern Britain and Ireland*, Vols 1 & 2, 2011

Bayliss, A. & Healy, F., 'Gathering Time: the social dynamics of change', in Whittle, A. et al. (eds), *Gathering Time: Dating the Early Neolithic Enclosures of Southern Britain and Ireland*, 2011

Bradley, R. & Edmonds, M., *Interpreting the Axe Trade: Production and exchange in Neolithic Britain*, 1993

Coles, S., Ford, S. & Taylor, A., 'White Swan Public House, Yabsley Street, Blackwall, Tower Hamlets: A Post Excavation/Assessment', 2003

Coles, S., Ford, S., Taylor, A., Anthony, S, Gale, R., Keith-Lucas, M., Raymond, F., Robinson, M. & Vince, A., 'An Early Neolithic Grave and Occupation, and an Early Bronze Age Hearth on the Thames Foreshore at Yabsley Street, Blackwall, London', *Proceedings of the Prehistoric Society* 764, 2008

Curry, A., 'The Neolithic Toolkit', in *Archaeology*, Nov/Dec 2014

Davies, S. R., 'Location, Location, Location: a landscape-based study of early Neolithic longhouses in Britain', *Rosetta* 7, 2009

Field, D., *Earthen Long Barrows: The Earliest Monuments in the British Isles*, 2006

Garbett, G., 'The Elm Decline: The Depletion of a Resource', *New Phytologist*, Vol. 88, No. 3, July 1981

Hayden, C. & Stafford, E., *The Prehistoric Landscape at White Horse Stone, Aylesford, Kent*, 2006

Hillam, J., Groves, C. M., Brown, D. M., Baillie, M. G. L., Coles, J. M. &

Coles, B. J., 'Dendrochronology of the English Neolithic', *Antiquity* Vol. 64, Issue 243, 1990

Konvalina, P., Capouchova, I., Stehno, Z., Moudry Jr, J. & Moudry, J., in *Romanian Agricultural Research*, No. 28, 2011

Lewis-Williams, D. & Pearce, D., *Inside the Neolithic Mind*, 2005

McCarter, S., *Neolithic*, 2007

Mithen S., *The Prehistory of the Mind, A Search for the Origins of Art, Religion and Science*, 1996

O'Connell, M. & Molloy, K., 'Farming and woodland dynamics in Ireland during the Neolithic', *Proceedings of the Royal Irish Academy* 101B, 2001

Parker, A. G., Goudie, A. S., Anderson, D. E., Robinson, M. A. & Bonsall, C., 'A review of the mid-Holocene elm decline in the British Isles', *Progress in Physical Geography*, Vol. 26, Issue 1, 2002

Rowley-Conway, P., 'How the West was Lost, A Reconsideration of Agricultural Origins in Britain, Ireland and Southern Scandinavia', *Current Anthropology* 45, August–October 2004

Rowley-Conway, P., 'Westward Ho! The Spread of Agriculture from Central Europe to the Atlantic', *Current Anthropology* 52, No. S4, October 2011

Russell, M., *Rough Quarries, Rocks and Hills: John Pull and the Neolithic Flint Mines of Sussex*, 2001

Sheridan, J. A., 'The Neolithization of Britain and Ireland: The "Big Picture"', in Finlayson, B. & Warren, G. (eds), *Landscapes in Transition*, Levant Supplementary Series, Vol. 8, 2010

Stevens, C. J. & Fuller, D. Q., 'Did Neolithic farming fail? The case for a Bronze Age agricultural revolution in the British Isles', *Antiquity*, Vol. 86, No. 333, 2003

Symonds, M., 'Exploring a prehistoric landscape at Kingsmead Quarry', *Current Archaeology* 292, 2014

Teather, A., 'Interpreting hidden chalk art in southern British Neolithic flint mines', *World Archaeology*, Vol. 43, Issue 2, 2011

Whittle, A., 'The Neolithic Period, *c.* 4000–2400 Cal BC', in Hunter, J. & Ralston, I. (eds), *The Archaeology of Britain: An Introduction from Earliest Times to the Twenty-First Century*, 2009

Whittle, A., Healy, F. & Bayliss, A. (eds), *Gathering Time: Dating the Early Neolithic Enclosures of Southern Britain and Ireland*, Vols 1 and 2, 2011

CHAPTER SIX

Castleden, R., *The Making of Stonehenge*, 1993
Chapman, H. P., 'Rudston "Cursus A" – Engaging with a Neolithic monument in its landscape setting using GIS', *Oxford Journal of Archaeology* 22 (4), 2003
Clark, P. A., *The Neolithic Ritual Landscape of Rudston*, 2004
Emberling, G., 'Urban Transformations and the Problem of the "First City"', in Smith, M. L. (ed.), *The Social Construction of Ancient Cities*, 2003
Loveday, R., *Inscribed Across the Landscape: The Cursus Enigma*, 2006
Oswald, A., *Causewayed Enclosures*, 2011
Schulting, R., Murphy, E., Jones, C. & Warren, G., 'New dates from the north and a proposed chronology for Irish court tombs', *Proceedings of the Royal Irish Academy* 112C, 2011
Simpson, D. D. A. (ed.), *Economy and settlement in Neolithic and early Bronze Age Britain and Europe*, 1971
Woodbridge, J., Fyfe, R. M., Roberts, N., Downey, S., Edinborough, K. & Shennan, S., 'The impact of the Neolithic agricultural transition in Britain: a comparison of pollen-based land-cover and archaeological ^{14}C date-inferred population change', *Journal of Archaeological Science*, 2012

CHAPTER SEVEN

Burl, A., *The Stone Circles of Britain, Ireland and Brittany*, 2000
Darvill, T., Marshall, P., Parker Pearson, M. & Wainwright, G., 'Stonehenge remodelled', *Antiquity* 86, 2012
French, C., Scaife, R., Allen, M. et al., 'Durrington Walls to West Amesbury

by way of Stonehenge: A major transformation of the Holocene land-scape', *The Antiquaries Journal* 92, 2012

Malone, C., *Avebury*, 1989

Parker Pearson, M., Chamberlain, A., Jay, M., Marshall, P. et al., 'Who was buried at Stonehenge?', *Antiquity* 83, 2009

Parker Pearson, M., Pollard, J., Richards, C., Thomas, J., Tilley, C. & Welham, K., 'The Stonehenge Riverside Project: Exploring the Neolithic landscape of Stonehenge', *Documenta Praehistorica* XXXV, 2008

CHAPTER EIGHT

Amesbury, M.J. et al., 'Bronze Age upland settlement decline in southwest England: Testing the climate change hypothesis', *Journal of Archaeological Science* 35, 2008

Clutton-Brock, J., *A Natural History of Domesticated Animals*, 1987

Darvill, T., Marshall, P., Parker Pearson, M. & Wainwright, G., 'Stonehenge remodelled', *Antiquity* 86, 2012

Field, D., 'The Development of an Agricultural Countryside', in Pollard, J. (ed.), *Prehistoric Britain*, 2008

Hansen, V. & Curtis, K. R., *Voyages in World History*, 2013

Harari, Y., *Sapiens: A Brief History of Humankind*, 2014

Ixer, R. A. & Budd, P., 'The Mineralogy of Bronze Age Copper Ores from the British Isles: Implications for the Composition of Early Metalwork', *Oxford Journal of Archaeology* 17 (1), 1998

Leary, J. & Field, D., *The Story of Silbury Hill*, 2013

Parker Pearson, M. et al., *The Age of Stonehenge*, 2007

Pitts, M., *Hengeworld: Life in Britain 2000 BC as revealed by the latest discoveries at Stonehenge, Avebury and Stanton Drew*, 2001

Stevens, C. J. & Fuller, D. Q., 'Did Neolithic farming fail? The case for a Bronze Age agricultural revolution in the British Isles', *Antiquity*, Vol. 86, No. 333, 2012

Strachan, D., *The Carpow Logboat: A Bronze Age vessel brought to life*, 2010

Topping, P., *Grimes Graves*, 2011

Williamson, T., *England's Landscapes: East Anglia*, 2006

CHAPTER NINE

Brown, T., 'The Bronze Age climate and environment of Britain', *Bronze Age Review* 1, November 2008

Clark, R., *Works of Man*, 2011

Downes, J. (ed.), *Chalcolithic and Bronze Age Scotland: ScARF Panel Report*, 2012

Green, M., *Celtic World*, 1996

Harding, D. W., *The Iron Age Round-house: Later Prehistoric Building in Britain and Beyond*, 2009

Lynch, A. H., Hamilton, J. & Hedges, R. E. M., 'Where the Wild Things Are: Aurochs and Cattle in England', *Antiquity* 82, 2008

Monbiot, G., *Feral*, 2013

Parker Pearson, M., *Bronze Age Britain*, 2005

Pitts, M., 'Heathrow Today, Tomorrow the World', *British Archaeology* 75, March 2004

Quartermaine, J. & Leech, R., *Cairns, Fields, and Cultivation: Archaeological landscapes of the Lake District uplands*, 2012

Yates, T. Y., *Land, Power and Prestige: Bronze Age Field Systems in Southern England*, 2007

CHAPTER TEN

Armit, I., *Towers in the North: The Brochs of Scotland*, 2003

Brown, T., 'The Bronze Age climate and environment of Britain', *Bronze Age Review* 1, November 2008

Cunliffe, B., *The Danebury Environs Programme: The Prehistory of a Wessex Landscape* 1, 2000

Cunliffe, B., *Danebury Hillfort*, 2011

Gelling, M., *Place-Names in the Landscape: The geographical roots of Britain's place-names*, 1984

Hall, P., *Cities in Civilization*, 1998

Herodotus, de Selincourt, A. (trans.), Marincola, J. (ed.), *The Histories*, 1962

McCormick, M., Büntgen, U. et al., 'Climate Change during and after the Roman Empire: Reconstructing the Past from Scientific and Historical Evidence', *Journal of Interdisciplinary History* XLIII:2, Autumn 2012

Miles, D., Palmer, S., Lock, G., Gosden, C. & Cromarty, A. M., *Uffington White Horse and its Landscape: Investigations at White Horse Hill Uffington, 1989–95 and Tower Hill Ashbury, 1993–4*, 2003

Payne, A., Corney, M. & Cunliffe, B., *The Wessex Hillforts Project: Extensive Survey of Hillforts in Central Southern England*, 2006

Sim, D., *The Roman Iron Industry in Britain*, 2012

Thomas, C., 'Claudius and the Roman Army Reforms', *Historia: Zeitschrift für Alte Geschichte*, Bd. 53, H. 4, 2004

Wainwright, G. J., 'The Excavation of an Iron Age Hillfort on Bathampton Down, Somerset', *Transactions of the Bristol and Gloucestershire Archaeological Society*, Vol. 86, 1967

Wanner, H. et al., 'Mid- to Late Holocene climate change: an overview', *Quaternary Science Reviews* 27, 2008

CHAPTER ELEVEN

de la Bédoyère, G., *Companion to Roman Britain*, 1999

Bowman, A. K., Champlin, E. & Lintott, A. (eds), *The Cambridge Ancient History* (2nd edn), Vol. X, 1996

Brady, K., Smith, A. & Laws, G., 'Excavations at Abingdon West Central Redevelopment: Iron Age, Roman, Medieval, and Post-medieval Activity in Abingdon', *Oxoniensia* LXXII, 2007

Cunliffe, B., *The Extraordinary Voyage of Pytheas the Greek: The man who discovered Britain*, 2001

Frere, S., *Britannia: A History of Roman Britain*, 1967

Frodsham, P., 'Forgetting *Gefrin*: Elements of the Past at Yeavering', *Northern Archaeology* 17/18, 1999

Fulford, M., *A Guide to Silchester: The Roman Town of Calleva Atrebatum*, 2002

Fulford, M., 'Nero and Britain: the Palace of the Client King at *Calleva* and Imperial Policy towards the Province after Boudicca', *Britannia* 39, 2008

Gascoyne, A. & Radford, D., with contributions from Crummy, P., Crummy, N., Niblett, R., Stenning, D., Benfield, S., Murphy, P. & Phillips, A., *Colchester, Fortress of the War God: An Archaeological Assessment*, 2012

Humphrey, J., *Roman Circuses: Arenas for Chariot Racing*, 1986

McOmish, D., *Oppida*, 2011

Markey, M., Wilkes, E. & Darvill, T., 'Poole Harbour: An Iron Age Port', *Current Archaeology* 182, 2002

Mattingly, D., *An Imperial Possession, Britain in the Roman Empire, 54 BC–AD 409*, 2006

Niblett. R., 'Roman Verulamium', in Niblett, R. & Thompson, I., *Alban's Buried Towns: An Assessment of St Albans' Archaeology up to AD 1600*, 2005

Owen, O., 'Eildon Hill North', in Rideout, J., Owen, O. & Halpin, E. (eds), *Hillforts of southern Scotland*, 1992

Palliser, D., *The Cambridge Urban History of Britain, Volume I: 600–1540*, 2000

Pitts, M., 'Re-thinking the southern British oppida: Networks, kingdoms and material culture', *European Journal of Archaeology*, Vol. 13, No. 1, 2010

Salway, P., *Roman Britain*, 1984

CHAPTER TWELVE

Adkins, L. & Adkins, R., *Handbook to Life in Ancient Rome*, 2004

Allen, J., 'The "Petit Appareil" Masonry Style in Roman Britain: Geology, Builders, Scale and Proportion', *Britannia* 41, 2010

Allen, J. & Fulford, M., 'Early Roman Mosaic Materials in Southern Britain, with Particular Reference to Silchester (Calleva Atrebatum): A Regional Geological Perspective', *Britannia* 35, 2004

Baggs, A. P., Board, B., Crummy, P., Dove, C., Durgan, S., Goose, N. R., Pugh, R. B., Studd, P. & Thornton, C. C., 'Iron-Age and Roman Colchester', in Cooper, J. & Elrington, C. R. (eds), *A History of the County of Essex: Volume 9, the Borough of Colchester*, 1994

Barker, J. & Downing, R. et al., 'Hydrogeothermal studies in the United Kingdom', *Quarterly Journal of Engineering Geology and Hydrogeology* 33, 2000

Beavis, J., 'Some aspects of the use of Purbeck Marble in Roman Britain', *Proceedings of the Dorset Natural History and Archaeological Society* 92, 1971

de la Bédoyère, G., *Roman Towns in Britain*, 1992

de la Bédoyère, G., *The Golden Age of Roman Britain*, 1999

de la Bédoyère, G., *Companion to Roman Britain*, 1999

Bidwell, P., *Roman Forts in Britain*, 2007

Breeze, D., *Hadrian's Wall*, 2006

Cassius Dio, Kilvert, I. S. (trans. & ed.), *The Roman History*, 1987

Champlin, E., 'The Suburbium of Rome', *American Journal of Ancient History*, 1982

Coates, R., 'A New Explanation of the Name of London', *Transactions of the Philological Society*, Vol. 96, No. 2, 1998

Crow, J., 'The Northern Frontier of Britain from Trajan to Antoninus Pius: Roman Builders and Native Britains' in Todd, M. (ed.), *A Companion to Roman Britain*, 2004

Cunliffe, B., *The Roman Baths at Bath*, 1993

Cunliffe, B., *Fishbourne Roman Palace*, 1998

Davies, H., *Roman Roads in Britain*, 2008

Davies, J., *Venta Icenorum: Caistor St Edmund Roman Town*, 2001

Fulford, M., *The Second Augustan Legion in the West of Britain*, 1996

Fulford, M., 'Nero and Britain: the Palace of the Client King at Calleva and Imperial Policy towards the Province after Boudicca', *Britannia* 39, 2008

Gascoyne, A. & Radford, D., *Colchester, Fortress of the War God: An Archaeological Assessment*, 2012

Gerrard, J., *The Ruin of Roman Britain: An Archaeological Perspective*, 2013

Halsall, T., 'Geological constraints on the siting of fortifications: examples from medieval Britain', in Rose, E. & Nathanail, C. (eds), *Geology and Warfare: Examples of the influence of terrain and geologists on military operations*, 2000

Hanson, W., 'The Roman Presence: Brief Interludes', in Edwards, K. & Ralston, I. (eds), *Scotland After the Ice Age: Environment, Archaeology and History, 8000 BC–AD 1000*, 2003

Hayward, K., 'A Geological Link between the Facilis Monument at Colchester and First-Century Army Tombstones from the Rhineland Frontier', *Britannia* 37, November 2006

Hind, J., 'A. Plautius' Campaign in Britain: An Alternative Reading of the Narrative in Cassius Dio (60.19.5–21.2)', *Britannia* 38, November 2007

Jones, E., *Towns & Cities*, 1966

Jones, R., *Roman Camps in Britain*, 2012

Jones, R., 'A False Start? The Roman Urbanization of Western Europe', *World Archaeology* 19, 1987

Josephus, F., Whiston, W. (trans.), *War of the Jews (the Works of Flavius Josephus, Book III, Chap. V)*, 1737

King, A., *Roman Gaul and Germany*, 1990

MacDonald, W., *The Architecture of the Roman Empire, Volume II: An Urban Appraisal*, 1986

MacKendrick, P., 'Roman Town Planning', *Archaeology*, Vol. 9, No. 2, June 1956

Magie, D. (trans.), 'The Life of Hadrian', *Historia Augusta*, 1921

Mandich, M., 'Re-defining the Roman "suburbium" from Republic to Empire: A Theoretical Approach', in Brindle, T., Allen, M., Durham, E. & Smith, A. (eds), *TRAC 2014: Proceedings of the Twenty Fourth Theoretical Roman Archaeology Conference*, 2015

Mazza, L., 'Plan and Constitution – Aristotle's Hippodamus: Towards an "Ostensive" Definition of Spatial Planning', *The Town Planning Review*, Vol. 80, No. 2, 2009

Morris, J., *Londinium: London in the Roman Empire*, 1998

Museum of London Archaeology, *Londinium: A new map and guide to Roman London*, 2011

Niblett, R. & Thompson, I., *Alban's Buried Towns: An Assessment of St Albans' Archaeology up to AD 1600*, 2005

Ordnance Survey, *Londinium: A descriptive map and guide to Roman London*, 1983

Poulter, J., *Further Discoveries about the Surveying and Planning of Roman Roads in Northern Britain*, 2014

Pritchard, F., 'Ornamental Stonework from Roman London', *Britannia* 17, 1986

Richmond, I., *Roman Britain*, 1955

Rivet, A., *Town and country in Roman Britain*, 1958

Rivet, A. & Smith, C., *The Place-names of Roman Britain*, 1979

Rogers, A., *Water and Roman Urbanism: Towns, Waterscapes, Land Transformation and Experience in Roman Britain*, 2013

Rose-Redwood, R., 'Genealogies of the Grid: Revisiting Stanislawski's Search for the Origin of the Grid-Pattern Town', *Geographical Review*, Vol. 98, No. 1, January 2008

Rotherham, I., *Roman Baths in Britain*, 2012

Royal Commission on Historical Monuments of England, 'Colchester' in *An Inventory of the Historical Monuments in Essex, Volume 3, North East*, 1922

Salway, P., *Roman Britain*, 1984

Sauer, E., 'Alchester: In Search of Vespasian', *Current Archaeology* 196, 2005

Sauer, E., 'Inscriptions from Alchester: Vespasian's Base of the Second Augustan Legion(?)', *Britannia* 36, 2005

Schrüfer-Kolb, I., *Roman Iron Production in Britain: Technological and socio-economic landscape development along the Jurassic Ridge*, 2004

Sealey, P., *The Boudican Revolt against Rome*, 2004

Sim, D., *The Roman Iron Industry in Britain*, 2012

Smith, W., *A Dictionary of Greek and Roman Antiquities*, 1875

Stanislawski, D., 'The Origin and Spread of the Grid-Pattern Town', *Geographical Review*, Vol. 36, No. 1, January 1946

Suetonius, Edwards C. (trans.), *Lives of the Caesars*, 2000

Tacitus, Mattingly, H. (trans.), revised by Handford, S., *Agricola*, 1970

Tacitus, Jackson, J. (trans.), *Annals* (Loeb Classical Library, Volume V), 1937

Vitruvius, Morgan, M. (trans.), *The Ten Books on Architecture*, 1914

Wacher, J., *The Towns of Roman Britain*, 1997

Wallace, L., *The Origin of Roman London*, 2014

Wilmott, T., *Richborough and Reculver*, 2012

Wilson, P., *Roman Forts and Fortresses*, 2011

Wilson, R. J. A., 'Urban Defences and Civic Status in Early Roman Britain', in Wilson, R. J. A. (ed.), *Romanitas: Essays on Roman archaeology in honour of Sheppard Frere on the occasion of his ninetieth birthday*, 2006

Witcher, R., 'The hinterlands of Rome: settlement diversity in the early imperial landscape of Regio VII Etruria', in *Papers in Italian archaeology VI: communities and settlements from the neolithic to the early medieval period*, 2005

Witcher, R., 'The extended metropolis: urbs, suburbium and population', *Journal of Roman Archaeology* 18, 2008

Wichter, R., '(Sub)urban Surroundings', in Erdkamp, P. (ed.), *The Cambridge Companion to Ancient Rome*, 2013

CHAPTER THIRTEEN

Ammianus Marcellinus, Rolfe, J. (trans.), *The Roman History*, 1939

Blair, P., *An Introduction to Anglo-Saxon England* (3rd edn), 2003

Campbell, J. (ed.), *The Anglo-Saxons*, 1982

Clarke, G., Rigby, V. & Shepherd, J., 'The Roman Villa at Woodchester', *Britannia* 13, 1982

Cotterill, J., 'Saxon Raiding and the Role of the Late Roman Coastal Forts of Britain', *Britannia* 24, 1993

Davies, H., *Design and Construction of Romans Roads in Britain*, 2001

Douglas, A., Gerrard, J. & Sudds, B., *A Roman Settlement and Bath House at Shadwell*, 2011

Drijvers, H., *Cults and Beliefs at Edessa*, 1980

Evans, J. A. A., *The Age of Justinian: The Circumstances of Imperial Power*, 1996

Fleming, R., *Britain after Rome: The Fall and Rise, 400–1070*, 2011

Gräslund, B. & Price, N., 'Twilight of the gods? The "dust veil event" of AD 536 in critical perspective', *Antiquity* 86, 2012

Gunn, J. (ed.), *The Years without Summer: Tracing A.D. 536 and its aftermath (British Archaeological Reports (BAR) International)*, 2000

Gurney, D., *Outposts of the Roman Empire. A Guide to Norfolk's Roman Forts at Burgh Castle, Caister-on-Sea and Brancaster*, 2002

Hamerow, H., Hinton, D. & Crawford, S. (eds), *The Oxford Handbook of Anglo-Saxon Archaeology*, 2011

Higham, N. & Ryan, M., *The Anglo-Saxon World*, 2013

Jones, M., *The End of Roman Britain*, 1996

Jones, R., *Roman Camps in Scotland*, 2011

Kondoleon, C., 'Signs of Privilege and Pleasure: Roman Domestic Mosaics' in Gazda, E. (ed.), *Roman Art in the Private Sphere: New Perspectives on the Architecture and Decor of the Domus, Villa, and Insula*, 1991

Little, L. L. (ed.), *Plague and the End of Antiquity: The Pandemic of 541–750*, 2007

McCormick, M. & Büntgen, U. et al., 'Climate Change during and after the Roman Empire: Reconstructing the Past from Scientific and Historical Evidence', *Journal of Interdisciplinary History*, XLIII:2, Autumn, 2012

Maddicott, J., 'Plague in Seventh-Century England', *Past & Present*, No. 156, August 1997

Oosthuizen, S., *Tradition and Transformation in Anglo-Saxon England: Archaeology, Common rights and Landscape*, 2013

Ottaway, P., *Roman York*, 2011

Pearson, A., 'Barbarian Piracy and the Saxon Shore: A Reappraisal', *Oxford Journal of Archaeology*, Vol. 24, No. 1, 2005

Pohl, W., 'Ethnic Names and Identities in the British Isles: A comparative perspective', in Hines, J. (ed.), *The Anglo-Saxons from the Migration Period to the Eighth century: An ethnographic perspective*, 1997

Powlesland, D., *25 Years of Archaeological Research on the Sands and Gravels of Heslerton*, 2003

Procopius, Dewing, H. (trans.) *Buildings*, 1940

Procopius, Dewing, H. (trans.), *History of the Wars*, 1914

Stenton, F., *Anglo-Saxon England*, 1971

Wagner, D. M., Klunk, J. et al., 'Yersinia pestis and the Plague of Justinian 541–543 AD: A genomic analysis', *The Lancet*, 2014

Wilson, R., *A Guide to the Roman Remains in Britain*, 2002

CHAPTER FOURTEEN

Abram, C., 'In Search of Lost Time, Aldhelm and The Ruin', *Quaestio (Selected Proceedings of the Cambridge Colloquium in Anglo-Saxon, Norse, and Celtic)*, Vol. 1, 2000

Alcock, L., *Arthur's Britain, History and Archaeology AD 367–634*, 1973

Asser, Keynes, S. (ed. & trans.), *Asser's Life of Alfred and Other Contemporary Sources*, 1983

Attenborough, F. (ed. & trans.), *The Laws of the Earliest English Kings*, 1922

Ayre, J. & Wroe-Brown, R., 'The Post-Roman Foreshore and the Origins of the Late Anglo-Saxon Waterfront and Dock of Æthelred's Hithe: Excavations at Bull Wharf, City of London', *Archaeological Journal*, 2015

Baker, J. & Brookes, S., *Beyond the Burghal Hidage: Anglo-Saxon Civil Defence in the Viking Age*, 2013

Bede, Sherley-Price, L., Latham, R. E. & Farmer, D. H. (trans.), *Ecclesiastical History of the English People*, 1990

Biddle, M. & Kjolbye-Biddle, B., 'Repton and the Vikings', *Antiquity*, Vol. 66, Issue 250, 1992

Blackmore, L., 'The origins and growth of Lundenwic, a mart of many nations', in Hårdh, B. and Larsson, L. (eds), *Central Places in the Migration and Merovingian Periods*, 2002

Blair, J., *The Church in Anglo-Saxon Society*, 2005

Brennan, N. & Hamerow, H., 'An Anglo-Saxon Great Hall Complex at Sutton Courtenay/Drayton, Oxfordshire: A Royal Centre of early Wessex?', *Archaeological Journal*, Vol. 172, No. 2, 2015

Brooks, N., 'The development of military obligations in eight- and ninth-century England', in Clemoes, P. & Hughes, K. (eds), *England Before the Conquest: Studies in Primary Sources presented to Dorothy Whitelock*, 1971

Carver, M., *An Iona of the East: The Early-medieval Monastery at Portmahomack, Tarbat Ness*, 2004

Carver, M., 'Early Scottish Monasteries and Prehistory: A Preliminary Dialogue', *The Scottish Historical Review*, Vol. LXXXVIII, Issue 2, 2009

Carver, M., *Portmahomack, Monastery of the Picts*, 2008

Carver, M. (ed.), *The Cross Goes North: Processes of conversion in northern Europe, AD 300–1300*, 2004

Frodsham, P., *Yeavering: People, Power & Place*, 2005

Fryde, N. & Reitz, D. (eds), *Walls, Ramparts, and Lines of Demarcation: Selected Studies from Antiquity to Modern Times*, 2009

Gardiner, M., 'An Early Medieval Tradition of Building in Britain', *Arqueología de la Arquitectura* 9, 2012

Garmonsway, G. N. (ed. & trans.), *The Anglo-Saxon Chronicle*, 1953

Hamerow, H., Hinton, D. & Crawford, S. (eds), *The Oxford Handbook of Anglo-Saxon Archaeology*, 2011

Haslam, J., 'Market and fortress in England in the Reign of Offa', *World Archaeology*, Vol. 19, No. 1, June 1987

Hooke, D., *The Anglo-Saxon Landscape: The Kingdom of the Hwicce*, 1985

Hope-Taylor, B., *Yeavering: An Anglo-British centre of early Northumbria*, 1977

Howe, N., *Writing the Map of Anglo-Saxon England: Essays in Cultural Geography*, 2008

James, S., Marshall, A. & Millett, M., 'An Early Medieval Building Tradition', in Karkov, C. (ed.), *The Archaeology of Anglo-Saxon England*, 1999

Loveluck, C., *Northwest Europe in the early Middle Ages, c. AD 600–1150: A comparative archaeology*, 2013

Maddicott, J., 'London and Droitwich, c. 650–750: Trade, industry and the rise of Mercia', *Anglo-Saxon England* 34, 2005

Marsden, R., *The Cambridge Old English Reader*, 2004

Miket, R. & Semple, S., *Yeavering: Rediscovering the Landscape of the Northumbrian Kings*, 2009

Millett, M., 'Early Medieval Walls and Roofs: A case study in interrogative excavation', *euroREA. Journal for (Re)construction and Experiment in Archaeology*, 2008

Millett, M. & James, S., 'Excavations at Cowdery's Down, Basingstoke, Hants. 1978–1981', *Archaeological Journal* 140, 1983

Oosthuizen, S., *Landscapes Decoded: The origins and development of Cambridgeshire's medieval fields*, 2006

Oosthuizen, S., 'The Anglo-Saxon Kingdom of Mercia and the Origins and Distribution of Common Fields', *The Agricultural History Review*, Vol. 55, No. 2, 2007

Oosthuizen, S., 'Medieval Field Systems and Settlement Nucleation: Common or separate origins?' in Higham, N. & Ryan, M. (eds), *The Landscape Archaeology of Anglo-Saxon England*, 2010

Oosthuizen, S., *Tradition and Transformation in Anglo-Saxon England: Archaeology, common rights and landscape*, 2013

Parks, G., *The English Traveler to Italy, Vol. 1: The Middle Ages (to 1525)*, 1954

Perry, D. & Munro, D., *Castle Park Dunbar: Two thousand years on a fortified headland*, 2000

Richards, J. D., *Viking Age England*, 1991

Rideout, J., Owen, O. & Halpin, E., *Hillforts of Southern Scotland*, 1992

Rippon, S., 'Landscape change during the "Long Eighth Century" in Southern England', in Higham, N. & Ryan, M. (eds), *The Landscape Archaeology of Anglo-Saxon England*, 2010

Spiegel, F., 'The *tabernacula* of Gregory the Great and the conversion of Anglo-Saxon England', *Anglo-Saxon England* 36, December 2007

Squatriti, P., 'Offa's Dyke between Nature and Culture', *Environmental History*, Vol. 9, No. 1, January 2004

Treharne, E. (ed.), *Old and Middle English c. 890–c. 1450: An Anthology* (3rd edn), 2010

Whitehouse, K., 'Early Fulham', *London Archaeology* 15, 1972

Wood, M., *In Search of the Dark Ages*, 2005

Wright, J., Powell, A. & Barclay, A., 'Excavation of Prehistoric and Romano-British Sites at Marnel Park and Merton Rise (Popley) Basingstoke, 2004–8', *Wessex Archaeology*, 2009

CHAPTER FIFTEEN

Baker, N. & Holt, R., *Urban Growth and the Medieval Church: Gloucester and Worcester*, 2004

Beresford, M., *New Towns of the Middle Ages: Town Plantation in England, Wales and Gascony*, 1988

Brooks, N. & Whittington, G., 'Planning and Growth in the Medieval Scottish Burgh: The example of St Andrews', *Transactions of the Institute of British Geographers*, Vol. 2, No. 3, 1977

Carpenter, D., *The Struggle for Mastery: The Penguin History of Britain 1066–1284*, 2003

Champ, J., *The English Pilgrimage to Rome: A Dwelling for the Soul*, 2000

Cooper, A., 'The King's Four Highways: legal fiction meets fictional law', *Journal of Medieval History*, Vol. 26, No. 4, 2000

Dyer, C., *Making a Living in the Middle Ages: The People of Britain 850–1520*, 2002

Fleming, R., *Kings and Lords in Conquest England*, 1991

Foot, S., 'The Making of Angelcynn: English Identity Before the Norman Conquest', in Liuzza, R. (ed.), *Old English Literature: Critical essays*, 2002

Garnett, G., *The Norman Conquest*, 2009

Gillingham, J. & Griffiths, R., *Medieval Britain: A Very Short Introduction*, 2000

Haslam, J., 'The Metrology of Anglo-Saxon Cricklade', *Medieval Archaeology* 30, 1986

Haslam, J., 'The second burh of Nottingham', *Landscape History* 9, 1987

Haslam, J., '*Domnoc* and Dunwich: A Reappraisal', *Anglo-Saxon Studies in Archaeology and History* 5, 1992

Haslam, J., 'Excavations at Cricklade, Wiltshire, 1975', *Internet Archaeology* 14, 2003

Haslam, J., 'King Alfred and the Vikings: Strategies and tactics 876–886 AD', *Anglo-Saxon Studies in Archaeology and History* 13, 2005

Haslam, J., 'The Development of Late-Saxon Christchurch, Dorset, and the Burghal Hidage', *Medieval Archaeology* 53, 2009

Haslam, J., 'The Development of London by King Alfred: A reassessment', *Transactions of the London and Middlesex Archaeological Society* 61, 2010

Haslam, J., 'The Two Anglo-Saxon *Burhs* of Oxford', *Oxoniensia* 75, 2010

Haslam, J., 'King Alfred, Mercia and London, 874–86: A reassessment', *Anglo-Saxon Studies in Archaeology and History* 17, 2011

Henry, Archdeacon of Huntingdon, Greenway, D. (ed.), *Historia Anglorum: The History of the English People*, 1996

Higham, N. J., *The English Conquest: Gildas and Britain in the fifth century*, 1994

Higham, N. & Ryan, M. (eds), *Landscape Archaeology of Anglo-Saxon Britain*, 2010

Higham, R., *The Making of Anglo-Saxon Devon: Emergence of a Shire*, 2008

Hill, D. & Rumble, A. (eds), *The Defence of Wessex: The Burghal Hidage and Anglo-Saxon fortifications*, 1996

Holt, R., 'The Urban Transformation in England, 900–1100', in Lewis, C. (ed.), *Anglo-Norman Studies* 32, 2010

Jones, S. R. H., 'Transaction costs, institutional change, and the emergence of a market economy in later Anglo-Saxon England', *Economic History Review*, Vol. 46, No. 4, 1993

Keynes, S., 'Anglo-Saxon Entries in the "Liber Vitae" of Brescia', in Roberts, J., Nelson, J. & Godden, M. (eds), *Alfred the Wise*, 1997

Keynes, S. & Lapidge, M. (trans.), *Alfred the Great: Asser's Life of King Alfred and Other Contemporary Sources*, 1983

La Regina, A. (ed.), *Archaeological Guide to Rome*, 2007

Langlands, A., 'Placing the burh in *Searobyrg*: rethinking the urban topography of early medieval Salisbury', *Wiltshire Archeological & Natural History Magazine* 107, 2014

Lankila, T., 'The Saracen Raid of Rome in 846: an example of maritime *ghazw*', in Akar, S., Hämeen-Anttila, J. & Nokso-Koivisto, I. (eds), *Travelling Through Time: Essays in honour of Kaj Öhrnberg*, 2013

Lilley, K. D., *Urban Life in the Middle Ages 1000–1450*, 2002

Lilley, K. D., *City and Cosmos: The Medieval World in Urban Form*, 2009

Miller, E. & Hatcher, J., *Medieval England: Towns, commerce, and crafts, 1086–1348*, 1995

Morris, M., *The Norman Conquest*, 2012

Orderic Vitalis, Forester, T. (trans.), *The Ecclesiastical History of England and Normandy*, 1842

Parks, G., *The English Traveler to Italy, Vol. 1: The Middle Ages (to 1525)*, 1954

Reuter, T. (ed.), *Alfred the Great*, 2003

Slater, T., 'The Analysis of Burgage Patterns in Medieval Towns', *Area*, Vol. 13, No. 3, 1981

Stell, G. & Tait, R., 'Framework and form: burgage plots, street lines and domestic architecture in early urban Scotland', *Urban History*, Vol. 43, No. 1, 2016

Sturgis, M., *When in Rome: 2000 Years of Roman Sightseeing*, 2011

Taylor, R., *Public Needs and Private Pleasures: Water Distribution, the River Tiber and the Urban Development of Ancient Rome*, 2000

White, G., *The Medieval English Landscape, 1000–1540*, 2012

CHAPTER SIXTEEN

Appleby, J. T. (ed.), *The Chronicle of Richard of Devizes of the Time of King Richard the First*, 1963

Baggs, A. et al., 'Woodstock: Introduction', in Crossley, A. & Elrington, C. (eds), *A History of the County of Oxford: Volume 12*, 1990

Bennett, H. S., *The Pastons and their England*, 1968

Broadberry, S., Campbell, B., Klein, A., Overton, M. & van Leeuwen, B., *British Economic Growth 1270–1870*, 2015

Brown, S., 'Exe Bridge, a history and guide', in Harvey, B., George, B. & Brown, S., *Devon Bridges*, Devon Buildings Group, 23rd Annual Conference, 2008

Davis, J., *Medieval Market Morality: Life, Law and Ethics in the English Marketplace, 1200–1500*, 2012

Fitz Stephen, W., *Descriptio noblissimae civitatis Londoniae, c. 1183*, in Logan, F. D. & Stenton, F. (eds), *Norman London*, 1990

Gerald of Wales, Thorpe, L. (ed. & trans.), *The Journey through Wales and The Description of Wales*, 1978

Gilchrist, R., *Medieval Life: Archaeology and the Life Course*, 2013

Hindle, P., *Medieval Roads and Tracks*, 2013

Hooker, J., Izacke et al., *The Ancient History and Description of the City of Exeter*, 1765

Millea, N., *The Gough Map: The earliest road map of Great Britain*, 2007

Page, W. (ed.), *A History of the County of Bedford: Volume 3*, 1912

Richards, P., *King's Lynn*, 1990

Shoesmith, R. & Johnson, A. (eds), *Ludlow Castle: Its history & buildings*, 2006

Stenton, F., *Norman London*, 1990

Stoyle, M., *Circled with Stone: Exeter's City Walls, 1485–1660*, 2003

Vaughan, R., *Matthew Paris*, 1958

CHAPTER SEVENTEEN

Alcock, N., *Cruck construction: An introduction and catalogue*, CBA Research Report No. 42, 1981

Brotton, J., *Great Maps: The world's masterpieces explored and explained*, 2014

Brunskill, R., *Illustrated Handbook of Vernacular Architecture*, 1971

Dyer, A., *Decline and Growth in English Towns 1400–1640*, 1991

Dyer, C., 'Deserted Medieval Villages in the West Midlands', *The Economic History Review*, New Series, Vol. 35, No. 1, February 1982

Fagan, B., *The Little Ice Age: How Climate Made History 1300–1850*, 2000

Febvre, L. & Martin, H. J., *The Coming of the Book*, 1976

Guy, J., *The Tudors: A Very Short Introduction*, 2000

Horrox, R. (ed. & trans.), *The Black Death*, 1994

Lavigne, F., Degeai, J.-P. et al., 'Source of the great AD 1257 mystery eruption unveiled, Samalas volcano, Rinjani Volcanic Complex, Indonesia', *Proceedings of the National Academy of Sciences of the United States of America*, Vol. 110, No. 42, 2013

Lobel, M. (ed.), 'Parishes: Tusmore', in *A History of the County of Oxford: Volume 6*, 1959

Miles, D. & Rowley, T., 'Tusmore Deserted Village', *Oxoniensia*, 1976

More, T., Baker-Smith, D. (ed. & trans.), *Utopia*, 2012

Parry, M., *Climatic Change, Agriculture and Settlement (Studies in Historical Geography)*, 1978

Rackwitz, M., *Travels to Terra Incognita: The Scottish Highlands and Hebrides in Early Modern Travellers' Accounts c. 1600 to 1800*, 2007

Rowley, T. & Wood, J., *Deserted Villages* (3rd edn), 2000

Rubin, M., 'After 1348', in *The Hollow Crown: A History of Britain in the Late Middle Ages*, 2006

Stevenson, K., *Power and Propaganda: Scotland 1306–1488*, 2014

Walford, E., 'Deptford', in *Old and New London: Volume 6*, 1878

Worsley, L., *If Walls Could Talk: An Intimate History of the Home*, 2011

CHAPTER EIGHTEEN

Ackroyd, P., *The Life of Thomas More*, 1998

Brigden, S., *New Worlds, Lost Worlds: The Rule of the Tudors 1485–1603*, 2000

Brimblecombe, P., 'Early urban climate and atmosphere', in Hall, A. & Kenward, H. (eds), *The Council for British Archaeology Research Report No. 43, Environmental Archaeology in the Urban Context*, 1982

Camden, W., Holland, P. (trans.), *Britain, or, a Chorographical Description of the most flourishing Kingdomes, England, Scotland, and Ireland*, 1610

Chandler, J., *John Leland's Itinerary. Travels in Tudor England*, 1993

Coward, B., *The Stuart Age: England 1603–1714* (2nd edn), 1994

Cunningham, I. (ed.), *The Nation Survey'd: Timothy Pont's Maps of Scotland*, 2001

Forbes, R., *Studies in Ancient Technology, Volume II* (3rd edn), 1993

Hall, A. & Kenward, H. (eds), *The Council for British Archaeology Research Report No. 43, Environmental Archaeology in the Urban Context*, 1982

Harbison, C., *The Art of the Northern Renaissance*, 1995

Harvey, P., *Maps in Tudor England*, 1993

Horace, Late Earl of Orford, Hentzner, P., (trans.), *Travels in England*, 1797

Houston., R. A., *The Population of Britain and Ireland 1550–1750*, 1992

Mackie, J. D., *The Earlier Tudors 1485–1558*, 1952

Rippon, S., 'Landscape Change during the "Long Eighth Century" in Southern England', in Higham, N. & Ryan, M. (eds), *The Landscape Archaeology of Anglo-Saxon England*, 2010

Saxton, C. & Ravenhill, W., *Christopher Saxton's 16th Century maps: The Counties of England and Wales*, 1992

Simmons, J. & Biddle, G., *The Oxford Companion to British Railway History from 1603 to the 1990s*, 1997

Toulmin Smith, L. (ed.), *The itinerary of John Leland: in or about the years 1535–1543 (5 volumes)*, 1906–1910

Vansittart, P., *A Literary Companion to London*, 1992

CHAPTER NINETEEN

Broodbank, J., 'The Howland Great Wet Dock', *The Mariner's Mirror*, Vol. 2, Issue 2, 1912.

Clark, P. & Slack, P., *English Towns in Transition 1500–1700*, 1976

Daly, A., *The History of Canvey Island and Surrounding Neighbourhood* (2nd edn), 1902

Defoe, D., *A Tour Through the Whole Island of Great Britain*, 1724–6.

Dietz, B., 'Dikes, Dockheads and Gates: English docks and sea power in the sixteenth and seventeenth centuries', *The Mariner's Mirror*, Vol. 88, No. 2, 2013

Forbes, R., *Studies in Ancient Technology: Volume II*, 1993

Gale, C. & Davidson, H., *Reports of Cases Argued and Determined in The Court of Queen's Bench, and upon . . . in Hilary, Easter, and Trinity Terms*, 1843

Green, H. & Wigram, R., *Chronicles of Blackwall Yard: Part I*, 1881

Harris, L., *Vermuyden and the Fens: A study of Sir Cornelius Vermuyden and the Great Level*, 1953

Harrison, S., Rowan, A. V., Glasser, N. F., Knight, J., Plummer, M. A. & Mills, S. C., 'Little Ice Age glaciers in Britain: Glacier climate-modelling in the Cairngorm Mountains', *The Holocene*, Vol. 24, No. 2, February 2014

Hobhouse, H. (ed.), 'Blackwall Yard: Development, to c. 1819' in *Survey of London, Volumes 43 and 44: Poplar, Blackwall and Isle of Dogs*, 1994

Jackman, W., *The Development of Transportation in Modern England* (3rd edn), 1966

Morrill, J., *Stuart Britain: A Very Short Introduction*, 2000

Morris, C. (ed.), *The Illustrated Journeys of Celia Fiennes 1685–c.1715*, 1995

Peet, H., *Thomas Steers, The engineer of Liverpool's first dock: A memoir*, 1932

Plot, R., *The Natural History of Oxford-shire, Being an Essay toward the Natural History of England*, 1677

Rotherham, I., *The Lost Fens, England's Greatest Ecological Disaster*, 2013

Sly, R., *From Punt to Plough: A History of the Fens*, 2003

Thirsk, J., 'The Isle of Axholme before Vermuyden', *The Agricultural History Review*, Vol. 1, No. 1, 1953

Thirsk, J. & Cooper, J. P. (eds), *Seventeenth-Century Economic Documents*, 1972

Wheatley, B., *London Past and Present: Its History, Associations, and Traditions, Volume 1: A–D*, 2011

White, J., *A Great and Monstrous Thing: London in the Eighteenth Century*, 2013

Wilkins, S. (ed.), *Sir Thomas Browne's World, including his Life and Correspondence*, 1836

Yarranton, A., *England's Improvement by Sea and Land*, 1698

CHAPTER TWENTY

Ashton, T., *An Economic History of England: The Eighteenth Century*, 1955

Barker, T. & Harris, J., *A Merseyside Town in the Industrial Revolution: St Helens 1750–1900*, 1954

Buchan, J., *Capitals of the Mind: How Edinburgh Changed the World*, 2003

Cooper, B., *Transformation of a Valley: Derbyshire Derwent*, 1991

Devine, T., *Scotland's Empire 1600–1815*, 2003

Devine, T., *Clearance and Improvement: Land, power and people in Scotland, 1700–1900*, 2006

Elliot, G., *Proposals for carrying on certain Public Works in the City of Edinburgh*, 1752

Houston, R. A., *The Population of Britain and Ireland 1550–1750*, 1995

Hughes, S., 'Memoir of James Brindley', in Weale, J. (ed.), *Quarterly Papers on Engineering* 1, 1844

Johnson, S., Levi, P. & Boswell, J., *A Journey to the Western Islands of Scotland*, 1984

McCutcheon, W., 'The Newry Navigation: The Earliest Inland Canal in the British Isles', *The Geographical Journal*, Vol. 129, No. 4, December 1963

Malet, H., *Bridgewater: The Canal Duke, 1736–1803*, 1977

Meade, M. K., 'Plans of the New Town of Edinburgh', *Architectural History* 14, 1971

Moritz, C., *Travels of Carl Philipp Moritz in England in 1782*, 1795

Osborne, R., *Iron, Steam and Money: The Making of the Industrial Revolution*, 2013

Overton, M., 'Re-establishing the English Agricultural Revolution', *The Agricultural History Review*, Vol. 44, No. 1, 1996

Peet, H. & Steers, T., *The Engineer of Liverpool's First Dock: A memoir*, 1932

Pennant, T. & Simmons, A., *A Tour in Scotland and Voyage to the Hebrides, 1772*, 1998

Plumptre, J., 'A Journal: Of a Tour through Part of North Wales, in the Year 1792', in Ousby, I. (ed.), *James Plumptre's Britain: The Journals of a Tourist in the 1790s*, 1992

Porter, R., '"In England's Green and Pleasant Land": The English Enlightenment and the Environment' in Flint, K. & Morphy, H. (eds), *Culture, Landscape, and the Environment, The Linacre Lectures 1997*, 2000

Priestley, J., *Historical Account of the Navigable Rivers, Canals, and Railways, of Great Britain*, 1831

Rawcliffe, C. & Wilson, R. (eds), *Medieval Norwich*, 2004

Rawcliffe, C. & Wilson, R. (eds), *Norwich since 1550*, 2004

Redmonds, G., *Huddersfield 1500–1800*, 1981

Richards, E., *The Highland Clearances*, 2002

Ruddock, T., *Arch Bridges and their Builders 1735–1835*, 1979

Searle, M., *Turnpikes and Toll-bars, Volume I*, 1930

Shoard, M., *A Right to Roam*, 1999

Smiles, S., *James Brindley and the Early Engineers*, 1864

Smiles, S., *Josiah Wedgwood F.R.S.: His Personal Story*, 1894

Smout, T., *A History of the Scottish People 1560–1830*, 1969

Walford, E., *Old and New London: Volume 5*, 1878

Wolmar, C., *Fire & Steam: How the Railways Transformed Britain*, 2007

Wood, C., *The Duke's Cut: The Bridgewater Canal*, 2009

Woods, R., *The Population of Britain in the Nineteenth Century*, 1995

Young, A., 'A Tour to Shropshire', *Tours in England and Wales, selected from the Annals of Agriculture*, 1932

CHAPTER twenty-ONE

The Farmer's Magazine, Volume the Fifth, January to June, MDCCCXLII April 1842

Anonymous review of Palmer, H., *Description of a Railway on a New Principle, The Philomathic Journal and Literary Review*, Vol. III, 1825

Bakewell, F., 'Centrifugal Pumps', in *Great Facts: A popular history and description of the most remarkable inventions during the present century*, 1859

Barratt, N., *Greater London: The Story of the Suburbs*, 2012

Blunden, J. & Curry, N. (eds), *The Changing Countryside*, 1985

Bonham-Carter, V., *The Survival of the English Countryside*, 1972

Brooke, D., 'The Railway Navvy – a reassessment', *Construction History*, Vol. 5, 1989

Clare, J., *The Village Minstrel and Other Poems*, 1821

Coleman, T., *The Railway Navvies*, 1968

Darwin, E., *Phytologia: or the Philosophy of Agriculture and Gardening*, 1800

Devine, V. & Clark, J., 'Cheshire Historic Towns Survey', *Crewe: An Archaeological Assessment*, 2003

Dockray, K., *The Victorian Peasant by Richard Heath*, 1989

Engels, F., *The Condition of the Working Class in England*, 1845

Evans, E. J., *The Forging of the Modern State: Early Industrial Britain*, 1993

Fowler, G., 'Shrinkage of the Peat-Covered Fenlands', *The Geographical Journal*, Vol. 81, No. 2, February 1933

Fussell, G., 'The Dawn of High Farming in England: Land Reclamation in Early Victorian Days', *Agricultural History*, Vol. 22, 1948

Geddes, P., *Cities in Evolution: An introduction to the town planning movement and to the study of civics*, 1915

Gibson, K. & Booth, A., *The Buildings of Huddersfield* (rev. edn), 2009

Hall, P. & Ward, C., *Sociable Cities: The Legacy of Ebenezer Howard*, 1998

Harvie, C. & Matthew, H. C. G., *Nineteenth-century Britain*, 2000

Herlihy, D., *Bicycle: The History*, 2004

Howard, E., *Garden Cities of Tomorrow*, 1902

Johnston, R. & Williams, M. (eds), *A Century of British Geography*, 2003

McAdam, J., *Remarks on the Present System of Road Making*, 1816

Morris, M., 'Towards an archaeology of navvy huts and settlements of the industrial revolution', *Antiquity*, Vol. 68, Issue 260, September 1994

Oakley, W., *Winged Wheel: The History of the First Hundred Years of the Cyclists' Touring Club*, 1977

Palmer, H., *Description of a Railway on a New Principle*, 1823

Paterson, D., *A New and Accurate Description of all the Direct and Principal Roads in England and Wales and Part of the Rods of Scotland*, 1803

Perry, P., 'High Farming in Victorian Britain: Prospect and retrospect', *Agricultural History*, Vol. 55, No. 2, April 1981

Phillips, G., *Walks Round Huddersfield*, 1848

Price, U., *An Essay on the Picturesque*, 1794

Prothero, R., *English Farming: Past and Present*, 1912

Rickman, J. (ed.), *Life of Thomas Telford, Civil Engineer*, 1838

Robinson, E., 'John Clare (1793–1864) and James Plumptre (1771–1832): "A Methodistical Parson"', *Transactions of the Cambridge Bibliographical Society*, Vol. 11, No. 1, 1996

Rocque, J., *An Exact Survey of the Citys of London, Westminster, ye Borough of Southwark, and the Country near Ten Miles round*, 1746

Rowley, T., *The English Landscape in the Twentieth Century*, 2006

Sinclair, J., *The Code of Agriculture: including Observations on Gardens, Orchards, Woods and Plantations etc.*, 1832

Smiles, S., *The Life of Thomas Telford*, 1867

Smith, J., *Remarks on Thorough Draining and Deep Ploughing* (6th edn), 1843

Stephens, W. B., 'Secular Architecture', in *A History of the County of Warwick, Volume 7: the City of Birmingham*, 1964

Stratton, M. & Trinder, B., *Twentieth Century Industrial Archaeology*, 2013

Vallance, J., *A Letter to M. Ricardo, Esq. in Reply to his Letter to Dr. Yates on the Proposed Method of Pneumatic Transmission or Conveyance by Atmospheric Pressure*, 1827

Webb, S. & Webb, B. P., *English Local Government: The Story of the King's Highway*, 1913

White, J., *Zeppelin Nights: London in the First World War*, 2015

Wolmar, C., *Fire & Steam: How the Railways Transformed Britain*, 2008

Wood, N., *A Practical Treatise on Rail-roads, and Interior Communication in General*, 1825

Woodforde, J., *The Story of the Bicycle*, 1970

Woods, R., *The Population of Britain in the Nineteenth Century*, 1995

CHAPTER twenty-TWO

Abercrombie, P., 'The Preservation of Rural England', *The Town Planning Review*, Vol. 12, No. 1, 1926

Abercrombie, P., 'Some Aspects of the County of London Plan', *The Geographical Journal*, Vol. 102, No. 5/6, November–December 1943

Arthur, M., *Last of the Few: The Battle of Britain in the Words of the Pilots who Won it*, 2010

Barratt, N., *Greater London: The Story of the Suburbs*, 2012

Bishop, P., *Battle of Britain: A Day-by-Day Chronicle, 10 July 1940 to 31 October 1940*, 2009

Blanchet, E., *Prefab Homes*, 2014

Bolton, D. K., King, H. P. F., Wyld, G. & Yaxley, D. C., 'Harrow, including Pinner: The growth of the hamlets', in Baker, T. F. T. et al. (eds), *A History of the County of Middlesex: Volume 4*, 1971

Clapson, M. & Larkham, P. (eds), *The Blitz and its Legacy: Wartime Destruction to Post-War Reconstruction*, 2013

Clunn, H., *The Face of London* (rev. edn), 1970

Cook, C. & Stevenson, J., *Britain Since 1945*, 1996

Cunliffe, B., Bartlett, R., Morrill, J., Briggs, A. & Bourke, J. (eds), *The Penguin*

Illustrated History of Britain & Ireland, 2001

Davies, C., *The Prefabricated Home*, 2005

Gardiner, J., *The Thirties: An Intimate History of Britain*, 2010

Gardiner, J., *The Blitz: The British Under Attack*, 2010

Glendinning, M., *Tower Block: Modern Public Housing in England, Scotland, Wales and Northern Ireland*, 1993

Golland, A. & Blake, R. (eds), *Housing Development. Theory: Process and Practice*, 2004

Gould, J., *Plymouth: Vision of a Modern City*, 2010

Hicks, J. & Allen, G., *A Century of Change: Trends in UK statistics since 1900*, Research Paper 99/111, 21 December 1999

Higham, R., *Bases of Air Strategy: Building Airfields for the RAF, 1914–1945*, 1998

Hobhouse, H. (ed.), 'Northern Millwall: Public housing in Northern Millwall', in *Survey of London, Volumes 43 and 44: Poplar, Blackwall and Isle of Dogs*, 1994

Jackson, A., *Semi-detached London: Suburban Development, Life and Transport, 1900–39*, 1973

Jackson, A., *London's Metro-Land: A unique British railway enterprise*, 2006

Kynaston, D., *Austerity Britain 1945–51*, 2007

Levine, J., *Forgotten Voices of the Blitz and the Battle of Britain*, 2006

Liddell Hart, B., *Paris; or the Future of War*, 1925

Longmate, N., *Island Fortress: The Defence of Great Britain 1603–1945*, 1991

Malden, H. (ed.), *A History of the County of Surrey: Volume 4*, 1912

Matless, D., 'Appropriate Geography: Patrick Abercrombie and the Energy of the World', *Journal of Design History*, Vol. 6, No. 3, 1993

Morton, H. V., *In Search of England*, 1927

Morton, H. V., *In Search of Scotland*, 1929

Overy, R., *The Bombing War: Europe 1939–1945*, 2013

Powers, A., *Britain: Modern architectures in history*, 2007

Saint-Amour, P., 'Air War Prophecy and Interwar Modernism', *Comparative Literature Studies*, Vol. 42, No. 2, 2005

Saint-Amour, P., 'On the Partiality of Total War', *Critical Inquiry* 40, Winter 2014

Saint-Amour, P., *Tense Future, Modernism, Total War, Encyclopedic Form*, 2015

Van Schaardenburgh, C., 'Shadow Factories of the WW2 British motor industry', *Engineering and Technology Magazine*, Vol. 8, Issue 10, 14 October 2013

Scott, P., *The Making of the Modern British Home: The suburban semi and family life between the wars*, 2013

Slavin, A., *The Development of the West of Scotland 1750–1960*, 1975

Smith, R., *National Parks of Britain*, 2008

Stratton, M. & Trinder, B., *Twentieth Century Industrial Archaeology*, 2013

Todman, D., *Britain's War: Into Battle, 1937–1941*, 2016

Tombs, R., *The English and their History*, 2014

Ullman, J., *Battersea Park*, 2016

Vale, B., *Prefabs: A History of the UK Temporary Housing Programme*, 1995

Ward, L., *The London County Council Bomb Damage Maps 1939–1945*, 2015

Ward, S., *Planning and Urban Change*, 1994

Wilkinson, J., *Select Views in Cumberland, Westmoreland and Lancashire*, 1810

Williams-Ellis, C., *England and the Octopus*, 1928

USEFUL WEBSITES

British History Online: http://www.british-history.ac.uk

Campaign to Protect Rural England: http://www.cpre.org.uk

A Vision of Britain through time: http://www.visionofbritain.org.uk

Office for National Statistics: http://www.ons.gov.uk

PastScape, Historic England: http://www.pastscape.org.uk/default.aspx

Historic England: https://historicengland.org.uk/listing/selection-criteria/listing-selection/

Royal Commission on the Ancient and Historical Monuments of Wales: http://www.coflein.gov.uk

Forestry Commission: http://www.forestry.gov.uk

Scottish Archaeological Research Framework: http://www.scottishheritagehub.com

Dictionary of Irish Architects: http://www.dia.ie

UK National Ecosystem Assessment: http:/uknea.unep-wcmc.org

INDEX